3,-

ABSTRACTION AND REPRESENTATION

BOSTON STUDIES IN THE PHILOSOPHY OF SCIENCE

Editor

ROBERT S. COHEN, *Boston University*

Editorial Advisory Board

THOMAS F. GLICK, *Boston University*
ADOLF GRÜNBAUM, *University of Pittsburgh*
SAHOTRA SARKAR, *McGill University*
SYLVAN S. SCHWEBER, *Brandeis University*
JOHN J. STACHEL, *Boston University*
MARX W. WARTOFSKY, *Baruch College of the City University of New York*

VOLUME 175

PETER DAMEROW

*Max Planck Institute for Human Development and Education,
Berlin*

ABSTRACTION AND REPRESENTATION

Essays on the Cultural Evolution of Thinking

With an Introduction by
WOLFGANG EDELSTEIN
*Max Planck Institute for Human Development and Education,
Berlin*

and

WOLFGANG LEFÈVRE
*Max Planck Institute for the History of Science,
Berlin*

Translated from the German by
RENATE HANAUER

KLUWER ACADEMIC PUBLISHERS
DORDRECHT / BOSTON / LONDON

Library of Congress Cataloging-in-Publication Data

```
Damerow, Peter.
   Abstracts and representation : essays on the cultural evolution of
 thinking / by Peter Damerow.
        p.   cm. -- (Boston studies in the philosophy of science ; v.
 175)
   Includes bibliographical references and index.
   ISBN 0-7923-3816-2 (hardcover : alk. paper)
   1. Abstraction.  2. Mental representation.  3. Thought and
 thinking.   I. Title.  II. Series.
 BF443.D36   1996
 153.4--dc20                                                 95-40140
```

ISBN 0-7923-3816-2

Published by Kluwer Academic Publishers,
P.O. Box 17, 3300 AA Dordrecht, The Netherlands.

Kluwer Academic Publishers incorporates
the publishing programmes of
D. Reidel, Martinus Nijhoff, Dr W. Junk and MTP Press.

Sold and distributed in the U.S.A. and Canada
by Kluwer Academic Publishers,
101 Philip Drive, Norwell, MA 02061, U.S.A.

In all other countries, sold and distributed
by Kluwer Academic Publishers Group,
P.O. Box 322, 3300 AH Dordrecht, The Netherlands.

Printed on acid-free paper

Translation edited by
Wolfgang Edelstein and Wolfgang Lefèvre

All Rights Reserved
© 1996 Kluwer Academic Publishers
No part of the material protected by this copyright notice may be reproduced or
utilized in any form or by any means, electronic or mechanical,
including photocopying, recording or by any information storage and
retrieval system, without written permission from the copyright owner.

Printed in the Netherlands

Bookkeeping with context-dependent numerals in early Mesopotamian civilization (3000 B.C.). See page 332.

TABLE OF CONTENTS

INTRODUCTION
by Wolfgang Edelstein and Wolfgang Lefèvre   xi

ON ACTION AND COGNITION

CHAPTER 1. ACTION AND COGNITION IN PIAGET'S GENETIC
EPISTEMOLOGY AND IN HEGEL'S LOGIC   1

Notes	10
Bibliography	27

CHAPTER 2. REPRESENTATION AND MEANING   29

Examples of Cognitive Structures	29
Cognitive Structures and Material Action	42
Cognitive Structures and Language	46
Notes	54
Bibliography	67

EDUCATION IN CONTEXT

CHAPTER 3. PHILOSOPHICAL AND PEDAGOGICAL REMARKS
ON THE CONCEPT "ABSTRACT"   71

The Common Usage of the Concept in the Didactics of Mathematics	71
Abstraction as the Isolation of Qualities	72
The Concept of Abstraction of Formal Logic	73
The Empiricist Concept of Abstraction	75
Models of Abstraction Based on Constitution Theory	75
The Dialectical Concept of Abstraction	76
The "Normal" Pattern of Teaching	78
On the Interpretation of Learning Processes in Cognitive Psychology	79
Understanding the Facts	80

The Function of Exercises	81
Abstraction and Concretization in Mathematics Teaching	83
Bibliography	86

CHAPTER 4. WHAT IS MATHEMATICAL ABILITY AND HOW DO ABILITY DIFFERENCES EMERGE IN MATHEMATICS EDUCATION? — 87

The Appearance of Mathematical Competence in Mathematics Education	87
Psychometric Constructs of Ability	93
Abilities as Cognitive Structures	98
Notes	108
Bibliography	110

CHAPTER 5. MATHEMATICS EDUCATION AND SOCIETY — 111

Introduction	111
The Vicinities of Mathematics— Mathematics of the Vicinities	113
Transmission of Knowledge Predetermined By Specific Ends	118
The Training of the Mind	122
Mathematics as a Profession	126
The Emergence of School Mathematics	131
Mathematics Education in the Context of Modern Institutional Conditions	138
Notes	144
Bibliography	145

CHAPTER 6. PRELIMINARY REMARKS ON THE RELATIONSHIP OF THE PRINCIPLES OF TEACHING ARITHMETIC TO THE EARLY HISTORY OFMATHEMATICS — 149

In Retrospect: Repetition in Unison	149
Action as the Starting Point for Mathematical Thinking?	151
On the Role of Historical Conditions of the Development of Mathematical Thinking	152
The Early History of Arithmetic as a Touchstone	153
Arithmetic as a System of Rules for Material	154
Representatives of Numbers	154
The Arithmetic of Constructive-Additive Sign Systems	156
Developmental Stages of Mesopotamian Arithmetic	159

Theoretical Perspective	167
Notes	169
Bibliography	170

CULTURAL EVOLUTION OF ARITHMETICAL THINKING

CHAPTER 7. THE DEVELOPMENT OF ARITHMETICAL THINKING: ON THE ROLE OF CALCULATING AIDS IN ANCIENT EGYPTIAN AND BABYLONIAN ARITHMETIC	173
Preliminary Remarks: Reckoning Board and Rod Numerals	173
Structural Characteristics of Ancient Egyptian Arithmetic	176
The Means of Calculation in Ancient Egypt	188
The Sources for Reconstructing the History of Old Babylonian Arithmetic	199
Structural Characteristics of Old Babylonian Arithmetic	204
The History of Mesopotamian Calculating Aids	211

CHAPTER 8. THE FIRST REPRESENTATIONS OF NUMBERS AND THE DEVELOPMENT OF THE NUMBER CONCEPT	275
Characteristics of the Numerical Signs in the Archaic Texts	275
Earlier Attempts at Interpretation	276
Rules for Using the Signs	279
Summations	281
Statistics as a Method of Decipherment	284
The Decipherment of Calendar Entries	286
Numerical Sign Systems with Specific Areas of Application	288
The Lack of an Abstract Number Concept	291
Number Analogues	293
From Number Analogues to the Abstract Number Concept	294
Bibliography	297

CHAPTER 9. ON THE RELATIONSHIP BETWEEN ONTOGENESIS AND HISTORIOGENESIS OF THE NUMBER CONCEPT	299
On the Preconditions of Piaget's Constructivist Conception of Numbers	301

Ontogenesis of the Number Concept in Piaget's Genetic Epistemology	302
Historiogenesis of the Number Concept in Piaget's Genetic Epistemology	308
Symbolic Representation and Historical Transmission	314
Protoarithmetic of a Stone Age Culture	321
Protoarithmetical Techniques in the Transition to a Literate Culture	329
Ontogenesis and Historiogenesis of the Number Concept	354
Notes	362
Bibliography	367

ON HISTORICAL EPISTEMOLOGY

CHAPTER 10. ABSTRACTION AND REPRESENTATION	371
Bibliography	381
CHAPTER 11. THE CONCEPT OF LABOR IN HISTORICAL MATERIALISM AND THE THEORY OF SOCIO-HISTORICAL DEVELOPMENT	383
Labor in Human History	383
Biological Evolution and Historical Development	387
CHAPTER 12. TOOLS OF SCIENCE	395
Labor and Cognition	395
The Emergence of Science	396
The Role of the Tools of Scientific Work	398
Continuity and Discontinuity of the Development of Science	401
LIST OF ORIGINAL PUBLICATIONS	405
NAME INDEX	407
SUBJECT INDEX	411

INTRODUCTION

The author of this book is affiliated with the Center for Development and Socialization of the Max Planck Institute for Human Development and Education in Berlin and heads its program on culture and cognition which devotes its labors to the reconstruction of scientific concepts through history in a perspective of what might be called "historical epistemology." He is also a member of a related research group in the newly founded Max Planck Institute for the History of Science in Berlin. Perhaps this double affiliation throws some light on the scope of Damerow's scientific interests. In any event it will explain why representatives of both these institutions join in an effort to introduce Peter Damerow's writings to an English speaking audience.

Damerow's scholarship ranges across widely different areas including philosophy and history of science, psychology, and education. Among his fields of expertise are the emergence of writing, early Babylonian mathematics, the history of arithmetic, the relationship between pure and applied mathematics, the theory and methods of mathematics instruction, the transition from preclassical to classical mechanics, and the history and theory of relativity.

Looking over this work, it is easy to see why it appears to contain a philosophical program of historical epistemology as the core of Damerow's intellectual pursuits. However, this is a picture that emerges only after the fact. When he started his work, there was no such program. No precise philosophical theory was later applied to various areas of research or illustrated by a set of examples. It probably began with an interest in processes of acquisition, development, and validation of knowledge. What now can be perceived as epistemological theory is the result of concrete projects. So the theory is saturated by a wealth of material derived from the sciences and their history. Nor is the theory separable from the study of concrete forms of the development of thought, a perspective hardly consistent with the formulation of universal propositions about human thinking.

The historical epistemology which Damerow has developed, and is continuing to develop, is based on philosophical epistemology, cognitive psychology, history of science, and education. But the theory itself is neither philosophical nor psychological nor historical in the sense defined by these academic disciplines. It is truly and strictly interdisciplinary—with the implication that it cannot be expressed in the conceptual framework of any of these disciplines.

Truly interdisciplinary scholarship is always subject to risk. On the one hand, it threatens to slide down the slippery slope of dilettantism, and thus will be accused of dilettantism whenever the results contradict positions of the disciplines. The times of universal scholarship belong irrevocably to the past. Therefore, an interdisciplinary enterprise must be grounded in collaboration among scholars from different disciplines. This has been Damerow's strategy throughout. But the other risk of interdisciplinary scholarship is not automatically averted thereby: the risk of being misunderstood because the results of interdisciplinary studies do not fit within the conceptual framework of any discipline. The reception of cross-disciplinary work also demands the spirit of interdisciplinarity in the reader.

For this reason we wish to add a few remarks about features of Damerow's work which may easily lend themselves to misunderstanding.

Peter Damerow's approach to the analysis of scientific thinking characteristically relies on a combination of the history of scientific (and extra-scientific) thought with philosophy and the cognitive sciences. Historians, therefore, may apprehend a reduction of the cultural history of thought to universal patterns, rules or laws, whereas philosophers or cognitive psychologists may entertain the obverse fears—that structures or results of thinking are subjected to a completely relativistic analysis.

Throughout this book Damerow frequently refers to Hegel and Piaget. This may arouse the suspicions of historians. Both Hegel and Piaget have elaborated theories of the development of general structures of thinking according to which the process of thinking and the development of the sciences are conceptualized as the epigenetic unfolding of structures. For Damerow, these theories are valuable because they propose that the development of thought be understood as a process of constructing and reconstructing cognitive structures. Thus, they generate insights into the developmental connection among those cognitive structures which order actions on objects and those structures from which emerge ideal objects such as the concept of number, concepts that appear to have *a priori* status.

The principal fault that Damerow finds with these theories consists in their neglect of the fact that cognitive activities depend on, and are determined by, material representations of prior knowledge and its structure. These representations range from tools to artificial symbol systems.

Damerow criticizes the unhistoric universalism of the theories of Hegel and Piaget that derives in part from this neglect. Taking these representations seriously as means of thought, thinking must be conceived not only as a social, but as a real historical process depending, as do these representations, on the historical development of societies and their material culture.

When Damerow highlights that thought depends on means embedded in the material culture of concrete historical societies, this focus can be traced back to his reception of Jerome Bruner's theory of external representations and, via Vygotsky and the so-called culture-historical school of Russian psychology, to Marx's theory of the means of production. It should be added, however, that originally this focus responded to Damerow's experience with the problems of mathematics teaching!

For Damerow, the history of thought and the sciences is not the mere appearance of the underlying epigenetic evolution of necessary structures of thinking. Rather it is a strictly historical process embedded in the history of societies and their entire material culture, and hence caught in the contingencies of that history. Cognitive structures and scientific theories cannot be ordered on a trajectory representing the path toward truth. Neither are we confronted with a chaotic emergence and fading of random structures of thinking. There is, in Damerow's view, a historical connection between major structures, each transmitted or replaced by the next. Damerow is particularly interested in these processes of replacement, both in ontogeny and in history, conceiving them as the restructuring of cognitive structures based on changes in available means of thought. The developmental patterns which he subjects to scrutiny are thus simultaneously intrinsically coherent and contingent due to their dependency on the historical development of material culture.

The present collection perhaps should be read starting with the last chapter which brings together many of the general features to which we have referred. The concrete studies of action and cognition of the history of arithmetical thinking and of the problems of sociohistorical development and education show this last chapter to be not an outline of an abstract theory of thinking, but a summary of results achieved by these studies. We hope the volume will convincingly demonstrate both the commanding conceptual framework of the author and the rich materials on which he draws to construct his historical epistemology.

A few remarks may be needed to explain the composition of this book. The individual chapters are based on original publications, in German, published over a range of years. While unified in their presentation, they represent the original line of argument without much change. Thus, they also document the progression of the author's thinking toward a more complete formulation of the historical epistemology to which the author has dedicated much of his endeavors.

Finally, we wish to gratefully acknowledge the support and assistance that the present project has received from various institutions and persons. Renate Hanauer has translated a difficult manuscript with a rare command of both language and subject matter and with unfailing dedication to the project. The Max Planck Institute for Human Development and Education has generously supported the translation. The technical staff has been extremely helpful at each stage of the growth of this book. Madeline Hoyt, in particular, has prepared both the layout and the type-setting. To the publishers we owe thanks for their care and patience with a book that has been very long in the making.

Berlin, September 1995 Wolfgang Edelstein
Wolfgang Lefèvre

CHAPTER 1

ACTION AND COGNITION IN PIAGET'S GENETIC EPISTEMOLOGY AND IN HEGEL'S LOGIC
[1977]

The following is a report on the results of a study of the connections between the genetic epistemology of the psychologist Piaget and Hegel's dialectical logic. A word about the reason for such an undertaking seems to be in order.

Piaget initiated a type of structuralism based on a theory of action which today represents the prevalent theoretical thinking in the area of psychology applied to problems of education and education policy. This structuralism mediates between psychological research and policy decisions in education. It must be seen as an attempt to recapture the view of the unity of science once exemplified by philosophy and its construction of systems. In my view, this attempt seems to be representative of current tendencies in the specialized scientific disciplines to transcend their boundaries and, by reflecting on the social practice of which they are a component, to obtain a theoretical frame of reference for the treatment of interdisciplinary problems generated by this practice.

Such attempts of philosophical reflection outside the theoretical tradition of a systematic philosophy are, however, always open to the charge of dilettantism. Even though the philosophy of today—barely tolerated by the sciences as the special science for radical questioning or as a substitute for religion—has lost all influence on the research strategies of the scientific disciplines and even though this development has not been without consequences for the relevance of its findings for these disciplines, this shattered tradition nevertheless assures a modicum of theoretical continuity which provides standards for the quality of philosophical thinking.

The attempt to relate Piaget's genetic epistemology—that is, the philosophical interpretation of the results of more than 50 years of research dedicated to the empirical investigation of the conceptual development of children—to the tradition of philosophical thought, is based on this consideration. Piaget himself relates his work to this philosophical tradition only eclectically—he only has superficial knowledge of Hegel's philosophy. The thesis implicit in the subject of my investigation—that is, that the developmental processes of the logical forms described and explained by Piaget have already been theoretically anticipated in the dialectical logic, must be shown by the results of the investigation.

Thus, the study under consideration has a twofold purpose. First, it provides an interpretation of Hegel's logic in terms of genetic epistemology as well as confronting it with the results of psychological investigations on the genesis of cognition. Second, the theoretical interpretation which Piaget provided for the results of his investigations is confronted with—in my opinion—a far more developed theory of their subject and judged in comparison with it.[1]

It is my intention to present results of this study by way of example. I am limiting myself to a comparison of Piaget's concept of reflective abstraction and Hegel's concept of reflection.

For Piaget reflective abstractions are the mark of a developmental stage of cognition. Reflective abstractions go beyond the sensory certainty to the necessity of thought. Human thinking is distinguished from animal intelligence by reflective abstractions.

If one defines intelligence, as does Piaget, as the control of behavior by means of cognitive systems, then reflective abstractions characterize that stage of the development of intelligence where the further development of the control of behavior has reached a particular structure. Piaget calls this developmental structure reflective abstraction. The result of reflective abstractions are the logico-mathematical structures which specifically distinguish human thinking from previous forms of intelligence. The process characterized by reflective abstractions is the process of the construction of these structures.[2]

The appearance of reflective abstractions can thus be identified in terms of developmental psychology: Reflective abstractions give rise to the transition from the stage of sensori-motor intelligence to the stage of concrete operations and, in a more developed form, to all subsequent transitions in the development of intelligence. In ontogeny it is the period between the ages of two and eight that corresponds to the transition to reflective abstractions. This is the period which begins with the acquisition of the semiotic skills (language, images, symbols) and the development of preoperative forms of the imagination and representation and ends with an extensive restructuring of the child's thinking resulting in the development of the basic logico-mathematical structures (structures of classification, seriation, number, space, time, etc.) in a concrete form not yet abstracted from the objects, that is to say, the transition to the stage of concrete operations.

In Piaget's view, the precondition for reflective abstractions is that the emerging mental activity focuses on itself, that mental activity becomes the subject of mental activity. To the extent that Piaget considers the cognitive systems instruments for controlling behavior and for this reason calls the structures of mental activity developing through the confrontation with the environment schemata of action, it means that the control of behavior itself becomes the subject of controlled acts of behavior. Thus, the mediated control of behavior acquires the characteristics of self-reference. Piaget calls the guidance of behavior in its most general form "regulation" and understands it as a process of feeding back the results of an act of behavior to the behavioral schema controlling this act. In biological terms, the process of becoming human proceeds from these "simple regulations" to "regulations of regulations" to "autoregulations" characterized by the self-organization of their inner structure ([13], p. 19).

In Piaget's view, these self-regulations of behavior constitute the basis of the reflective structure of the corresponding cognitive systems. At the same time they establish the relative independence of the developmental process of these systems from its biological preconditions, a process characterized by reflective abstractions.

Piaget describes the process of reflective abstraction as a process that begins with becoming conscious of acts of behavior and their cognitive equivalents. To be more precise: In view of the fact that these acts of behavior and the cognitive systems guiding them are themselves already instruments of cognition, the process begins with differentiating the contents of cognition into the components of direct sense impressions and the mediating activity. Reflective abstraction is the process of the constructive modification of acts of behavior and of cognitive activities initiated by the emerging consciousness.

With regard to the structure of the acts of behavior and/or the cognitive systems guiding them reflective abstraction encompasses two components that cannot be separated: the projection of the existing structure onto a higher level (which Piaget called "reflection in the physical sense of representation") and the reconstruction and reorganization of that which was transferred in this way with the structural elements of this level (reflection in the psychological sense "as the mental processing of matter provided earlier in its raw or unprocessed state"). The unity of the two components of reflective abstraction is ensured by the fact that this process does not take place in already existing hierarchial structural stages, but in structural stages in which the particular expanded structure represents precisely the result of the self-reference of reflective abstraction. In other words, the relations generated in the reflective process in the physical sense are at one and the same time the structural elements of the expanded structure into which the earlier structure is projected by this reflection. Piaget, therefore, calls the developmental process of reflective abstraction "endogenous development."[3]

However, when reflection is characterized as a developed form of behavior regulation, its differentiation with regard to prior forms poses problems. On Piaget's view, reflection is nothing but "the prototype of a regulation of regulations"; the self-reference does not define the specific difference of reflective abstraction with regard to its prior forms of organic self-regulation. Rather, Piaget tries to capture the specific difference by contrasting "automatic regulation" and "active regulation." The distinction concerns the means of regulation. In the case of automatic regulation, the means are barely varied; only their application and adaptation are specified. Active regulation, on the other hand, is characterized by the possibility of selecting various means. Piaget argues: "... automatic regulations do not, by themselves, lead to conscious awareness. Active regulations, on the other hand, do; and because they do, they lie at the source of the representation or conceptualization of material actions." ([13], p. 18)[4]

Let us now compare Piaget's conception of reflective abstraction as the structure of the process of the development of logico-mathematical structures of thinking in terms of genetic epistemology with Hegel's logic of reflection. In Hegel's system the term "reflection" characterizes the transition from the "logic of being" to the "logic of essence"; reflection is the term used for the structure of the process which distinguishes the logic of essence from the logic of being. Together the logic of being and the logic of essence constitute the "objective logic," "the genetic exposition of the notion" ([5], p. 577)—that is, in their totality they correspond to Piaget's "genetic epistemology."[5] In a comparison of the structure of his logic with the organization of Kant's philosophy Hegel characterizes "being" as the sphere of intuition, "notion" as the sphere of reason; "essence" is the sphere of the development of being into "notion," of intuition into concept—a sphere that is characterized by reflection.

What this means, however, is that Hegel's reflection and Piaget's reflective abstraction are of equal significance within the system.[6] On Hegel's view, reflection characterizes the transition from sense-certainty to the necessity of thought as well. "The sensible appearance is individual and evanescent: the permanent in it is discovered by reflection." ([3], p. 42) Reflective thinking "has to deal with thoughts as thoughts, and brings them into consciousness." ([3], p. 5) On Hegel's view, it is also reflection that distinguishes human thinking from animal intelligence. "It is in knowing what he is and what he does, that man is distinguished from the brutes." ([3], p. 34)

Man is a thinker, and is universal: but he is a thinker only because he feels his own universality. The animal too is by implication universal, but the universal is not consciously felt by it to be universal: it feels only the individual. The animal sees a singular object, for instance, its food, or a man. For the animal all this never goes beyond an individual thing. Similarly, sensation has to do with nothing but singulars, such as *this* pain or *this* sweet taste. Nature does not bring its [thought in its objective meaning] into consciousness: it is man who first makes himself double so as to be a universal for a universal. ([3], p. 47f.)

Hegel uses the term "determining" for the constructive moment of cognition and in particular also for the constructive moment of reflective thinking based on its self-reference. Like Piaget, he discerns in the structure of reflection the wellspring for a constructive process of "self-determination" resulting in the logical forms: "The form ... will have within itself the capacity to *determine* itself, that is, to give itself a content, and that a *necessarily* explicated content—in the form of a system of determinations of thought." ([5], p. 63) Consequently, Hegel perceives the doctrine of essence, the genetic exposition of the notion by means of reflection, as the doctrine of "the Notion as a system of *reflected determinations*" ([5], p. 61).[7]

The constructive process of "determining" by means of reflection is a necessary process. Thus, in Hegel reflection—just as the reflective abstraction in Piaget—is sharply delineated with regard to formal abstraction and the process of arbitrary concept formations resulting from it. In the introduction to the logic he comments on the structure of the process of reflection as follows:

... we are speaking of a determining of an object in so far as the elements of its content do *not* belong to feeling and intuition. Such an object is a *thought,* and to determine it means partly, first to produce it, partly, in so far as it is something presupposed, to have further thoughts about it, to develop it further by thought. ([5], p. 62)

In this description the analogy to Piaget's reflective abstraction is clearly evident.

In the Logic itself the structure of reflection is more carefully defined by means of the strict terminology employed there. The concept of "negation" is a key concept for the theory of determining developed in the logic. It defines the most elementary act of determining.

In the logic of being, the logic of the sphere of intuition, negation produces the "transition," the specific form of processing in this first part of

Hegel's logic. In the "transition" from "something" to an "other" the something vanishes; this is the hallmark of transition. Initially, this specific form of processing is not changed by the self-reference peculiar to negation, the negation of negation either. While it is true that the negation of negation, elevates being to a new stage respectively, this, nevertheless, remains a transition; in the outcome of the process its stages are not contained as internal moments.

In the logic of essence the structure of the developmental process is the reflection. It is developed from the self-reference of negation in the form of the "negation of negativity." Being as the negation of negativity is being not mediated by negation—that is, immediate being. However, because this immediate being emerges from negativity by means of negation, it is connected with negativity by dint of this emergence by means of negation. Thus, being as negation of negativity is at one and the same time immediate and mediated, no longer, however, as an immediate unity. In the negation of negativity immediacy and mediation are present as internal, contradictory moments, and the development of this contradiction moves the constructive process described as the determination of reflection forward. Thus, the immediate being becomes essence "which is Being coming into mediation with itself through the negativity of itself" ([3], p. 207).

According to this theoretical construction the transition from being to essence therefore necessarily entails an internal duplication. The external relationship between being and essence is by necessity also an internal relationship of essence designated by Hegel as "illusory being":

Essence accordingly is [illusory being, i.e.,] Being thus reflecting light into itself. ... That reflection, or light thrown into itself, constitutes the distinction between Essence and immediate Being, and is the peculiar characteristic of Essence itself. ([3], p. 207f.)[8]

At this point a further agreement of the theoretical foundations of Piaget's reflective abstraction and Hegel's reflection becomes apparent. For Piaget as well as Hegel the differentiation of the cognitive content into the components of the immediate sensory given and mediating activity is connected with reflection and is essential for the process of constructing or determining the logical forms respectively.[9] At the same time, however, the fundamental difference between Piaget's and Hegel's concepts of reflection emerges in the formula that reflection "is the showing of this illusory being within essence itself" ([5], p. 394).

Piaget understands reflective abstractions as a result of becoming conscious of the internal conditions of the coordination of behavioral acts and the cognitive activity determined by reflective abstractions as a unique process based on the biological conditions of the coordination of behavior which is not influenced by the cognitive activity that is directed to external objects, guided by sensory factors and determined by formal abstractions (see [11], p. 305ff.). Piaget writes:

> Starting from these primitive interactions in which the parts played by internal and external factors are indistinguishable (as well as subjectively fused), knowledge is then built up in two complementary directions, while still based on actions and action schemata outside which it has no hold either on the exterior world or on internal analysis.
>
> The first of these directions, which develops much earlier in the animal kingdom because it is the most essential for adaptation to environment, is the conquest of objects or knowledge of environmental data, which will eventually enable the subject to comprehend the exterior world objectively. ... The second direction, which is almost certainly confined to human intelligence, involves becoming conscious of the internal conditions of these coordinations, a development which leads, by means of "reflection," to the making of logico-mathematical constructions. ([11], p. 28; see also note 2)

On Piaget's view the differentiation of the cognitive contents associated with reflection into the moments of the sensory given and the mediating activity is thus a complete bifurcation resulting in two autonomous, unassociated, complementary cognitive modes with their own structures of developmental processes complementing each other only externally.[10]

In Hegel's theory, on the other hand, the concept of reflection is more broadly defined from the outset. Here reflection also begins with the differentiation of the cognitive content into the moments of the sensory given and the mediating activity. According to his theory, reflection does, however, not just refer to the mediating activity, to the internal conditions of the co-ordination of acts of behavior, but to the entire relationship of the sensory given and the mediating activity. It is reflection that mediates between these two components, because reflection as the negation of negativity defines the cognitive content as both immediate and mediated. In the process of reflection mediation and immediacy blend into each other. Immediacy is not only the beginning of mediation, but at the same time its outcome, mediated immediacy, always at a higher stage of cognitive development.

Thus, reflection is not understood here, as it is in Piaget's theory, as a form of abstraction; rather, reflection is the evolving unity of the sensory given and mediating activity. The logico-mathematical forms, the determinations of reflection, are the result of the cognitive contents and, conversely, these contents appear to be constituted by the logico-mathematical forms by means of which they are related to each other. The unity of these forms and their concrete contents—which, in Piaget's theory, can only be established by means of a particular kind of cognition—that is, in terms of the empirical cognition by means of formal abstractions—is for Hegel as much a result of reflection as the determination—that is, the construction, of these forms themselves.

This concept of reflection becomes all the more convincing, if we consider its origin: The concept was introduced in the *Jenaer Realphilosophie* to distinguish labor as "reflective activity" from activity as "pure mediation," as the mere satisfaction of a desire by means of the destruction of its object. What distinguishes labor as "reflective activity" from activity as "pure mediation" is the endurance of its material means, of its tools, in which activity as the unity of ideal purpose and material object has materially objectified itself materially (see [4], p. 216ff.). The unity of the sensory given and mediating activity constructed in Hegel's logic, the mediated immediacy as the result of reflection, not only constitutes a hypothetico-theoretical construct, but this unity is actually created in the material means of the objective activity in a myriad of forms. The constructive moment of reflection, the creation of the determinations of reflection in Hegel's logic, of the logico-mathematical structures in Piaget's genetic epistemology, corresponds to ever more growing abundance of material activities in the process of the production, which is, at one and the same time, condition and product of reflection (on this point, see also the publication [2] on "labor and reflection," which is a result of investigations carried out in connection with the present study).

Thus, what ultimately distinguishes Piaget's concept of reflective abstraction and Hegel's concept of abstraction is this: Piaget bases reflective abstraction on abstract activity, the realization of a purpose without further consequences, the activity in other words which Hegel calls "pure mediation." The structure of Hegel's reflection, on the other hand, is related to productive labor; logical "determining" need not be understood as a self-governing process, since material production corresponds to it. Reflection becomes determining—that is, constructively creating the determinations of reflection, as the unity of two opposing moments: of the "positing reflection," which assumes the "immediacy as the sublating of the negative" ([5], p. 401), and "external reflection," which "forms the starting point of the presupposed immediate" ([5], p. 400). To this corresponds the unity of positing the ends and pre-supposing the material object of labor which is represented by the means of labor.

This much on the comparison between the concept of reflective abstraction in Piaget's genetic epistemology and the concept of reflection in Hegel's dialectical logic. Out of this emerges the possibility of an alternative interpretation of the results of Piaget's investigations on the development of cognitive ability. If the interpretation is based on Hegel's concept of reflection, then the dichotomy of knowledge as empirical knowledge, to which the logico-mathematical forms are merely external, and logico-mathematical knowledge that is independent of the contents can be avoided.

Although Piaget considers the unity of cognition and action a necessary premise of any theory of knowledge which takes into account the actual genesis of cognition—a premise he developed primarily on the basis of the biological preconditions of cognition—, cognition and action, just as in the theoretical tradition of Kantianism to which Piaget refers critically in a variety of ways, ultimately diverge again in the dichotomy of reality-based empirical knowledge and action-based logico-mathematical knowledge. It became evident that this dichotomy in Piaget's theory is a consequence of the abstract concept of action on which the concept of reflective abstraction is based. Piaget's concept of action encompasses the biological foundations of action and the control of action by means of cognitive structures. This concept of action may suffice for an understanding of animal behavior and the forms of intelligence that become evident in this behavior; applied to human behavior, it necessarily leads to an undervaluation of the significance of the objective means of action for the development of cognitive ability.

If, however, action is understood as material activity, then the result is a concept of reflection which displays the structure of developmental processes elaborated by Hegel. In this case, reflection on action includes the reflection on its material means, and this moment of reflection—in Hegel: the determining reflection—becomes the source of both empirical knowledge and the development of its logico-mathematical forms.

The consequences of such a reinterpretation of the results of Piaget's investigations mainly concern the generalizing conclusions—that is, Piaget's attempt to regain the perspective of a unity of the sciences by means of philosophical reflection and, consequently, the relevance of the results of psychological investigations for problems of education and education policy.

If the material means of the actions on which cognitive activity is based are irrelevant for the development of cognitive abilities, as is implied by the concept of reflective abstraction, the development of the cognitive abilities of the child can be interpreted as the direct continuation of the development of animal intelligence. This is the reason why Piaget thinks he can explain the development of cognition biologically, why he understands the relationship of the individual to society in analogy to the relationship of the individual phenotype to the genetic population and hopes to be able to explain the history of the species in terms of the history of the individual, why the significance of the production of tools for this transition—hardly in dispute any longer—is almost totally neglected in his theory of the transition from animal to human intelligence.

If, however, the material means of actions, on which cognitive activity is based, do have the significance for the development of cognitive abilities implicitly attributed to them in Hegel's concept of reflection, then individual history must be interpreted in light of the history of the species. Hegel's concept of reflection opens up the possibility of explaining the development of cognitive structures as well as the development of cognitive abilities in terms of the material heritage of history and its acquisition in the context of the individual cognitive process.

Such an explanation would, however, dash the hope, nurtured in a variety of ways by the structuralism of Piagetian origin, that the development of cognitive structures could be achieved by an abstract competence training, that the deficits of intellectual competence resulting from social privation could be redressed by compensatory educational programs without changing the social conditions which caused them. In terms of educational policy this implies the following: The reality of learning practiced in educational institutions must not be organized as an isolated domain of action, but must be integrated into the social practice of the learners. The cognitive abilities of the learners can only be developed effectively by reflecting on this practice. However, even then the real meaning which the learning successes have for the learner poses a social barrier which can only be overcome collectively.

NOTES

[1] *On the Relationship of Hegel and Piaget*
The comparability of Piaget's genetic epistemology and Hegel's dialectical logic is not accidental. There is a historical connection between the problems of the last great scientific systems of idealist philosophy and the structuralism which grew out of research in the scientific disciplines.

With the development of the bourgeois mode of production, the principles of the division of labor increasingly penetrated science as well. The detachment of subject areas from philosophy and their establishment as independent sciences is indicative of this process. The growing importance of science for social reproduction was expressed in this process of the Taylorization of science. The introduction of the principles of the division of labor into science was an element of scientific progress which considerably increased the effectiveness of science. Philosophical reflection soon fell behind the results of research in specialized scientific disciplines. Hegel's misjudgment with regard to the number of the planets and his defense of Goethe's theory of colors are often cited as proof that it is no accident that his construction of a system was the last, that Hegel could not even give an adequate account of the scientific problems of his own time. Within the specialized scientific disciplines the conviction prevailed that their emancipation from philosophy and the repudiation of philosophical reflection were tantamount to the transition of prescientific speculation to scientific thought based on experience.

The accomplishments of philosophy without which the productivity of, for example, Galileo, Descartes, or Newton would have been unthinkable, were forgotten in this process. Initially, the problems associated with the establishment of the scientific disciplines were negligible. However, with the introduction of the division of labor into science, the unity of science was lost to an increasing extent until, ultimately, even the social communication linkages between the disciplines broke down. Increasingly, research goals and cognitive interests were presented to the sciences as accomplished facts no longer in need of examination. Social practice as a criterion of truth for science was reduced to technical possibilities for realization, abstract feasibility, usability under the conditions of the mode of production giving the sciences their immense social importance. The relationship of ends and means, the object of reason realizing itself in the reflection on this relationship, became the object of political decisions with priority over science. The disciplines were left with a neutral, instrumental rationality. In the dualism of norms and facts of the positivist theory of science this development was finally hypostatized into a postulate.

However, what once was a moment of scientific progress, has turned into an impediment today. Especially the ecological crisis has given rise to doubts about abstract productivity, about the reduction of scientific reason, successful in practice, to an instrumental rationality. For this reason the hope was nurtured in philosophy that, in bypassing the scientific disciplines as it were, philosophy could attain a function for the justification of political decisions as a "philosophy of practice," thus giving it renewed relevance. However, the effects of these doubts on the particular sciences themselves seem to be far more significant, as they have become increasingly insecure in their understanding of science due to the recognition of their immanent limitations. That is to say, the more inclusive the practical tasks of science are becoming, the more frequently problems emerge as problems of interdisciplinary research no longer capable of being solved by specialized scientific disciplines alone.

In this way, the attempts—largely inspired by structuralism—to regain the perspective of a theoretical unity of science by means of the interdisciplinary development of theories are necessarily linked to the problems of the philosophical construction of systems, and in this tradition Hegel's system occupies a preeminent position. Not only does it mark the end of this tradition, but within it the entire array of problems of this tradition was addressed opening perspectives for the future. It is clear that the theoretical level of this tradition which found its expression in Hegel's system has not, in any way, been reached by structuralism.

The position of Piaget's genetic epistemology is of similar preeminence with regard to the context of the structuralist attempts at interdisciplinary theory development in the context of which these problems are taken up again today. None of the comparable attempts of this kind succeeded in theoretically integrating the achievements of the disciplines to a similar extent. The proposed theoretical system presented in the three volumes of his introduction to genetic epistemology ([8]) in particular may serve as evidence. However, such a feat would not have been possible without breaking with the positivist analytical theory of science which considered the theoretical state of the sciences, just as they were, to be the norm. Piaget argues: "Against this viewpoint, we argue that the study of science *as it is,* as a fact, is fundamentally irrelevant. Genetic epistemology, as we see it, tries, to the contrary, to overcome this separation of norm and fact, of valuation and description. We believe that only in the real development of the sciences can we discover the implicit values and norms that guide, inspire, and regulate them." ([9], p. 4; translation corrected). This was precisely Hegel's position developed extensively in the *Phenomenology of Mind,* the "science of manifested spirit," which he considered to be the precondition of his system proceeding from the "pure knowledge" of dialectical logic: "Logic, then, has for its presupposition the science of manifested spirit, which contains and demonstrates the necessity, and so the truth, of the standpoint occupied by pure knowing and of its mediation." ([5], p. 68f.)

Despite the continuity of how this problem presents itself, a continuity that could serve as the basis for a comparison between Piaget's genetic epistemology and Hegel's dialectical logic, such a comparison is beset with considerable difficulties due to the considerable historical distance between the systems. Even if the problem is the same, the level of scientific knowledge to be integrated has drastically changed.

To start with, there is the development of mathematics. Piaget does not have to develop the logico-mathematical structures of thinking in metatheoretical terms in attempting to define their importance in epistemological terms; they are already provided by mathematical research which has a tendency toward structuralism, and their importance for the methodology of the scientific disciplines—in any case of mathematics—is already elaborated in detail by the Bourbaki system. Piaget adopted these accomplishments and identified his concept of logico-mathematical structure with the concept of structure of the Bourbaki group (see [10], p. 17ff.; [9], p. 23ff.). This enables him to begin his critique of formalism where the latter has reached its limit—that is, in terms of its merely inductive self-definition—and in conjunction with the criticism of formalism to prove the relevance of his genetic concept of reflective abstraction acquired through psychological research for the self-definition of mathematics as well.

Hegel, on the other hand, had to develop his genetic conception of logico-mathematical structures out of a critique of formalism at a time when this formalism was to a great extent merely programmatic: The most productive branch of mathematical research at the time, calculus, kept defying noncontradictory formalization, although its productivity was based on the very originality of its formalism. On the one hand, this situation made the basic critique of formalism easier, and it was relatively easy to identify its definite limits; on the other hand, however, it was impossible to assess the importance of this critique for scientific research, and any proposed system was forced to accommodate the results in the scientific disciplines. Only an interpretation taking this background into account allows us to reconstruct Hegel's theory of formalism and to compare Hegel's and Piaget's concepts of abstraction.

Far greater than the problems created by the disparate developmental level of mathematical science are the problems resulting from the fact that the level of knowledge in the scientific disciplines basically did not permit a systematic approach which explained the coherence of reality in terms of its development. To understand the difficulties encountered by Hegel's systematic approach, it suffices to remember how speculative Piaget's genetic epistemology would have to be if, in determining the stages of intelligent behavior, he could not have drawn on the development in biology initiated by Darwinism. In this sense the assertion is correct that Hegel was far ahead of the sciences and could, therefore, receive support from them only indirectly (see [1], p. 26ff.). Thus, Hegel paid a steep price for the boldness of his approach: He was forced to use idealist patterns of explanation with the utmost consistency and, ultimately, to reconstruct any kind of development as conceptual development in order to achieve the integration of the scientific knowledge of his time into his system. If the comparison with Piaget's genetic epistemology is to do more than to confirm once again that Hegel turns idea and reality upside down, a reinterpretation of all of Hegel's concept formations becomes necessary.

To be more precise, in the context of the topic under consideration, the following basic difference between Hegel and Piaget emerges which cannot be considered here. In interpreting theoretical behavior Piaget proceeds from practical behavior. He interprets the developing logical forms as instruments for the control of behavior at a high level of biological development. Hegel, on the other hand, proceeding from theoretical behavior, interprets practical behavior as a certain form of the realization of the developing mind and dialectical logic as the most general form of this development.

For the sake of comparison, Hegel's logic is therefore interpreted contrary to the intention of its author. The parts of the system which, in Hegel, represent conclusions from his logic, are considered its assumptions in this context. Examples used by Hegel to clarify logical concepts, in particular the addenda of the "Encyclopedia" which were compiled from lecture notes, are used in reconstructing their origin and meaning. The success of such a procedure will, however, necessarily remain limited. There are models for interpreting Hegel's logic in this fashion. I am referring in particular to Erdei's reconstruction of the first chapter ([1]). In the end it can only be justified in terms of the intended result: in terms of the theoretical value of the concepts of a genetic epistemology and of a dialectical logic obtained in the course of reconstruction.

2   *On the Developmental Stages of the Concept of "Reflective Abstraction"*

The concept of reflective abstraction acquired its central theoretical position in Piaget's theory relatively late. His book *The Psychology of Intelligence,* first published in French in 1947, arguably the best known summary of his theory of intelligence, does not yet contain the concept; it is, however, essentially adumbrated in this work: "In order to understand the mechanism of the formation of operations, it is first of all important to realise what it is that has to be constructed, i.e. what must be added to sensori-motor intelligence for it to be extended into conceptual thought. ...

In the first place, acts of sensori-motor intelligence, which consist solely in co-ordinating successive perceptions and (also successive) overt movements, can themselves only be reduced to a succession of states, linked by brief anticipations and reconstruction, but never arriving at an all-embracing representation ... sensori-motor intelligence acts like a slow-motion film, in which all the pictures are seen in succession but without fusion, and so without the continuous vision necessary for understanding the whole.

In the second place, and for the same reason, an act of sensori-motor intelligence leads only to practical satisfaction, i.e. to the success of the action, and not to knowledge as such. It does not aim at explanation or classification or taking note of facts for their own sake; it links causally and classifies and takes note of facts only in relation to a subjective goal which is foreign to the pursuit of truth. Sensori-motor intelligence is thus an intelligence in action and in no way reflective.

As regards its scope, sensori-motor intelligence deals only with real entities, and each of its actions thus involves only very short distances between subject and objects. ... There are thus three essential conditions for the transition from the sensori-motor level to the reflective level. Firstly, an increase in speed allowing the knowledge of the successive phases of an action to be moulded into one simultaneous whole. Next, an awareness, not simply of the desired results of action, but its actual mechanisms, thus enabling the search for the solution to be combined with a consciousness of its nature. Finally, an increase in distances, enabling actions affecting real entities to be extended by symbolic actions affecting symbolic representations and thus going beyond the limits of near space and time.

We see then that thought can neither be a translation nor even a simple continuation of sensori-motor processes in a symbolic form. It is much more than a matter of formulating or following up work already started; it is necessary from the start to reconstruct everything on a new plane. Perception and overt responses by themselves will continue to function in the same way, except for being charged with new meanings and integrated into new systems. But the structures of intelligence have to be entirely rebuilt before they can be completed." ([7], p. 120ff.)

In his three-volume work *Introduction à l'Epistémologie Génétique* published in 1950, Piaget used the term "reflective abstraction" for this constructive process. In the genesis of mathematical theories he believed to have found a model for demonstrating how to conceive of this process. He developed the concept of reflective abstraction using the example of the progressive construction of more and more inclusive number concepts and extended it to mathematical theory formation in general: "The infinitely productive construction of mathematics depends on a twofold movement of operative generalization which creates new structures by means of previous elements and on reflective abstraction or differentiation which obtains its elements from the proper action of the lower stages. The practical coordinations that are at the source of thought may be rudimentary and approximative at the outset; however, they progress to ever more formalized and abstract coordinations, since the abstraction characterizing them is an abstraction that takes its departure form previous operations as well as actions and not from the object." ([8], Vol. I, p. 333)

However, in the beginning Piaget was very cautious in drawing conclusions for epistemology. In his final theoretical conclusions he put the concept of reflection on a par with the concept of construction and stressed the interdependence of the processes of reflection and construction: "Thus, the growth of knowledge consists in a progressive structuration with or without a trend towards stable forms of equilibrium. ... Thus, and this is of the essence, it incessantly vacillates between two extremes which, however, never present themselves in a pure state. The first amounts to a sequence of superimposed constructions which is precisely what causes the whole problem of the connection between the old and the new. However, on the other hand, a point upon which L. Brunschvicg in particular has insisted, 'progress is reflective' and also consists in changing the points of departure by extending the initial structures more and more. Thus, these two processes are not opposed to each other at all, for every construction is more or less reflective and every 'reflection' is constructive to varying degrees. Thus, the genetic mechanisms are situated between these two extremes, naturally with a great variety of forms of construction in turn vacillating themselves between the free or deductive construction and the construction imposed by an empirical discovery; and a great variety of forms of 'reflection' that again vacillate between simply 'becoming concious' of a previous condition that had gone unnoticed and the axiomatic change of the whole." ([8], Vol. III, p. 297). "The abstraction resulting from action and the operative composition provide the key to the double feature of continuity and novelty that is part of the genetic process only on condition that we recognize their reciprocity based on the steady interdependence of reflection and construction. Indeed, the abstraction resulting from previous actions or operations is determined by the new construction and is of significance only in connection with this new structuring that constitutes the form of equilibrium towards which it tends. But in return the new composition soon mirrors that which existed before, and it is in this reflection that the reflective process consists. It is for this reason that the reflection is responsible for the construction and the construction itself permits a reflective perspective which continues the abstraction described before. Consequently, reflection confers a new reality on the elements abstracted from their preceding system: by causing the old element to pass from the unreflected state embedded in its previous context to a reflected and abstract state, the reflection elaborates it by putting it on a different level and attributes a form to it which it did not have until then, because it is the result of new relations and their overall equilibrium. Due to reflection every new construction structures the elements retroactively by relying on those elements that preceded it." ([8], Vol. III, p. 303f.) "... it is not at all absurd to admit that the schematism of the logico-mathematical operations, the conclusion of the explication of the intuitive coordinations, has already been functionally prepared by the sensorimotor schematism without being 'contained' in it as a finished structure. ... The deeper reason for this continuity is that such a steady creation of new forms reflecting on the preceding elements in itself expresses the essential characteristics appropriate for any biological development (physical or mental): the complementary differentiation and integration. Indeed, if reflection is already constructive, it is because it differentiates the structures on which it is based, and if the corresponding construction is in turn reflective, it is because it integrates the preceding elements that had been thus differentiated." ([8], Vol. III, p. 305f.)

In the 1960s Piaget incorporated the concept of reflective abstraction into his "biological epistemology": "... knowledge does not start in the subject ... or in the object ... but rather in interactions between subject and object and in other interactions originally set off by the spontaneous activity of the organ as much as by external stimuli. Starting from these primitive interactions in which the parts played by internal and external factors are indistinguishable (as well as subjectively fused), knowledge is then built up in two complementary directions, while still based on actions and action schemata outside which it has no hold either on the exterior world or on internal analysis.

The first of these directions, which develops much earlier in the animal kingdom because it is the most essential for adaptation to environment, is the conquest of objects or knowledge of environmental data, which will eventually enable the subject to comprehend the exterior world objectively. ... The second direction, which is almost certainly confined to human intelligence, involves becoming conscious of the internal conditions of these coordinations, a development which leads, by means of 'reflection,' to the making of logico-mathematical constructions. These, in the human child, are even to be observed in elementary form before there is any systematized physical knowledge.

From the point of view of the regulatory functions of the nervous system, is there a connection between this second line of development and the general autoregulations of organic life? None whatever, it would seem. And yet if we think of cybernetic models, which are so far, the only ones throwing any light on the nature of autoregulatory mechanisms, we notice at once that they all bring some sort of logic into play (or at least some sort of binary arithmetic, which amounts to the same thing). Moreover, the essential function of logical operations, if they are to function effectively and in a living way, is to set up systems of control and autocorrection. Since logic, in psychological terms, is abstracted by reflection, not about objects, but about general coordinations of action, it would not be very daring to form the opinion that there is in existence a common fund of regulatory mechanisms which belong to the nervous regulatory mechanism which belong to the nervous regulatory systems in all their forms, and of which the general coordinations of action are only one manifestation among others. And since the nervous system is not a State within a State but the differentiated product of organic and morphogenetic coordinations, there is no a priori reason to set an initial limit to regressive analysis.

To sum up: to suppose, as our ruling hypothesis does, that cognitive functions are reflections of the essential mechanisms of organic autoregulation is a perfectly valid proposition." ([11], pp. 27–29)

The present comparison of Piaget's reflective abstraction and Hegel's reflection is based on the theoretically developed concept of reflective abstraction which Piaget outlined in his works *Biology and Knowledge* ([11], first published in 1967) and *The Equilibration of Cognitive Structures* ([13], first published in 1975) in particular. In the meantime a work announced by Piaget in the latter publication has been published posthumously in which he presents his complete theory of reflective abstraction once again ([12]). Since this work did not appear in time for the present essay, it could not be taken into account here.

3   *On the Structure of the Process of Reflective Abstraction*
Following are the two most important, somewhat abbreviated references providing the context for Piaget's theory of the structure of the process of reflective abstraction: "We are ... compelled to think of the construction of logico-mathematical structures in the form, not of a development that is integrated unpredictably with external elements, but as a kind of endogenous evolution going forward in stages. These stages are of such a kind that the combinations characteristic of any one of them will be new as combinations, yet based entirely upon the elements already present in the preceding stage." ([11], p. 318f.) "From the psychological angle ... the abstraction process is very characteristic of logico-mathematical thought and differs from simple or Aristotelian abstraction. In the latter, given some external object, such as a crystal and its shape, substance, and color, the subject simply separates the different qualities and retains one of them—the shape, maybe—rejecting the rest. In the case of logico-mathematical abstraction, on the other hand, what is given is an agglomeration of actions or operations previously made by the subject himself, with their results. In this case, abstraction consists first of taking cognizance of the existence of one of these actions or operations, that is to say, noting its possible interest, having neglected it so far ... Second, the action noted has to be 'reflected' (in the physical sense of the term) by being projected onto another plane—for example, the plane of thought as opposed to that of practical action, or the plane of abstract systematization as opposed to that of concrete thought (say, algebra versus arithmetic). Third, it has to be integrated into a new structure, which means that a new structure has to be set up, but this is only possible if two conditions are fulfilled: (a) the new structure must first of all be a reconstruction of the preceding one if it is not to lack coherence and congruity; it will thus be the product of the preceding one on a plane chosen by it; (b) it must also, however, widen the scope of the preceding one, making it general by combining it with the elements proper to the new place of thought; otherwise there will be nothing new about it. These, then, are the characteristics of a 'reflection,' but now we are taking the term in the psychological sense, to mean a rearrangement, by means of thought, of some matter previously presented to the subject in a rough or immediate form. The name I propose to give this process of reconstruction with new combinations, which allows for any operational structure at any previous stage or level to be integrated into a richer structure at a higher level, is 'reflective abstraction' *[abstraction réfléchissante]*.

This will explain why logico-mathematical construction is, properly speaking, neither invention nor discovery; proceeding by means of reflective abstractions, it is a construction in the full sense of the word; that is, it is productive of new combinations. But such combinations can only be brought about by a combinatorial attainable by calculation at the levels below and previous to the construction of the new structure, for the latter, by retroactive effect ... demands a reflective rearrangement of the preceding elements, and achieves a synthesis which outstrips the original structures and thereby enriches them." ([11], p. 320f.) "Reflective abstraction includes two indissociable activities. One is 'reflecting' or projecting onto a higher level something borrowed from a lower level. ... The other is more or less conscious 'reflexion' in the sense of cognitive reconstruction or reorganization of what is transferred. Reflective abstraction does not, however, just utilize a succession of hierarchical levels organized by some other means. Abstraction itself engenders the levels by means of alternating reflecting and reflexion so intimately linked to the perfecting of regulations that only a single general mechanism is involved.

a. Two points must be kept in mind. To begin with, every regulation moves toward both retroactive and anticipatory effects. That is why the amplitude of corrections and reinforcements varies. Second, anticipations depend on indexes announcing what is to come. The first forms of such indexes are very precocious and can even be appreciated in the neonate's nursing during the first week of life. They are coordinated according to a law of 'transference' or, better yet, 'recurrence.' For example, *a* announces *x,* then *b* preceding *a* announces *a* and *x,* then *c,* even earlier, announces *b, a,* and *x,* and so on. (Cf. the auditory indexes announcing a baby's feeding.) This organization of indexes constitutes a new level in relation to the initial regulations operating as corrections or reinforcements after the fact. For example, in the development of the logic of relations, seriations are first accomplished by putting in order several groups of two or three elements. Eventually these evolve into systematic operatory seriation of a whole set of elements. Between the two extremes there occurs a level where success is achieved only through trial and error. At that level, corrections after the fact are gradually coordinated, thanks to anticipatory and retroactive progress that continues up to the point that such corrections become rarer and rarer and finally become unnecessary. This is the significance of the intermediary representative level seen between simple trial-and-error behavior and programmed operation. Regulations obviously play a role in reflection onto a level brought about by their coordination.

b. On each new level, what we have called 'reflexion' gives rise to new equilibrations through regulation of indexes, and so forth. These regulations of slightly higher rank naturally extend the regulations of the starting level through 'reflective abstraction.'

c. It is a matter of course that the higher system constitutes a regulator exercising control over the regulations of lower level. That is the case wherever reflexion occurs, because reflexion is a reflexion 'on' what has been acquired previously. Reflexion thus represents the prototype of a regulation of regulations, since it is itself a regulator and takes control over whatever is inadequately controlled by previous regulations. That is what happens with active regulations or when conceptionalizations begin to direct action. Moreover, it occurs anew at every stage, every developmental step being characterized by a new reflecting and a new reflexion.

d. The formation of regulations of regulations, whether expressed in those terms or in terms of the reflectings and reflexions belonging to reflective abstraction, constitutes a very general and, apparently, paradoxical process. Whatever language is used, the fact is that every cognitive system depends on the system that follows it for control and for completion of its own regulatory function. It is in this peculiar situation that autoregulation gradually comes into being. Apparently such dependence is a necessary condition for the formation of systems where there is an interplay of differentiations and integrations such that totalities serve as regulators through their action on subsystems and particular schemes, as was described in section 4. ...

This collaboration, if not identity, of regulations and reflective abstraction, both evolving level by level, accounts for the central process of cognitive development, which is to say, for the indefinite formation of operations on operations. If, as we have just suggested, regulations of regulations exist and if, as we have shown elsewhere, there likewise exist reflexions to different powers, it goes without saying that it will always be possible to apply new operations to any given operatory system. These new operations may be drawn from other systems or may represent operations from the same system that have been raised to a higher power." ([13], pp. 29–31)

[4] *On the Differentiation of Reflective Abstraction and Other Forms of Self-Reference*
This interpretation is based on the following reference: "We must distinguish quasi-automatic regulations from active ones. The first are seen in simple sensorimotor actions where means are subject to little variation except in details of accommodation or adjustment. For example, when trying to grasp an object, the subject has to take account of the distance to the object or of how far his hand must open. By contrast, active regulation comes into play when the subject has to change the means he uses or when he hesitates among several different ways of doing something. Such behavior is seen, for example,

when a child constructs a house of cards. Whether changing a decision already made or trying to decide what to do next, the child must make a choice. Although the frontier between automatic and active regulations is difficult to define, it is important to distinguish between them because automatic regulations do not, by themselves, lead to conscious awareness. Active regulations, on the other hand, do; and because they do, they lie at the source of the representation or conceptualization of material actions. Eventually they will be subordinated to a control system of higher order, which is the beginning of regulation to the second degree." ([13], p. 18)

By according an important role to the material means of actions as far as reflective abstraction is concerned, Piaget is coming closer to materialistic explanations which assume the close association of the use of tools and the development of consciousness. In Piaget's earlier publications this argumentation is, however, not to be found. In his *Biology and Knowledge* he limits himself exclusively to emphasizing the correlations between organic and cognitive regulations and to describing, in phenomenological terms, the transition from the irreversibility of sensori-motor schemata to the reversibility of logico-mathematical operations and to the conceptual identity based on this reversibility. The corresponding passage in the book mentioned above reads as follows (abbreviated): "Whereas elementary cognitive regulations can thus be shown to be of the same type as organic ones, the higher kinds of regulation, which are in fact operations, are of a different form, though they constitute the result of a complete transition from ordinary regulations ... Before going any farther, we should point out that there is nothing sudden about this transition and that the earliest representational, though preoperational, regulations make a transition between sensorimotor regulations and operations. ... But we must go even farther and see a higher form of regulation in the operation as such—regulation in which retroactive control has become complete, rigorous reversibility. ... It may be said that such a regulation presupposes conservation. But that is not true at all, for it is precisely the reversibility which brings about the conservation, as can easily be demonstrated in the psychological field: reversibility is the very process from which conservation is produced, and this process varies in degrees of approximation as long as it remains in the state of a regulation in the usual sense of the word. This explains the intermediary responses obtained between nonconservation and conservation. On the other hand, the child's arguments when he is trying to justify some conservation that may have become self-evident to him are of the reversibility type ... or of the identity type. ... Identity teaches the child nothing new; he knew all the time, whatever his age, that nothing had been added, but that did not prevent his concluding that there was nonconservation. In fact, identity actually becomes an argument in itself as soon as it is subjected to reversibility. Thus, it is certainly reversibility that entails conservation and not the reverse ... However, since this interpretation consists of viewing operations as regulations of the higher type, hence, as the final state that can be reached by ordinary regulations when their approximating retroactions lead to complete reversibility, it truly holds a deep biological significance, going much farther than the limited scope of formal isomorphisms. If regulation of the lower or ordinary type is a process for correcting or modifying errors, then operational regulation is seen as a process for precorrecting, avoiding, or eliminating errors, which is something much greater. Indeed, an operational deduction is not subject to any error if it is in conformity with the laws of its structure. ... An error in logic or in mathematics is the result of an individual slip, of factors of attention, memory, and the like, which have nothing to do with the structure being used, whereas a perceptual structure, for example, has a probabilistic aspect that excludes any kind of composition other than by approximate regulations.

Thus, it can be seen that during the gradual extension of the knowledge 'environment,' that is, the sum total of external objects on which the intelligence is brought to bear, and during the gradual dissociation of forms and their contents, that is, the elaboration of abstract and conceptual 'forms' as opposed to perceptual or sensorimotor forms, and, a fortiori, as opposed to the material forms of the organism, the regulations whose task it is to control cognitive exchanges with this environment, that is, to organize experience in terms of deductive frames, will reach a level of precision never found in elementary regulations. Instead of being restricted to corrections after the event that bear only on the results of processes or behaviors, or to an approximate guidance of anticipations that are never more than probabilistic, they carry out a function of precorrection in the proper sense of the word." ([11], pp. 208–211)

5   *On the Teleological Mystification of Reflection in Hegel and Piaget*
In its entirety the quote reads as follows: "... the *Notion* is to be regarded in the first instance simply as the third to *being* and *essence,* to the *immediate* and to *reflection.* Being and essence are so far the moments of its *becoming;* but it is their *foundation* and *truth* as the identity in which they are submerged and contained. They are contained in it because it is their *result,* but no longer as *being* and *essence.* That determination they possess only in so far as they have not withdrawn into this their unity.

Objective logic therefore, which treats of *being* and *essence* constitutes properly the *genetic exposition of the Notion.*" ([5], p. 577)

Similarly, if Piaget's genetic epistemology is considered to be the genetic exposition of the developed intelligence and its logico-mathematical forms of thought, a further agreement emerges. In Hegel the "objective logic" is a logic of the becoming of the notion. In the "subjective logic," the "doctrine of the notion," the result of this becoming is described. However, this result, the notion, is to be simultaneously the "foundation and truth" of this becoming. Hegel expresses this by means of a seemingly paradoxical formulation. The genesis of the notion is described as a "withdrawal into unity." This phrasing is reminiscent of Piaget's regulation of the constructive process of reflective abstraction by the result of that process: "The formation of regulations, whether expressed in those terms or in terms of the reflectings and reflexions belonging to reflective abstraction, constitutes a very general and, apparently, paradoxical process. Whatever language is used, the fact is that every cognitive system depends on the system that follows it for control and for completion of its own regulatory function." ([13], p. 31)

In my view this idea shared by Piaget and Hegel represents an idealistic mystification of the relationship between relative and absolute truth. For as soon as the process of the development of cognitive structures is conceived as independent of its material foundation and as soon as the existence of an autonomous logic of development of this process is assumed—and, as a matter of fact, both Piaget and Hegel do this—then the result of this cognitive process, the concept reflecting its material object, must at the same time appear to be the foundation of this process. The process of the development of cognitive structures acquires an immanent a priori criterion of truth not attained until the end of the cognitive process: "Now observation and experiment show as clearly as can be that logical structures *are* constructed, and that it takes a good dozen years before they are fully elaborated; further, that this construction is governed by special laws, laws which do not apply to any and every sort of learning. Through the interplay of reflective abstraction ... which furnishes increasingly complex 'materials' for construction, and of equilibration (self-regulation) mechanisms, which make for internal reversibility, structures—*in being constructed*—give rise to that nescessity which a priorist theories have always thought it necessary to posit at the outset. Necessity, instead of being the prior *condition* for learning, is its *outcome.*" ([10], p. 62)

6   *On Relating Hegel's System to the Phases of Cognitive Development*
However, the process of relating parts of the system of Hegel's logic to phases of the genesis of cognition which can be identified in the context of developmental psychology, on

which I am basing the thesis of the comparability of Piaget's reflective abstraction and Hegel's reflection, must be considered merely a limited interpretation strategy. This necessarily follows from the fact that Hegel conceived of his dialectical logic as an integrated approach to both, genesis of cognition and ontology. For example, Hegel asserts: "Being is the immediate. Since knowing has for its goal knowledge of the true, knowledge of what being is *in and for itself*, it does not stop at the immediate and its determinations, but penetrates it on the supposition that at the back of this being there is something else, something other than being itself, that this background constitutes the truth of being. ... When this movement is pictured as the path of knowing, then this beginning with being, and the development that sublates it, reaching essence as a mediated result, appears to be an activity of knowing external to being and irrelevant to being's own nature. But this path is the movement of being itself." ([5] p. 389) Thus, Hegel's objective logic must not merely be understood as the theoretical representation of the genesis of the logical forms of cognition, but must simultaneously be interpreted as a system of logical categories as well. Therefore, the implicit assumptions concerning the relationship of the cognitive process and the real movement of the object also require an explication.

If, however, an interpretation of the objective logic in terms of genetic epistemology is considered permissible at all, then the alleged correlation goes beyond the connections between reflective abstraction as the structure of the process of the development of logico-mathematical structures and reflection as the most general processing structure of Hegel's logic of essence detailed above and provides numerous interpretive aids.

Thus, the structure of the process in the logic of being which Hegel defines as "transition," that is, as the structure of a development of categories determined by external moments, corresponds to the development of sensori-motor schemata as theoretically reconstructed by Piaget as the result of the "assimilation" of new objects to these schemata and the "accommodation" of the schemata to these new objects. Hegel differentiates the structure of the processes in the logic of being, the "transition," and the structure of the processes in the logic of essence, the "reflection," as follows: "In the sphere of Essence one category does not pass into another, but refers to another merely. In Being, the form of reference is purely due to our reflection on what takes place: but it is the special and proper characteristic of Essence. In the sphere of Being, when somewhat becomes another, the somewhat has vanished. Not so in Essence: here there is no real other, but only diversity, reference of the one to *its* other. The transition of Essence is therefore at the same time no transition: for in the passage of different into different, the different does not vanish: the different terms remain in their relation. ... In the sphere of Being the reference of one term to another is only implicit; in Essence on the contrary it is explicit. And this in general is the distinction between the forms of Being and Essence: in Being everything is immediate, in Essence everything is relative." ([3], p. 206) In the Hegelian concept of transition a mediating activity, albeit unconscious, merges with the immediate certainty of sensible qualities into a unity. Thus, the concept of transition corresponds to the activity of sensori-motor intelligence which consists in "co-ordinating successive perceptions and (also successive) overt movements," ([7], p. 120) and which, within this process, develops not merely cumulatively, but in its categorial form (cf. the lengthy quotation in note 2).

Furthermore, categories of the logic of being correspond to numerous concepts of Piaget's theory related to figurative and sensori-motor aspects of cognition; and the categories of the logic of essence correspond to logico-mathematical concepts in the nonformal form not yet abstracted from reality as characterized by Piaget for the stage of concrete operations.

The parallel between the systematic structure of Piaget's genetic epistemology and Hegel's objective logic is, however, not consistent. The following is an important difference: Hegel includes three basic categories in the logic of being—that is, "quality," "quantity," and "measure." According to Piaget's theory, however, quantity and measure are results of reflective abstraction. As such they ought to be part of the logic of essence. In Hegel's classification which, incidentally, is not based on empirical considerations concerning the ontogenetic sequence of the categories, the theoretical tradition persists of perceiving space and time as categories of intuition and of basing mathematics on the forms of intuition (cf. the Kantian system). Hegel claims: "Mathematics is concerned with the abstractions of time and space. But these are still the object of sense, although the sensible is abstract and idealised." ([3], p. 34) According to Piaget's experiments, however, only a "practical concept of space" and a "practical concept of time" are developed at the stage of sensori-motor intelligence represented by the material actions and their sensori-motor coordination and lacking precisely the structural details of the developed concepts of space and time characterized by quantity and measure.

7 *On the Concepts of "Determination" and "Negation"*
The concept of "determining" as an operational concept closely associated with the dialectical development in Hegel's system is, however, of more general significance and really demands a more thorough explication of its relationship to the concept of construction in Piaget's theory of the development of cognitive systems. The following comment may serve as an illustration.

The concept of "determining" applies to the various parts of Hegel's logic. It was already pointed out in note 6 that the determining in the logic of being conceived as "transition" is distinguished from the determining in the logic of essence conceived as "reflection." The transition of the logic of being to the logic of essence is, however, not characterized by the addition of thought to intuition; being determines itself as essence. The process of determining changes itself by means of itself. It is conceived as self-determination, and in this way passes through the processing forms of "transition" and "reflection." This, therefore, represents a theory of the transition to thought and its logical forms similar to Piaget's theory in which thought—based on the biological theory of development—is understood as the highest developmental stage of differentiating forms of self-regulation.

Hegel designates the most general form of transition in the logic of being as "becoming." "*Transition* is the same as becoming except that in the former one tends to think of the two terms, from one of which transition is made to the other, as at rest, apart from each other, the transition taking place *between* them. Now wherever and in whatever form being and nothing are in question, this third must be present; for the two terms have no separate subsistence of their own but *are* only in becoming, in this third." ([5], p. 93) Becoming as the most general form of transition might be compared with the transition of one sense perception to another, and Hegel (like Piaget with his concept of "sensori-motor schemata" and their character essential for the organization of perceptions) already perceives this process as an active process, an activity of thought (albeit not yet conscious of itself). Hegel, therefore, further elaborates on the concept of becoming: "Becoming is the first concrete thought, and therefore the first notion." ([3], p. 167) In this sense becoming is not just a given fact, but is subject to development and changes in the process of its own unfolding; becoming is simultaneously "determination" in the above sense of the term: "As the first concrete thought-term [i.e., determination by thought], Becoming is the first adequate vehicle of truth" ([3], p. 168), a determination that brings thought closer to the truth.

Hegel designates the transition as an act of determining as "negation" saying: "The foundation of all determinateness is negation." ([3], p. 171) Thus, negation is the most elementary precondition of any form of mediation. Hegel conceives of mediation as follows: "To mediate is to take something as a beginning and to go onward to a second thing; so that the existence of this second thing depends on our having reached it from something else contradistinguished from it." ([3], p. 20) Thus, following Hegel's logic, the quality of a sense perception is already the result of a mediating process and "the relationship between 'immediacy' and 'mediation' in consciousness" is such "that, though the two 'moments' or factors present themselves as distinct, still neither of them can be absent, nor can one exist apart from the other." ([3], p. 20) For what Hegel calls "being-for-self," the "completed Quality," ([3], p. 178) is the being that as "negation of the negation, is restored again." ([3], p. 176) Without this mediation by the negation of the negation the determination of quality would merely be "a finite determinateness—a somewhat in distinction from an other" and not a determinateness that "contains distinction absorbed and annulled in itself." ([3], p. 179) However, because immediacy and mediation are not separated on the level of the "being-for-self," a self-consciousness is associated with the "completed Quality" of sense perception which is not yet separated from the latter. "The readiest instance of Being-for-self is found in the 'I.' We know ourselves as existents, distinguished in the first place from other existents, and with certain relations thereto. But we also come to know this expansion of existence (in these relations) reduced, as it were, to a point in the simple form of being-for-self. When we say 'I,' we express the reference-to-self which is infinite, and at the same time negative. Man, it may be said, is distinguished from the animal world, and in that way from nature altogether, by knowing himself as 'I': which amounts to saying that natural things never attain a free Being-for-self, but as limited to Being-there-and-then, are always and only Being for an other." ([3], p. 179) Thus, sense perception as the source of cognition is understood by Hegel as active perception unfolding in terms of "negation" and "negation of the negation," and in this developmental process which Hegel constructively conceives as "determination," the egocentrism of the sensori-motor period of intellectual development appears to follow by necessity from the as yet undivided unity of the sensory given and mediating activity.

Thus, in Hegel's view negation is not just a logical category in the usual sense. It defines a more elementary act of determination as the determination by means of reflection. In his late book *The Equilibration of Cognitive Structures,* Piaget operates with a similarly generalized concept of negation. In Piaget "assimilation" to a cognitive schema is the precondition for all cognition, and at the stage of sensori-motor intelligence the development of these schemata still takes place exclusively by means of "accommodation" to the changing external conditions (analogous to Hegel's concept of "transition" as a processing determined by external events). In order to give this process of "accommodation" a more precise theoretical definition, Piaget introduces a generalized concept of negation which is not identical with the logical negation arrived at by means of reflective abstraction. "Negation" designates the negative feedback by means of which a schema of a regulation is accommodated to the conditions modified by a disturbance. In contrast, Piaget considers positive feedback which reinforces a certain perceptual quality as more complicated and understands it as the "negation of a negation," similar to Hegel's definition of the being-for-self: "As for positive feedback, it is a form of reinforcement. For that reason it appears not to involve negation of any sort. That is not entirely the case, however. In the cognitive domain, positive reinforcement differs from simple assimilatory activity in the following way. The latter aims at generalizing what can be assimilated into a scheme ... whereas the former tends to reinforce simple assimilations by filling in lacunae (weaknesses, etc.) that prevent a goal from being reached or a system from being stabilized. A lacuna is a negative factor, and to eradicate it is to suppress something inadequate. It is, then, not just playing with words to see the negation of a negation in positive feedback. For example, suppressing the spatiotemporal distance separating the subject from a goal negates something that is negative vis-à-vis achieving that goal." ([13], p. 21)

ACTION AND COGNITION 23

A comparison between Hegel and Piaget not limited to the logic of reflection, but including the logic of being would have to proceed from this concept of determination or construction that can be characterized in terms of "negation."

8   *On the Metaphor of the Mirror for the Concept of Reflection*
In explaining this duplication in the reflection Hegel, just like Piaget, uses the metaphor of the mirror, but in a different way, and it is in this difference that the dissimilarities of Piaget's and Hegel's concepts of reflection become visible. Piaget calls attention to the double meaning of the term reflection: reflection as mirroring and as contemplation. In reflection as mirroring he sees the metaphor for one of the aspects of reflection, that of the duplication of reality in the reflecting consciousness. In the reflection as contemplation, on the other hand, he sees the metaphor for the constructive aspect of this duplication in consciousness (see also note 3). Piaget believes that both aspects of reflection—they might be defined as the empirical and the constitutive aspects of reflection in the cognitive process—can be separated from each other and are defining distinct cognitive modes: empirical knowledge and logico-mathematical knowledge.

Hegel, on the other hand, believes that the double meaning of the term reflection stands for the same thing, that mirroring as well as contemplation are characterized by the evolving unity of immediacy and mediation which is his theoretical characterization of reflection: "The point of view given by the Essence is in general the standpoint of 'Reflection.' This word 'reflection' is originally applied, when a ray of light in a straight line impinging upon the surface of a mirror is thrown back from it. In this phenomenon we have two things,—first an immediate fact which is, and secondly the deputed, derivated, or transmitted phase of the same.—Something of this sort takes place when we reflect, or think upon an object; for here we want to know the object, not in its immediacy, but as derivative or mediated." ([3], p. 208)

9   *On the Differentiation of the Cognitive Content into the Components of the Sensory Given and the Mediating Activity*
To say that reflection is associated with a differentiation of the cognitive content into the moments of the sensory given and the mediating activity, represents a considerable simplification of both Hegel's logic and Piaget's genetic epistemology. This simplification can only be justified in terms of the proposed comparison: This aspect, relevant for both theories but trivial in and of itself, makes the difference between the "reflection" of Hegel's logic and the "reflective abstraction" of the genetic epistemology particularly clear. However, both concepts of reflection are only inadequately described by the assertion that the differentiation into the sensory given and mediating activity is associated with reflection. Some further comments may be useful.

As was already pointed out in note 1, an interpretation proceeding from the real actions of the subjects and referring thought operations as forms of consciousness back to them, is unacceptable to Hegel. Accordingly, Hegel does not identify the immediacy with its sensori-material origin and mediation with sensori-material activity. For that reason there is no comparable formulation in his logic except in comments and explanatory passages. However, leaving aside this specific difference of the formulation I chose with regard to Hegel's argumentation, it still does not do justice to Hegel's concept of reflection in another respect. With the pair of concepts undifferentiated versus differentiated applied to the sensory givens and mediating activity the process of development is captured only externally and the decisive moment of self-determination in the development of being into essence is omitted. As a matter of fact, sensory given and mediating activities already operate in the logic of being in a number of different ways; only this difference cannot yet be reflected in any categories. Thus, it can be said that the repressed relationship of material being and ideal consciousness reappears in a mystified form in the relationship of the concepts of Hegel's logic operating throughout (e.g., the concept of negation) to the gradually developed categories. By means of "external reflection," Hegel himself has to distinguish that which, at the level of the logic of being, is purported to still be undifferentiated, for example, when he asserts the lack of differentiation of "being and negation," in order to develop the categories of "determinate being" and "limit": "In [determinate] Being (determinate there and then), the determinateness is one with Being; yet at the same time, when explicitly made a negation, it is a Limit, a Barrier. ... In Being-there-and-then, the negation is still directly one with the Being, and this negation is what we call a Limit (Boundary). A thing is what it is, only in and by reason of its limit" ([3], p. 172f.); the same is true for the lack of differentiation of "being" and the "negation of the negation" in the development of the concept of the "being-for-self": "... essentially relative to another, somewhat is virtually an other against it ... what is altered is the other, it becomes the other of the other. Thus Being, but as negation of the negation, is restored again: it is now Being-for-self." ([3], p. 176) In this case, being and negation must therefore be assumed to be simultaneously differentiated and undifferentiated: differentiated by external reflection and undifferentiated in the process of the development of the logical categories. This makes sense, if it is taken as the theoretical expression of the contradiction between actions with which a subject adjusts his or her own actions to the objects to a considerable degree, and a consciousness reflecting these actions in cognitive schemata, that is, a consciousness in which these actions and objects are, nevertheless, not yet separated into logical categories and which accordingly, is not capable itself of providing a theoretical explanation of this contradiction.

The lack of differentiation of the sensory given and the mediating activity in the sensori-motor period is expressed in Piaget's theory with a comparable degree of sophistication. The theory is, however, more simple in the sense that Piaget consistently makes a conceptual distinction between the material actions and their ideal correlates in the cognitive schemata of consciousness.

The following is a brief summary of the general role of "reflective abstraction" in his theory of intelligent behavior: According to Piaget, cognitive activity as an active process based on object-related action encompasses two inseparably linked, opposing moments which he calls "assimilation" and "accommodation." The concept of assimilation "... expresses the fundamental fact that any piece of knowledge is connected with an action and that to know an object or a happening is to make use of it by assimilation into an action schema" ([11], p. 6). "We shall apply the term 'accommodation' (by analogy with 'accommodates' in biology) to any modification produced on assimilation schemata by the influence of environment to which they are attached." ([11], p. 8) In the successful process of cognition assimilation and accommodation constitute a "dynamic equilibrium," in which the stabilization and generalization of the behavioral schema by means of the assimilation of ever new objects and the differentiation and expansion of the behavioral schema by means of accommodation to new circumstances maintain their equilibrium. A preponderance of assimilation leads to the stagnation of the cognitive process, a preponderance of accommodation results in the loss of the identity of the behavioral schema, a dissolution of existing cognitive structures. At the developmental stage of sensori-motor intelligence the equilibrium is achieved by means of regulations, with accommodation taking place by means of "compensations" in the form of negative and positive feedback. At the developmental stage of concrete operations and at the subsequent developmental stages, that is, at the stages of conscious mental activity, the reorganizations participate in the creation of the equilibrium by means of reflective abstractions.

Thus, the total lack of differentiation of the sensory given and mediating activity is true only for the first innate schemata of assimilation of sensori-motor intelligence. "It seems ... that only global observables exist at the initial levels of sensorimotor development. ... They are not object-observables, since objects are not dissociated from the body proper, for example, as objects to suck. Nor are they observables relative only to the subject's actions, since the subject does not understand them as such, does not perceive them in detail, and is completely ignorant of the self as well as of his body as something that belongs to him. In sum, he does not conceive himself to be a subject any more than he conceives objects to be permanent or localizable." ([13], p. 70) To the extent, however, to which the compensatory regulations of sensori-motor intelligence are developed, a separation of object and action and a coordination of ends and means occurs at the level of practical behavior. "The structure of compensatory regulations makes them tools for forming negations. ... Nothing about such mechanisms is conscious, however, since at the beginning they involve only the negative dimensions of action, and what is observed of action is conceived in terms of differences rather than in terms of opposites or negations. Nevertheless, these in some way motoric, practical negations have great importance because they are the source of the conceptualized negations that come later. ... This reflecting of practical into conceptual negations is an essential aspect of the constructive process tied so closely to the interplay of regulations. We have called that process, studied elsewhere, reflective abstraction." ([13], p. 28f.)

Thus, the lack of differentiation of the sensory given and mediating activity in the phase of sensori-motor intelligence only holds for the logical forms of the cognitive structures of consciousness which guide behavior. "Sensorimotor intelligence consists of the direct coordination of actions without any representation or thought." ([11], p. 7) The transition to logical thinking takes place presupposing the contradiction between the separation of object and action in terms of practical activity and their lack of differentiation in terms of the conceptualization of this activity in consciousness. It goes without saying that this contradiction itself is reflected in consciousness. According to the Piagetian theory, too, the sensory given and mediating action must, in this sense, be understood as both differentiated and undifferentiated in consciousness at the developmental level of sensori-motor intelligence. External reflection on the thinking of the child can perceive them as differentiated and realize that, in this thinking, they are undifferentiated in terms of logical categories. Piaget describes this contradictory situation of consciousness as a situation of the repression of facts for logical reasons: "In the first place, the part of an action the subject retains in his conceptualization of it, or in other words what he centers on and overestimates, is what he was able to assimilate and to understand from the beginning. ... In the second place, however, an essential part of the action, although well executed, escapes this taking of consciousness. ... elements are not neglected simply because the child cannot pay attention to everything at once. There is a better reason for what happens. The missing facts have been set aside because they contradict the child's habitual way of conceptualizing some part of the situation. ... In the third place, it must be pointed out that the elements left out when consciousness is taken are subject to a kind of cognitive repression. ... In such situations the subject has, in fact, understood something of the idea that he refuses to accept in his conceptualization, but his understanding has been achieved in terms of action. In other words, it proceeds from a sensorimotor scheme and not from an idea. We are obliged to conclude, therefore, that the scheme, whose existence as a scheme of action is undeniable, is eliminated from the conscious conceptualization by a sort of active rejection or repression because it is incompatible with other concepts the subject believes in." ([13], p. 118f.)

[10] *On Piaget's Thesis of the Fundamental Difference Between Empirical and Logico-Mathematical Cognition*

The Piagetian concept of reflective abstraction inevitably results in stipulating a cognitive activity other than that determined by reflective abstraction, that is, one characterized by formal abstraction. The assumption, however, that both cognitive modes are independent of each other without which Piaget's biological definition of reflective abstraction would be untenable, leads to theoretical difficulties that can hardly be resolved; for Piaget must postulate the unity of action and cognition, if his most general theoretical assumptions are to be true, of empirical cognition as well. Thus, Piaget conceives of the formal abstraction for empirical cognition more specifically as a simple act of assimilation to the logico-mathematical structures of reflective abstraction: "The third main type of knowledge is that which begins by learning and wich achieves its highest expression in what is commonly known as experimental knowledge. In this context we are referring to 'physical' experiment as against logico-mathematical experiment ... simply in order to express the fact that the information, in this case, is obtained from the object and not by action ... Experimental knowledge constitutes a considerable part of man's cognitive work, and it is quite as important as logico-mathematical knowledge. Being exogenous in origin ... it is something quite distinct from logico-mathematical knowledge and yet is always inextricably bound up in it ... when we turn to physical experiments at however primitive a level, the necessity for such a framework is extremely significant, since it demonstrates the impossibility of 'pure' experiment in the sense of a direct and immediate contact between subject and objects. To put it another way: any kind of knowledge about an object is always an assimilation into schemata, and these schemata contain an organization, however elementary, which may be logical or mathematical." ([11], pp. 333–335, shortened) This approach, however, inevitably leads to the well-known epistemological aporia that there is no reasonable explanation for the congruity of concrete sensory contents and abstract logico-mathematical forms of cognition.

Consequently, in his use of language Piaget vacillates between the extremes of assuming that the natural material environment of consciousness, whose existence Piaget does not question, is chaotic or determined by natural laws. On the one hand he argues: "... the union of mathematics and physics is not one of sign and meaning but one of structuring activity and datum, which, without this activity, would remain chaotic." However, in the same context he characterizes the cognitive activity of the physicist: "... the physicist constantly acts, and the first thing he does is to transform objects and phenomena in order to get at the laws validating these transformations" ([11], p. 338). And elsewhere: "To know is to asssimilate reality into systems of transformations. To know is to transform reality in order to understand how a certain state is brought about." ([9], p. 15) However, how can this be possible if the transformation system is not derived from laws which determine the change of state? (On this point see the criticism of [6].)

Piaget himself has pointed out another theoretical possibility without convincingly refuting it. Because of the significance I attribute to it, it is cited in full: "Since we have just seen that physical and experimental knowledge cannot possibly be established without some structuration and logico-mathematical framework, the simplest way to explain the harmony between these frameworks and their contents is, of course, to say that the contents act in return upon the frameworks and hence that the adaptation under discussion is carried out by means of progressive gropings, in other words, by an equilibration between the assimilation of the contents into the frameworks and the differentiating accommodations of the frameworks to the contents.

Such a concept, however, is tantamount to saying that logico-mathematical structures are not derived solely from the action of the subject upon objects but also from the objects themselves, since physical experiment will gradually bring about modifications in them. This is, of course, possible, and, if it were so, my interpretation would need some rather basic revision. It would just be one of those unfortunate things that happen, but that is not the point. Only we must weigh the consequences carefully: to say that logico-mathematical structures undergo modification under the influence of physical experiment would simply mean that there is no difference, except a basic one of degree, between physics and mathematics, in which case both would be assimilated into a general knowledge which might be called logico-experimental.

Therefore, the only proper way to solve this problem is by epistemological analysis, making use especially of the historico-critical method, which is the only one capable of judging the real relationship between physicists and mathematicians (on condition, of course, that the task is entrusted to professionals in the respective disciplines and not to philosophers, who think themselves capable of judging such cases without any technical training, either logico-mathematical or in physics)." ([9], p. 242f.)

BIBLIOGRAPHY

[1] Erdei, L.: *Der Anfang der Erkenntnis. Kritische Analyse des ersten Kapitels der Hegelschen Logik.* Budapest: Akademía Kiadó, 1964.
[2] Furth, P.: "Arbeit und Reflexion." In Furth, P. (Ed.): *Arbeit und Reflexion. Zur materialistischen Theorie der Dialektik—Perspektiven der Hegelschen Logik.* Köln: Pahl-Rugenstein, 1980, pp. 70–80.
[3] Hegel, G. W. F.: *The Logic of Hegel.* Trans. from The Encyclopedia of the Philosophical Sciences by W. Wallace. 2nd ed., Oxford: Clarendon Press, 1892.
[4] Hegel, G. W. F.: *Frühe politische Systeme.* Frankfurt a.M.: Ullstein, 1974.
[5] Hegel, G. W. F.: *Hegel's Science of Logic.* Trans. by A. V. Miller. Reprint ed., Atlantic Highlands, NJ: Humanities Press International, 1989.
[6] Jahnke, N., Mies, T., Otte, M. & Schubring, G.: "Zu einigen Hauptaspekten der Mathematikdidaktik." In *Schriftenreihe des IDM, 1,* 1974, pp. 5–84.
[7] Piaget, J.: *The Psychology of Intelligence.* London: Routledge & Kegan Paul, 1950a.
[8] Piaget, J.: *Introduction à l'Epistémologie Génétique.* 3 vols., Paris: Presses Universitaires de France, 1950b.
[9] Piaget, J.: *Genetic Epistemology.* New York: Columbia University Press, 1970a.
[10] Piaget, J.: *Structuralism.* New York: Basic Books, 1970b.
[11] Piaget, J.: *Biology and Knowledge: An Essay on the Relations Between Organic Regulations and Cognitive Processes.* Chicago, IL: University of Chicago Press, 1971.
[12] Piaget, J.: *Recherches sur l'Abstraction Réfléchissante.* 2 vols., Paris: Presses Universitaires de France, 1977.
[13] Piaget, J.: *The Equilibration of Cognitive Structures.* Chicago, IL: University of Chicago Press, 1985.

CHAPTER 2

REPRESENTATION AND MEANING
[1980]

EXAMPLES OF COGNITIVE STRUCTURES

The concept of cognitive structure is central to those theories of thought that emerged as a result of the critical evaluation of the studies on developmental psychology and epistemology of the Geneva group around Jean Piaget. It denotes developmental stages of logico-mathematical structures of thinking which—based on empirical observations and experiments—were genetically reconstructed in this area of investigation.[1] The considerations to be presented here deal with a particularly controversial question: What is the importance of the material representation of cognitive structures with regard to their ontogenetic and phylogenetic development and how should the representation be defined empirically and theoretically.

Since the concept of cognitive structure is used ambiguously, I consider it to be necessary to preface my remarks on this topic with a clarification of this concept. To this end, I shall illustrate this concept with the help of some of Piaget's well-known experiments, explain its explanatory force, and emphasize specific aspects which, from my perspective, appear to be essential. Furthermore, some theoretical conclusions will be drawn on which I am basing my reflections on the representation of cognitive structures.

Let us first turn to an especially simple experiment, the so-called seriation experiment. In this experiment children of varying ages are asked to arrange objects according to some ordinal characteristic, for example different types of sticks according to their lengths. The attempts to solve this problem serve to reconstruct problem-solving strategies. The results allow inferences as to the thought processes involved in the problem-solving process.[2]

The task of sorting sticks according to their lengths is relatively simple, although it does require logical thinking. It is tempting to consider the ability to solve this problem as a cognitive universal of the human species, that is to say, to attribute the required logical operations to a general, possibly biologically predetermined basic endowment of human beings. The results of the seriation experiment, however, prove such an assumption to be highly questionable. The problem presents children of up to about age five with insurmountable difficulties, and there is every indication that these difficulties are not merely superficial; rather, they are due to the lack of the required "logic" itself. In view of this, the question arises as to what it might be that enables an adult to solve the problem without the slightest difficulty.

First, there is the ability to anticipate the result, the ordered series, mentally. In arranging the sticks, the *goal* to be realized can already be envisioned. Further, mention must be made of the physical ability to order the sticks. The *means* for realizing the goal must be available—in this case in the shape of biologically highly developed physical organs trained by means of learning processes. And finally—one might want to add—solving such problems had to be learned.

This kind of learning is special. Let us assume for a moment that the thought processes accompanying the problem-solving process can be completely described as functionally dependent on a problem-solving process in which acquired knowledge about how objects are ordered is associatively recalled and applied to a given case, and in which, guided by this knowledge, the correct sequence of actions to be performed is controlled. In other words, let us assume that the ability to order the sticks according to their lengths is the result of a simple learning process. Now, somebody with the ability to solve the problem without difficulty is convinced in advance that he or she will be able to solve the problem. Thus, if the ability to solve the problem were the result of a simple learning process as described above, then there could only be an empirical argument to support this conviction: In recent years he or she never failed to solve such a problem (although, according to the results of the seriation experiment, he or she had difficulties with such a task as a child). His or her confidence in his or her ability to solve the problem would be the result of an inductive inference from a large, but finite number of successful attempts to the general effectiveness of the acquired problem-solving method. And it is precisely such an inductive inference from the particular to the general that produces that which, and only that which, we call *empirical certainty*.

It certainly cannot be denied, however, that nobody can claim absolutely that he or she will successfully solve the problem posed by the seriation experiment. The earth could open up and swallow the table with the sticks to be ordered. A heart attack could prevent the intention from being executed. However, these objections obviously miss the point. Even if we assume that the actual preconditions for solving the problem exist (whatever may be part and parcel of these preconditions that we neglected to remember when we presented the problem), the fact will remain that the success of the acquired problem-solving method can only have empirical certainty.

At this point the real problem with the seriation experiment emerges. That is to say, if we assume the actual preconditions for solving the problem as given (an assumption which, applied to the actual case, always implies an infinite number of empirical assumptions), then it is not only the empirical certainty concerning the possibility of solving the problem that increases for a subject equipped with sufficient logical means. For under these preconditions only two possibilities are conceivable: either the problem can be solved or there is a possibility that, in an actual case, it cannot be solved in principle, that nobody can solve it—and it can be proved that this case cannot occur. A method can be cited by means of which the sticks can effectively be ordered in a sequence in a finite number of steps. The sequence can be shown to possess the characteristics of an ordered series with each element of this series dividing it into two parts such that one part is made up only of larger sticks and the other only of smaller sticks. Assuming that all the sticks differ in length, it can even be proved that the place of each stick in the ordered series is uniquely determined, that, therefore, *any* conceivable method for arranging the sticks necessarily leads to the same result.

This does not mean much, however, as long as the methods of proof are not considered. With regard to the problem raised here it is, however, evident that the conviction that the sticks can be ordered according to size has a different epistemological status than that of empirical certainty. If the actual conditions for solving the problem are given (if, for example, there is empirical certainty that no earthquake is to be expected that could open up the earth, that the heart is healthy and a heart attack thus unlikely, etc.), then the confidence that the problem can be solved is based primarily on the *logical necessity* by means of which certain procedures, as long as they can be realized at all, lead to the desired result. The theory of cognitive structures provides an answer with regard to the nature and the conditions for the development of this form of knowledge which, in this context, is distinguished from empirical certainty by the concept of logical necessity.

For the purpose of delineating this answer, let us take a closer look at the strategies that might be used in arranging the sticks! If some of the sticks have already been ordered according to size, then a few comparisons will suffice to pinpoint the place where any of the remaining sticks will fit into the ordered series. The fact that the procedures which we employ in this situation must lead to success can be proved with the help of the formal characteristics of the ordering relation, antisymmetry and transitivity.

If, for instance, we have verified that stick A is larger than a given stick B from the series already in place, and if the sticks of the ordered series that are smaller than B are to the left of it, then it follows from the transitivity of the ordering relation, that stick A is also larger than all the sticks to the left of B, and that, consequently, it must undoubtedly be inserted into the series to the right of stick B which was used for the empirical comparison. Once the empirical comparison between stick A and stick B has been carried out, the strategy of subsequent comparisons cannot only be based on the result of this comparison, but at the same time on the knowledge that stick A is also larger than any stick C to the left of stick B. The empirical comparison between A and C no longer needs to be carried out. The knowledge that A is larger than C can be generated from of the outcome of the comparisons carried out between A and B which makes use of a system of rules based on the ordering structure.

Knowledge generated in this fashion does not basically differ from empirical knowledge acquired directly by the comparison operation. It is empirical as well, for it can always be acquired or verified by a comparison which can be carried out directly. Knowledge generated in this fashion is distinguished solely by the *means* employed in obtaining it. While the knowledge that A is larger than B was acquired by an isolated comparison, the knowledge that, consequently, A is larger than all C which are smaller than B follows from an operation which is part of a system of rules relating a variety of empirical experiences to each other. The pieces of information thus gained from the isolated instance that A is larger than B are no longer isolated; rather, taken together, they constitute a system which assigns a special place to each particular information in the overall context. Such an informational system including the system of rules generating the linkages shall be designated as *cognitive structure*.

The special cognitive structure illustrated by the seriation experiment confers a dual character on the quantitative relations between the sticks to be arranged in order: on the one hand, they are empirical facts and, as such, capable of being obtained empirically; on the other hand, they are necessary and as such identifiable elements of the system of the quantitative relations between the sticks combined in this cognitive structure. The question why it is not only empirically certain, but logically necessary that stick A can be inserted into the existing series of ordered sticks and that there is exactly one place for this stick can be answered by showing that the place is contained in this cognitive structure and can be found in it regardless of whether or not the actual conditions for realizing the task of ordering the sticks obtain. In the process of solving the problem of arranging the sticks, only that which is already given as a possibility in the cognitive structure is realized.

Let me summarize these reflections: A person confronted with the problem of the seriation experiment and equipped with the cognitive structure described, not only draws isolated pieces of information from the sense data, but simultaneously assimilates this information to a cognitive structure. This structure is generative in the following sense: It completes the perceptible quantitative relations thereby forming a closed structure and in this process defines a logically consistent system of possibilities of action which, if necessary, can actually be performed with the sticks and the outcome of which is already anticipated in the cognitive structure. The realization depends, however, on the assumption that the conditions for the realization presumed to exist in the course of the assimilation to the cognitive structure (e.g., no earthquake, sound heart), which are external to this structure, actually obtain.

The concept of cognitive structure is not identical with the mathematical concept of structure, although they are obviously closely related. I would like to point out two important differences in particular.

The first difference concerns the abstractness of mathematics. Mathematically speaking, there is no difference between the order relations between sticks of varying lengths, corresponding order relations between varying quantities of liquids related to the lengths of the sticks by a monotone mapping, or corresponding order relations between points in time. For the cognitive structures, however, the assimilated contents are essential, and it is conceivable and, with reference to the examples mentioned, for a certain phase of child development even to be assumed that different cognitive structures are used in solving the problem posed by the seriation experiment, if it is conducted with the various contents mentioned above.[3]

Cognitive structures consequently change in the process of assimilating new contents although, in general, this change only affects the contents and does not alter the formal relationships between contents assimilated earlier—that is, the relationships generated by the cognitive structure. I propose to call this change which is brought about by the assimilation of new contents *extensive expansion*. This expansion of a cognitive structure consists of two contravening processes. On the one hand, the extension of the structure expands; it loses its specificity and its close association with particular contents, and in this sense the cognitive structure becomes more abstract as well with regard to one of its essential characteristics—that is, the system of generative rules. On the other hand, however, the existing structure is enhanced by the assimilation of new contents, the possibility of assimilating additional contents increases; it becomes more differentiated and is linked to other cognitive structures without inconsistency or related to them in a contradictory way. Thus, in this instance, the process of abstraction is not independent of the contents; rather, it is a contravening aspect of a process of concretization which ultimately decides the further developmental destiny of the structure.

The second difference which I would especially like to emphasize concerns the possibility of formalizing mathematical structures. Structures corresponding to the mathematical concept of structure possess an identity that can be described and analyzed in metamathematical terms. It is true that formal mathematical deductions and metamathematical reflections cannot be separated in heuristic considerations, especially when thinking about the substance of mathematical objects; however, with regard to the form of the presentation of the results—and it is to these that the concept of structure is referring—a strict distinction is made between formal conclusions on the basis of a form remaining identical with itself and substantial considerations relating to the level of knowledge attained.[4] For cognitive structures, on the other hand, the reflectiveness is essential, and the relative identity and stability of formal relationships is a phenomenon generated by means of reflectiveness. The identity of a cognitive structure is merely an aspect—albeit an essential aspect—of an inclusive developmental process including nonidentity. The reflectiveness of cognitive structures provides us with the key for understanding their development into more and more highly organized forms as well as for understanding the relative stability of these forms.

With regard to the example of arranging sticks of unequal lengths, the assumption of the reflectiveness of the cognitive structure to which the perceptions of the situation in question are assimilated solves an important theoretical problem. It is conceivable to counter the view presented here—that is, the view that a generative process which can be described mathematically is associated with this assimilation and that this process expands the perceptible order relations into a closed system. The objection to this view would be something like this: This process cannot be observed empirically in the thought processes actually taking place in solving a problem and must, therefore, be considered to be nothing more than the external reflective interpretation of the problem-solving strategies realized in these thought processes. Indeed, those who have no trouble with the problem presented by the seriation experiment will—in answer to the question what they thought they were doing in the process of solving the problem—probably respond that they did not think about it at all or that they were wondering why they were presented with such a trivial problem. Thus, the problem before us is the following: If we assume that the ability to order the sticks is indeed due to the generative processes of a cognitive structure that is, processes based on the transitivity of the order relation, why is it not possible to show these processes in the thought processes actually taking place while arranging the sticks?

An answer to this question presents itself when we assume that cognitive structures not only *expand* extensively, but also *reflectively*. For what this indicates is that each assimilation of new contents to a structure not only generates knowledge about the object of cognition, but simultaneously also about the means employed in this process.

With respect to the example of arranging the sticks this means that the problem-solving strategy regarding the problem at hand, which was shaped in the course of the generation of the system of quantitative relations out of the sense data, itself becomes a fact that can be remembered and as such can be assimilated to the cognitive structure. Thus, not even the simple repetition of solving a problem takes place under the same conditions, and the more the extension of a structure expands, the more varied the system of generative rules completing the perceptible relations to the system becomes. With regard to the particular problem this implies a foreshortening of the operations of thought necessary for solving it; and, in order to be explicated, the latent logico-mathematical basis of these operations in turn itself requires an act of external reflection. In the case of our example, the ordering of sticks, the course of thinking has been foreshortened to such an extent that we can hardly imagine the elaborate thought processes we ourselves once had to perform as children in order to solve such a problem.

This train of thought also clarifies the question how cognitive structures can be investigated empirically and why the theory of cognitive structures arose in the context of research in developmental psychology. The most important method for investigating cognitive structures—a method exemplified by the seriation experiment—resembles the method of the biologist who seeks to confirm hypotheses on relationships concerning their historical development and who, to this end, searches for the connecting links in the chain of development that can be theoretically identified.[5] The basic functional modes of cognitive structures transformed into a status of latency by means of reflectiveness are clearly evident in the developmental stage in which they arise. If it is true, for example, that solutions of problems like that of our example are based on the generative processes of a cognitive structure, then it must be possible to find a transitional phase in the development of the child in which these generative processes are not yet reflectively foreshortened. That is to say, there must be a transitional phase between the phase in which the problem cannot yet be solved and the phase in which it is mastered without any problem. In this phase the preconditions of the anticipated goal and the availability of the means for solving the problem are already in existence. The cognitive structure is, however, not yet fully developed so that the generative processes to be demonstrated have to be executed in their entirety. In the attempts to solve problems they appear directly in the form of thought processes. Their results possess empirical certainty only insofar as the system of the developing cognitive structure is not yet closed so that the thought operations are still capable of transcending the system.

With respect to the problem of arranging sticks of unequal lengths, we generally find this transitional phase in children that grew up under our cultural conditions around the ages of four or five. In the most distinct case the phase is characterized as follows: The child is already in a position of anticipating the objective of the problem and of remaining focused on it during the attempt of solving the problem. Given favorable circumstances, this can be shown by the fact that the child can even draw the ordered series of sticks although it cannot, or at best with difficulties, actually produce it. Furthermore, the child is able to deal with the comparison of size between individual sticks without the slightest difficulty with the result that the discernible quantitative relations are assimilated to a cognitive structure by means of which two sticks at a time are associated with each other in a fixed quantitative relation. It is, however, characteristic of this transitional phase that a logico-mathematical correlation between the comparison operations and the anticipated, completely ordered series of sticks does not yet exist for the child. Thus, in this phase the attempts at solving the problem of ordering the unarranged sticks are characterized by the fact that each individual action—for example, the insertion of a new stick into an already existing series or the correction of an obvious mistake—can only be checked empirically after the fact. The approximation to the anticipated image of the ordered series takes place step by step on the basis of trial and error. The child can never be quite certain of the influence an intended operation will have on the total arrangement of the sticks to be ordered as long as it has not actually carried out this operation. It is not capable of determining why the inaccurately constructed sequence does not correspond to the anticipated arrangement of the ordered series except by trial and error.

Since the example I have chosen to illustrate the concept of the cognitive structure is one of the best examined cognitive structures and the problem chosen has by now become a classical experiment of developmental psychology, there are numerous individual observations in the scientific literature that confirm the regular occurrence of the transitional phase described above.

For instance, at the beginning of this phase, when the child essentially limits itself to bringing its attempts of solving the problem into line with the anticipated goal and is hardly guided by the means of comparison operations and the possibilities inherent in that means, the following problem-solving strategy is frequently encountered: The child tries to bring the incorrectly arranged sticks optically into line with the anticipated image of the ordered series ascending in step-like fashion by moving them vertically while disregarding the irregularities at the other end of the sticks or in the futile hope that this operation will not only solve the problem at one end of the sticks, but at the same time at the other end as well.

It is, furthermore, typical of the transitional phase that the children are by and large successful in constructing an ascending or descending series correctly for some part of the unarranged sticks. To this end the comparison operation simply needs to be reiterated, and the gradual construction of the series can be guided step by step by the anticipated total picture of the ordered series. In contrast, the attempt of fitting in the remaining sticks afterwards creates considerable difficulties, is accomplished only after several attempts have been made, or does not result in a correct arrangement of the sticks at all. Likewise, as a rule, correcting an incorrect series presents great difficulties. Sometimes attempts to correct the situation lead to more mistakes than are eliminated.

What all these observations indicate is the following: In this stage of transition, the quantitative relations capable of being ascertained empirically and the image of the ordered series are not parts of a common structure; they are not part of a generatively created context, but still remain unconnected and can only be linked by means of empirically controlled actions. In other words: The anticipated image of the ordered series has a different logical structure than it does after it has been assimilated to the cognitive structure which generates the relationships of the ordered series from the comparison operations. Thus, in this phase, there remains only one possibility of solving the problem of ordering the sticks—that is, by means of logically uncoordinated, empirically controlled operations. As soon as the number of the sticks to be ordered exceeds a certain number that can still be apprehended visually, such a procedure is bound to fail.

The cognitive structure corresponding to the seriation experiment which I have used to explain the concept, is a particularly simple example of a cognitive structure. In order to disassociate the concept somewhat from this example, a few more examples will be added.

We obtain a second example for a cognitive structure as soon as we pose the problem of the first example in a slightly different form. Instead of ordering sticks of unequal lengths, the task now consists in ordering a variety of containers according to the amount of liquid contained in them.[6] Obviously the problem has now taken on a completely different quality. As a matter of fact the problem does not make any sense, unless it is assumed that the liquids in the containers have a property which remains unchanged when they are poured into different containers or divided among several containers, yes, even when they are spilled, that is to say, their quantity according to which the containers are to be ordered. There are, however, hardly any clues at all with regard to the perception that, when a liquid is poured from one glass into another of a different shape, anything other than color, smell, etc. remains identical. For the goal of ordering the containers to be anticipated at all, the existence of such an "extrasensory" property remaining identical with itself and invariant with regard to the manipulations mentioned earlier must be assumed. Our conviction that there is such a property as quantity obviously is as indirect a piece of knowledge as our conviction examined at the beginning that any number of sticks of unequal lengths can unambiguously be ordered according to size. That is to say: Our certainty of the identity of the quantity of the liquid results from the system of generative rules of a cognitive structure.

At this point I am disregarding the details of this structure. I would, however, like to point out that compositions and decompositions (pouring together and dividing liquids) must be assimilated to this structure. Thus, the system of rules of this cognitive structure must generate structural relationships with forms to be attributed to algebraic structures. Therefore, this cognitive structure is no doubt more complex than the structure of the seriation experiment discussed earlier. The concept of quantity in which identity is represented only seems simple, because in it the unity of this structure is reflectively expressed.

We encounter the transitional phase within which this cognitive structure is constructed at age five or six. The given age in turn only applies to our cultural environment. Intercultural comparisons indicate that the speed with which cognitive structures develop depends on specific environmental conditions despite their essentially universal nature concerning the human species as is true of the examples under discussion. Thus, a study of children in Senegal raised without the benefit of schooling in the bush, for example, showed that 50 percent of the eleven- to 13-year-olds did not believe in the invariance of a certain liquid when poured into different containers. This can be taken as a sure sign that they did not presume the existence of a quantity identical with itself at all. In contrast, in children of the same village who attended school the cognitive structure was already fully developed in this age group.[7]

In this second example the transitional phase of a cognitive structure is in turn characterized by the fact that the goal of an activity devised to produce certain quantities is already anticipated. Furthermore, the means for attaining this goal are presumably available, but the necessary logical connection between the end and the means which would allow them to be related to each other reciprocally through their definitions does not yet exist. In concrete terms this means that children already have an understanding of the quantity of a certain liquid which makes it appear meaningful to them to try and obtain a certain quantity not identical with its outward appearance by decreasing or increasing it. Furthermore, they are capable of handling liquids properly, of using the proper utensils correctly, and of making inferences from the change of external characteristics to changes in quantity. The manipulations with the liquid and their perceptible results, on the one hand, and the anticipated outcome, on the other, are not yet, however, logically related to each other in such a way as to operationalize the end by the means—that is, of attributing the criterion of reaching the goal by way of definition to the possible manipulations with the liquid. Thus, the criterion of reaching the goal can only be empirical which means that the quantity is not yet sufficiently distinguished from its manifestations. The difficulties to which this can lead become palpable in the following description of a problem-solving process adopted from one of Piaget's protocols:[8]

A seven-year-old boy watches as the experimenter fills about a fifth of a glass with some red juice. The experimenter then selects a narrower, but taller glass from among the other glasses and asks: 'Can you pour as much juice into this one as your Mom has in hers?' The boy affirms: 'Just as much,' and pours juice into the narrower glass up to the same level. The experimenter asks: 'Is that the same amount?' The boy responds: 'No,' and pours in some more until the glass is about half full, compares the levels, says: 'No, that is too much,' and pours back enough until the equality of the levels has been restored. The experimenter asks: 'Who gets more?' The boy replies: 'Mom, because her glass is wider,' and again pours some more into the narrower glass. The experimenter asks: 'Do both of you now have the same amount?' The boy replies: 'No, I have more,' and takes away the excess in the narrower glass. The experimenter asks again: 'Do you have the same amount now or does someone have more?' The boy replies: 'Mom, because she has a wider glass.' He pours more into the narrow glass, hesitates, says: 'No, now I have more,' pours the liquid back again and restores the level and then says: 'No, Mom has more.' In the course of the experiment he does not succeed in finding a solution that satisfies him. This problem-solving behavior in which the means are used properly and the envisioned goal is already anticipated so as to be capable of serving as a criterion for judging the success of the problem-solving attempts, is typical of the transitional phase in which the cognitive structure necessary for recognizing the invariance of the quantity of liquids with regard to certain routine operations takes shape.

In order to obtain a third example, let us once again vary the problem by substituting a number of distinct objects, let us say, large beads, for the liquid the amount of which can be changed in a continuous manner and which can be divided into any number of components.[9] If the same experiments as with the liquids are conducted with this experimental set-up, the same or similar results are obtained. The misjudgments based on the lack of identity concerning quantity are evident with beads as well.

This result raises the question whether or not, in this case, counting would solve any given problem. This is not the case, however. In the transitional phase the result of counting is one empirical information among others and logically consistent with changing the number of beads when they are poured into a different container. Thus, the concept of number in its most elementary form is based on a cognitive structure as well. The latter generates the correlations between the perceptions of discrete objects which indirectly results in a recognition of a quality of these objects that is invariant with regard to changes of their location in space: their number.

The last example I want to give is another thoroughly investigated cognitive structure which is the basis of the classificatory use of concepts.[10] In the course of the assimilation to this structure, associations are generated between the sense perceptions by means of which they are integrated into systems of extensionally determined, hierarchically organized concepts. On account of this cognitive structure each analytical act of an empirical classification results in generating a system of indirect knowledge which can be represented as classificatory statements, statements constructed with general and subordinate terms in a direct and indirect, ascending and descending line as well as with the connectives (negation, disjunction, conjunction, formal implication, etc.). The examination of this cognitive structure has shown that the formation of the extension of a concept from its intension is already based on very complicated processes. At age eight many children do not yet comprehend the logical relation of the inclusion of two concepts. They use intensional, but not extensional relations between concepts, and the inclusion is a relation between the extensions of two concepts to which a material implication of the intensions rarely corresponds.

The examination of this cognitive structure yields a specific interpretation of Russell's and comparable antinomies. If we proceed from any predicate statement—that is, a statement that assigns a predicate to an object, it can be asked whether this statement is to be interpreted as a classifying statement independent of its content. If we were to affirm this question, this would mean that any similar statement of our language can be assimilated to a cognitive structure the system of rules of which generates a nonprocessing extension of the predicate identical with itself. However, in this case the system of rules would have precisely the effect of the universal axiom of set formation (for any predicate P there exists the set of all x for which 'x is P' is true) which, as we know, results in Russell's antinomy.[11]

The reflective expansion of a cognitive structure is, however, a cognitive process which leads to the restructuring of cognitive structures if this becomes necessary in the course of assimilating new contents, and this process is expressed in the language which functions as the means of thought. In language reflective propositions are permitted without restriction provided they are substantially sound. This situation can be illuminated with the help of a simple example. The proposition: "No answer is still an answer" expresses a contradiction. As a reflective statement it may nevertheless be meaningful depending on what it refers to. In this case it signifies a logical restructuring, that is to say, a common process of thought. Thus, with regard to the theory of cognitive structures presented here, the assertion that each predicate proposition can be interpreted as a classifying statement is false.

In mathematics as well there are relations that can be interpreted reflectively. An example is the axiom $x = x$ which, as a spoken sentence, only makes sense, if it is interpreted reflectively, for an identity and a difference is stated at one and the same time. Identity can only be claimed for objects of thought that were previously considered to be dissimilar. However, in mathematics, when relations capable of being interpreted reflectively are introduced into a system of sign transformations, there is always an attempt to make sure (implicitly or explicitly) in metamathematical terms that the given logico-mathematical structure is not restructured in this process. The universal set formation axiom by means of which (without metamathematical control) a classifying meaning can even be given to any reflective statements by attributing an extension to the predicate, violates this tacit rule which consists of nothing more than the reflective meaning of Aristotles' Law of Contradiction, unless this set formation axiom is appropriately restricted on the basis of metamathematical considerations (e.g., by means of Zermelo's 'axiom of choice'). Russell's antinomy is obviously based on a reflective statement the predicate of which is assigned a fixed extension by means of the set formation axiom, that is to say, the statement: x is not a member of x. If the predicate of this statement is interpreted extensionally and applied to itself as the object, the contradiction of Russell's antinomy occurs.[12]

COGNITIVE STRUCTURES AND MATERIAL ACTION

After illustrating the concept of cognitive structure with the help of some examples and developing basic features of a theory of cognitive structures, we can now move on to the question of the origin of cognitive structures.[13]

Let us once more consider the transitional phases in which the cognitive structures are developed and the generative systems of rules emerge. As we have seen, these transitional phases can be described by the fact that the goal of an action or a sequence of actions is already anticipated and that furthermore the means for carrying out the action are available, while the cognitive structure has not yet or has only incompletely emerged. Thus, the anticipated goal cannot yet be assimilated to this cognitive structure. As a result, the course of action can only be checked empirically by means of the anticipated goal (trial and error); a logical inference of a guiding schema of action is not yet possible. The existence of such a preliminary genetic stage of any cognitive structure led to the unanimous conclusion that the cognitive structures result from experience through action. To cite a pithy characterization: cognitive structures are internalized coordinating schemata of actions. They are based on experiences gained in the context of concrete action.

Incompatible with this interpretation of the results of studies in the field of developmental psychology concerning the genesis of cognitive structures are those philosophical systems which hold logico-mathematical structures of thought to be given *a priori* (i.e., prior to any experience). In this context, the transcendental philosophy of Kant comes to mind, but also Platonism. Cognitive structures are considered to be the results of experience.

However, more can be said. The *goals* of actions and of sequences of actions anticipated in the transitional phases can hardly have much influence on the system of the generative rules of the cognitive structure as it emerges; for as long as they themselves have not yet been assimilated to the already existing structure, they do not have any influence on the course of the concrete actions in the problem-solving process except for the empirical control of the success of an action after the fact. The anticipated goals determine primarily the definition of the problem in the concrete context of action—that is, the contents that are the focus of interest, and thus to a large degree the extension of the emerging cognitive structure, however not its system of generative rules.

The relationship of the available *means* employed to perform the actions to the generative rules of the emerging structure is, on the other hand, quite different. For if these means are regarded not as particular concrete objects and aids for action, but in terms of their general abstract function for the action capable of being transferred to other contexts of action as well, then the system of the results of action that appear conceivable corresponds exactly to the system of the logico-mathematical operations generated by means of the generative rules of the emerging cognitive structure. In the process of their objective use the means represent that capability of productive action that is only reproduced in the generative rules of the cognitive structure. In the reflective expansion and reconstruction of those cognitive structures to which the experiences acquired in the course of employing the means are assimilated in the transitional phase of the development of a new structure, the generative rules are ascertained empirically in terms of the means that effect the reconstruction.

This thesis has far-reaching consequences. Let me give an example from the area of phylogeny.[14] It is a fair assumption that the development of the first cognitive structures as defined by the theory presented here and thus the very development of thought is associated with anthropogenesis. We know, for instance, that there is a close historical connection between the biological anthropogenesis and the use of tools; to be specific, the first forms of the use of tools precede the last stage of the biological anthropogenesis in which the cerebrum in particular attained the form characteristic of homo sapiens. Thus, the primitive use of tools cannot be viewed as a mere outcome of the improved capacity of thought associated with the enlargement of the cerebrum. However, in view of the thesis that the systems of generative rules that make thought possible originate in the material means of actions, the antecedence of the primitive use of tools seems plausible. The capacity of thought based on this use of tools subsequently created the very superiority of the hominids in the struggle for survival which turned the enlargement of the brain into a mutation with a selective advantage.

The relationship between the material means of the realization of goals for action and the cognitive structures is not a one-sided relationship, even though the formula that cognitive structures are internalized schemata of action suggests such a view. Although it is true that, concerning the process of the development of cognitive structures, the actual material actions are considered to be primary, each developed cognitive structure, however, affects the actions. Once a cognitive structure has been developed, there is a striking increase of competence in terms of the system of action corresponding to the structure. With regard to the problem of ordering sticks of unequal lengths according to size, for example, the transition from the empirically controlled problem solution to the logical deduction in the context of the generative system of a cognitive structure takes place within a relatively short time period; and once the cognitive structure has been developed, the probability of making mistakes is almost zero. This distinguishes the development of a cognitive structure from the acquisition of "skills." With the latter the problem is not with comprehension; rather, it is with the physical and psychological limits concerning the correct application of the means necessary for solving a given problem. Cognitive structures structure and coordinate such skills by developing stable, albeit inflexible schemata for action for particular goals by means of shortening the necessary reflection. Concrete learning processes encompass both forms of the cognitive competence for action. The development of math skills is a typical example for the interaction of the development of cognitive structures and the development of cognitive skills. The acquisition of math skills is therefore a complicated process with erratic and steady developmental phases and with different individual courses of development.

The most important type of effect of cognitive structures on material actions is the effect with regard to those productive actions in which the means for material actions are generated or improved upon on the basis of a change in the anticipated goal. With regard to the example of the sticks this means devising a measuring rod, in the case of the invariance of the amount of liquid it means creating measuring cups as well as producing a useful design for containers for certain purposes and adjusting size and form to the intended use. In the case of the development of the number concept it means producing calculating devices which, as we know, played a decisive role in the history of arithmetic.

With the help of all these devices, cited as examples, relationships can be created empirically that go beyond those generated by the cognitive structure which is necessary for the development of these devices. In the first example it concerns the metrics of the concept of length. In the second example it concerns the metrics of the concept of the amount of a liquid, the differentiation of weight and volume of the liquid as well as the correlations of relationships with regard to length, form, and volumes, etc. In the third example it concerns the augmentation of the natural numbers by zero, the introduction of place-value systems, the expansion of the number concept by number theoretical considerations, written arithmetical procedures, etc., which may be represented in the simple calculating devices so that, in using them, the generative rules of the corresponding cognitive structures can be deduced empirically.[15]

Let me sum up the results of these deliberations on the relationship of cognitive structures to the material actions from which they arise:

The *cognitive structures* give the anticipated goals of material actions a logico-mathematical form which serves to establish abstract relationships between the conditions of action and the anticipated goals. At the same time they confer upon the perceptible conditions of action a reference to a totality of possible anticipated goals transcending the perceptual certainty of these perceptions. In the present context this reference shall be characterized as the *subjective meaning* of these conditions of action for the acting subject equipped with the cognitive structures.

When the *material means* are taken as means and especially when they have been intentionally established to be used in certain contexts of action (i.e., when they are themselves the results of the realization of subjective meanings), they represent abstract-general relationships not identical with the subjective meaning in the form of possibilities for action not yet anticipated which may actually be realized when the conditions for their realization are given. In this context, these objective possibilities for action shall be designated as *objective meanings*.

In the concrete context of material actions, means established with subjective meaning are therefore *representatives of objective meaning*.[16] The process of the individual development of cognitive structures in the context of material action is therefore to be essentially understood as a process of acquiring objective meanings of means provided by society as subjective meanings. As a process concerning the history of the species the development of cognitive structures is, however, the result of the production of new means and thus a process of the interaction of subjective and objective meanings.

COGNITIVE STRUCTURES AND LANGUAGE

So far, the reflections on the representation of meanings embedded in the material means have neglected the role of language. The possibility of intersubjective communication concerning the proper use of the means of material actions was attributed to the conditions for realization which were tacitly assumed as given. However, this communication is itself objectively mediated, that is to say, by means of the spoken and the written word. The representation of meanings embedded in these means will be discussed below.

Modern linguistics distinguishes between linguistic competence and linguistic performance.[17] *Linguistic competence* is the ability of a speaker or listener (writer or reader) to generate formally correct linguistic products or to recognize them as such by mentally reproducing their formal structure and by analyzing them in terms of their formal elements, if the conditions for the realization of concrete linguistic performance permit this (i.e., abstractions are derived from the conditions governing linguistic performance).

Linguistic performance, on the other hand, is the actual use of the spoken or written word in concrete situations. Linguistic production—that is, a material action or sequence of actions in which concrete linguistic products are purposefully generated and used as means, constitutes the core of linguistic performance. Thus, linguistic competence is only one of the requirements governing linguistic performance. Linguistic performance can only be adequately understood in terms of the totality of the goals, means, and requirements for the realization of linguistic actions.

In linguistic terms linguistic competence is defined as generative grammar—that is, as a system of rules which generates structurally possible forms for each linguistic performance and subjects the concrete or anticipated linguistic products to these forms. We can, therefore, apply the concept of linguistic competence directly to the theory developed in these pages: Linguistic competence is a cognitive structure to which linguistic products are assimilated during the process of actual linguistic performance.

As with all the cognitive structures mentioned up to now, there is a parallel in mathematics in the present case as well. If we take the system of generative rules of a generative grammar without the contents assimilated to this system—that is, if we abstract this system of rules from the cognitive structure developing in the course of linguistic performance, and if we assume that the rules cannot be changed reflectively, then this system of rules corresponds to the semiotic systems of mathematical logic and model theory.[18] There is, however, a difference between formal languages and the systems of rules of the linguistic competence of actual linguistic performances in that the latter are much more complex than the former.

Thus, in the concrete linguistic performance the anticipated goal of the speech act is assimilated to the cognitive structure of linguistic competence in the process of *speaking*. In this way this goal is matched up with a possible—that is, a formally correct, linguistic product generated by the generative system of rules of the cognitive structure making use of the basic phonetic components of language, the phonemes. Subsequently, this goal, assimilated to the cognitive structure of linguistic competence, is the basis for the direction of the speech act in which the actual linguistic product is generated under the prevailing conditions for the realization of the speech situation. In the process of *listening,* on the other hand, the actual product of linguistic performance is assimilated to linguistic competence and as such is identified as a linguistic product with a particular grammatical structure.

Linguistic competence understood in this way obviously explains only a technical aspect of language.[19] We have to ask ourselves how thoughts and information can become the content of the linguistic product and how they can be communicated intersubjectively by means of language.

To clarify this question theoretically on the basis of the above explanation of linguistic competence as a special cognitive structure, an attempt will be made to examine the processes taking place in the context of linguistic performance in situations where linguistic performance is closely associated with goal-directed material actions.

If, in such a situation, the actor hears somebody speak and if he or she perceives the speaker to be located in the same context of action and that the words spoken are related to this context, then the linguistic product perceived is not only assimilated to the cognitive structure of linguistic competence, but simultaneously to that cognitive structure which is the basis of his or her own goal-directed action. If the linguistic product proved to be a goal-oriented linguistic product associated with the context of action which was competently created, then it will be possible to assimilate at least parts of the structural relationships generated in the process of assimilation of the linguistic product by means of the linguistic competence of the listener, specifically to the structural relationships of the cognitive structure which is the basis of action. In this way, the linguistic product is not merely related to a system of formally conceivable linguistic products by means of the system of generative rules, but, at the same time, some are specifically identified and related to the possibilities of action by means of the system of generative rules of the cognitive structure associated with the action.

It therefore follows from the assimilation of the linguistic product to the cognitive structure of the context of action that, conversely, cognitively structured perceptions associated with the context of action are assimilated to the cognitive structure of linguistic competence as well. These are structured perceptions of a nonlinguistic character to the extent that, with regard to their structure, they can neither be created by means of the generative rules of linguistic competence, nor can they be identified by this competence as belonging to the linguistic product given their sensible form (which does not represent a weakness of linguistic competence, since they were in fact not created by means of speech and they did not get their structures from a generative linguistic grammar however understood). Thus, by means of linguistic performance in concrete contexts of action and with reference to them, the non-linguistic cognitive structures are related to linguistic competence in an inclusive cognitive structure of linguistic performance to be defined in the present context as *subjective semantics*.

It is thus the specific achievement of subjective semantics as the cognitive structure underlying the specific linguistic performance in connection with concrete actions to be generating a system of relationships between the grammatically structured linguistic products (in the broader sense of modern linguistics) and the cognitively structured conditions of action. This means, using the concepts that were introduced with regard to the relationship of cognitive structures to material actions, that subjective semantics links the linguistic products with the subjective meanings that include the conditions of action for the subject equipped with cognitive structures. Thus, with regard to concrete contexts of action the linguistic products become concrete *representatives of subjective meanings* by means of linguistic performance, although they are not related to the conditions of action in any way other than that of linguistic performance within the context of collective material action.

This latter fact is often expressed by stating that the meaning of linguistic signs and symbols is established by convention while, for instance, sounds and other nonlinguistic perceptions, which convey information to us, derive their meaning from their natural contexts. This led to the conclusion that the interpretability of *natural* signs is assured by means of material actions, especially of work, while the interpretability of *artificial* signs is established by the structures of interaction between individuals and must be explained in terms of the discourse between them. Keeping in mind the theory which I have presented above, the concept of convention must be considered a rather vague description of the development of the meaning of linguistic products. In addition, it invariably presupposes the ability to communicate by means of language.

On the basis of the reflections presented above a simple explanation emerges for the possibility of intersubjective communication by means of language. If language is to be a means of communication between subjects, then it must have objective semantics. At the very least there must be partial agreement of subjective meanings generated by different individuals in a concrete situation of material action by means of their subjective semantics concerning the linguistic products. As we have seen, the subjective meanings are acquisitions of the objective meanings represented in the material means of the actions. They are generated by the subjects by means of their cognitive structures in relation to the conditions for action common to those subjects. The subjective meanings differ from each other only in that they are the result of different individual processes of the development of cognitive structures by means of which the objective meanings represented by the material means of action are reconstructed individually.

In light of the theory presented here, we may think of the process of intersubjective communication in situations of concrete collective actions in the following terms: Person A is attempting to solve a problem by means of goal-oriented action. Person B is observing the situation and assimilates it to the cognitive structure that she herself would base her actions on if she were in person A's position. In the context of this cognitive structure, person B constructs the external course of action required by the solution of the problem by reflecting on her own possible action and translates the result relating directly to perceptible conditions for action into the form of the representation within the context of a linguistic product. Person A generates the subjective meaning of the linguistic product by means of the subjective semantics with which she (person A) is equipped and is in a position to realize the result as an action, provided the actions to be performed are within the purview of her actions at all and, thus, in the area of subjective meaning that the conditions for action have for her. However, it is obviously not necessary for person A alone to have been in a position to find the solution. It is not even necessary for person A to be at all capable of executing operations comparable to those used by person B in order to chart the course of action for solving the problem.

Intersubjective communication by means of language in the sense of an effective intersubjective transfer of competence related to action therefore has its basis in an objective semantics of language. This is based on partial agreements of subjective meanings generated by the communicating subjects in the context of collective material action by means of their subjective semantics, agreements that can be attributed to the foundation of subjective meanings within the objective meanings of the material means of the actions.

This argument can be summarized as follows: Linguistic products are indirectly not merely representatives of subjective meanings, but also representatives of objective meanings, just as the means of the material actions themselves. The generation of linguistic products supplies the means of thought, the cognitive structures, with a material appearance and turns the results of the anticipation of goals of action by means of cognitive structures themselves into a perceptible condition of action. Language becomes a material means of actions. The objective meaning of linguistic products, however, remains dependent on the material means of action having an immediate effectiveness. There is no process in language comparable to the creation of new or improved means of material actions by means of which new objective meanings of conditions of action are actually generated.

The function of language of representing objective meanings indirectly in the linguistic products nevertheless has far-reaching consequences for the development of the possibilities for action inherent in the nonlinguistic means of material actions as objective meanings. Mental operations based on systems of generative rules of cognitive structures in the problem-solving process within the actual context of action divide the actions into two distinct, but inseparable components: planning and execution. The two components cannot be separated because they are united in one person. However, the representation of objective meanings in linguistic products makes it possible to actually separate these components and to organize the division of labor between individuals in the actual context of action which initially develops as a specialization into basically similar, linguistically coordinated partial actions, as a division of labor between planning and executive actions—that is, between material and mental activity. Outside the context of material activities, speaking also becomes a meaningful activity of the generative creation of linguistic products with objective meaning.

In order to develop this potential inherent in language, a medium of representation other than that of ephemeral sound is needed. Thus, it is not surprising that the development of early forms of the division of labor into planning and executive activities—the rise of the state on the basis the emergence of classes—is intimately connected with the development of writing.[20] In writing, some of the formal structures of the linguistic products are isomorphically represented in a permanent medium. This increases the possibilities of referring thought processes to their products to an extent hardly to be overestimated. In particular, the possibility of transforming the generative processes of a cognitive structure into rules for treating the material representatives in writing emerges—that is, of *mechanically* executing linguistically mediated thought processes abstracted from their meaning with regard to the material representatives of meanings. In more simple terms it can be stated that together with writing the prerequisite has been created for that technology of manipulating symbolic systems on which mathematics is based. This, however, does not take into account the existence also of direct connections between the means of material actions and mathematical technology that did not come into being via language. In this context, the importance of calculi for arithmetic and of the straightedge and compasses for geometry in Greek mathematics must be remembered. The Euclidean postulates, for example, by means of which the figures are generated to which the axiomatic system refers, are (with the exception of postulate 5 referring to the existence of parallels) functional descriptions of the straightedge and compasses. Euclid's methods of proof, however, are based on the linguistic mediation of thought processes and the material representation of structures of linguistic competence in writing.[21]

The investigation of the material means of thinking in language, writing, and symbolic mathematical systems directs attention to a problem so far disregarded which, however, does at least deserve to be briefly mentioned. The generative rules of linguistic competence and the generative rules of cognitive structures are integrated into an inclusive cognitive structure by means of the cognitive structures of semantics. Since semantics, just like any other cognitive structure, develops not only extensively, but reflectively as well, it is to be expected that, at higher stages of development, rules of linguistic competence are themselves already produced with reference to semantic functions. Rules governing the generation of linguistic products and in particular grammatical rules of writing and rules of mathematical technology may have a formal function as rules of linguistic competence as well as a substantial function within the framework of nonlinguistic cognitive structures related to material action. Due to the possibility for abstraction from meaning inherent in the material representation and the relative autonomy of language entailed by this it is often difficult to identify these two functions of generative rules accurately.

Thus, with regard to Central European languages, for example, some rules for the formation of syllables and words have meaning and some do not; in general, purely syntactic functions and functions with semantic content are easily distinguished. The combination of syllables into words, for instance, is in part purely lexical in nature (in the sense of the "lexicon" of Chomsky's theory of linguistic competence the entries of which represent a syntactic classification of formatives), in part (as in the use of prefixes for negating concepts) a logical operation within a cognitive structure. It is also evident that the system of rules of linguistic competence regulating the formal linguistic treatment of comparative concepts does not contain any rule corresponding to the transitivity of order described as a generative rule within the context of the cognitive structure in the example above which forms the basis for the arrangement of the sticks. It is, therefore, possible to make correct linguistic use of comparative concepts without comprehending their meaning. On the other hand, for instance, it is a matter of some controversy whether a specific semantic function is to be ascribed to the syntactic categories of subject and predicate and their relationship to each other and how it might possibly be interpreted as part of a cognitive structure related to material action as objective meaning of the subject-predicate relationship.[22]

These reflections on the relationship of subjective and objective semantics within the framework of the theory of cognitive structures presented above do not only explain the possibility of intersubjective communication and the development of language towards independence with regard to the formation of theories and to discourse implied by it, but they also indicate the limits of intersubjective communication which cannot be transcended by means of language. For the prerequisite for communication is, first, the existence of collective experiences concerning the means of material actions the objective meaning of which makes the partial agreement of subjective meanings possible; secondly, the development of the cognitive structures of the subjects participating in the communication must have been developed to the point that the objective meanings indirectly represented in the linguistic products which are the object of communication can be subjectively reconstructed by the participating subjects. The first prerequisite, for instance, becomes a problem in the reconstruction of the meaning of historical sources, if the knowledge of the actual conditions of action of the period from which the source originates is limited. The second prerequisite mainly concerns linguistic communication with children about objects that do not yet lie within the purview of their material actions and thus not in the purview of the subjective meanings that can be generated by means of the existing cognitive structures.

With this we have come full circle. If we ask a four-year-old to arrange sticks of unequal lengths in order, it is more than likely that this child has not yet or only incompletely developed the cognitive structure for intelligently solving this problem. This same child has most certainly already been using the terms "big" and "small" to denote relationships of size for some time. And it has probably been using the comparatives "bigger" and "smaller" for at least close to a year. If we are casual in drawing conclusions from this use of language to thinking, we fail to see that these terms are used with a subjective meaning other than that which we unintentionally attribute to these terms. We easily misunderstand what the child wants to say, and for this reason alone we will not necessarily succeed in teaching the child how to order the sticks according to their lengths.

If we pay attention to the context of action the child uses to articulate the problem of arranging the sticks, we will soon realize that the child is talking about something other than we are. We may, for instance, notice that the terms "bigger" and "smaller" are often used without reference to the comparative object, and that they are used especially in the following way: to emphasize the given attribute. We may also notice that the superlatives "the greatest" and "the smallest" are used not at all or incorrectly. Finally, an experimental arrangement like the one under discussion reveals that the comparatives are not at all used in terms of a transitive relationship.

My reflections on a theory of cognitive structures show why this would be so. The development of linguistic competence is, as it were, ahead of the development of semantics, and the grammatical structure of the comparison of adjectives generated by linguistic competence simply does not contain the transitivity of the comparative.

These reflections do, however, have further implications. The words are not just used with a different subjective meaning; they cannot even be understood with regard to their objective meaning as long as the corresponding cognitive structure has not been developed and linked to linguistic competence by means of semantics. At this stage of cognitive development language cannot yet replace direct experience. It attains its dominating role as a material means only much later. For the time being it remains itself an object of thought and action to be cognitively explored indistinguishable from the other conditions of action.

To illustrate the resulting logic, the psychologist Vygotsky quotes an anecdote by Humboldt about a simple-minded peasant who supposedly said:[23] "I understand that people have measured the distance from the Earth to the most distant stars with these instruments, that they have identified their distribution and movement. What I want to know is how they learned their names." According to Vygotsky a child that has just learned to speak argues in a similar way: "A cow is called 'cow' because it has horns, a calf 'calf' because his horns are still small, a horse 'horse' because it has no horns, a dog 'dog' because it has no horns and is small, and an automobile 'automobile' because it is not alive at all." This empirically corroborated logic of the early stages of language acquisition attests to the correctness of the above reflections.

NOTES

[1] The concept of cognitive structure used here has been borrowed from the field of psychology. Its use in the present context demands an explanation.

The influence of Piaget's work on the psychology of our century is considerable. The field of "cognitive psychology" based on his work, became an influential alternative to the psychology inspired by behaviorism and committed to psychometric methods which emerged when psychology first established itself as an empirical science following its emancipation from philosophy. It is emphatically viewed as a "cognitive turning point" of Western psychology. This is accurate at least insofar as psychological research today focuses to a considerable extent on questions which, due to the empiricist scientific understanding and its consequences for the conception of the subject matter of psychological inquiry, had been resolved onesidedly.

These questions are not new, and the answers are only partially new as well. Many of them are an integral part of the traditional topics of philosophical inquiry, and since they are comprehensive and cannot be reduced to psychology alone, they can, in different contexts, also be found in other sciences. The concept of cognitive structure in particular has more or less adequate equivalents in philosophy and other sciences. Stated without reference to any special discipline: When cognitive structures are under discussion, the reference is to the forms of thinking and of consciousness, their function, origin, and development.

By way of illustration the following example may suffice. The controversial theoretical discussion concerning the concept of meaning as a basic concept of sociology addresses, albeit in a different theoretical context, the same problems raised once again by cognitive psychology as it is rejecting predetermined, empiricist conclusions. Thus, the concept of cognitive structure occurs in Luhmann's systems theory as a concept of the system constituting meaning, in the writings of Habermas as a concept of the cultural interpretive system based on interaction, in Heidtmann's theory in Marxist terms using Hegel's terminology as a concept of being as totality, as a unity of immediacy and mediation conveyed by means of reflection. The fact that all these concept formations arise in close association with the terminologically identical concept of reflection may serve as evidence for this assertion (cf. [25]; [17]; [19]).

The objective of the present investigation is limited. Relationships between the treatment of cognitive structures within different theoretical contexts are acknowledged implicitly, but not systematically discussed. Rather, the discussion is guided by the theoretical discussion in psychology, even though the arguments are primarily drawn from the philosophical discussion concerning the materialist concept of labor and related to the current discussion within the discipline. This one-sided treatment of the topic imposed solely by limitations of space is expressed by the use of the concept of cognitive structure as the central concept of my discussion.

[2] Piaget repeatedly used this experimental arrangement in his investigations (cf. [34], p. 96ff.; [39]; [40], p. 29ff.). He uses it with particular frequency to explicate his concept of cognitive structure (see, e.g., [37]).

[3] At this point I am deviating from the general use of the concept of cognitive structure in one crucial aspect. An explanation can be found in my essay *Action and Knowledge in Piaget's Genetic Epistemology and in Hegel's Logic* reprinted in this volume (Chap. 1). The concept of cognitive structure presented here emerges within Piaget's theory, if the assumption of the existence of two independent forms of abstraction is revised: the abstraction from the action and the abstraction from the object of the action. If we proceed instead from the mutual mediation of both forms of abstraction, content and form become the elements of a uniform cognitive process which takes the place of the two processes postulated by Piaget, that is to say that of empirical knowledge and that of mathematical knowledge. Without getting into a detailed discussion, I would like to call attention to the fact that the problem of the "décalages," the dependence of the level of thought on the contents of thought (see [48]) often cited by proponents of empiricism as an argument for denying the very existence of cognitive structures, takes on a completely different meaning, if the development of the forms of thinking is from the very outset considered to be determined by the contents rather than by an autonomous developmental logic. The course suggested here for overcoming the abstract dichotomy of the form and content of thinking in Piaget's theory (and simultaneously the conclusion following from it that the development of human thought attains a final structural stage on reaching the stage of formal thought) must be distinguished from a widespread practice of revising the Piagetian theory on this point. This approach is based on a reinterpretation of the concepts of assimilation and accommodation. Assimilation is viewed as the structuring of environmental data by means of existing cognitive structures and accommodation as a complementary process of adapting these structures to the environment (see, e.g., [15], p. 147f.; [23]; [22], p. 41ff.). First, it must be remembered that Piaget explicitly rejects such an interpretation of the concepts concerning the logico-mathematical structures of thinking (see,

e.g., [38], p. 342ff.). His objection that those who interpret the cognitive structures as the result of a "simple progressive accommodation" to the environment take the easy way out, stems from the recognition that his theory would indeed not adequately explain the logico-mathematical structures in terms of the progressive accommodation even if it were to be assumed that they are prefigured in the structure of the objects, because the coordination of actions, formally understood, excludes the mediation between object structure and cognitive structure.

However, the problem presents itself in a different light if, on the basis of interactionist theories, the social constitution of cognitive structures and social reality are taken into account equally. In this case object structure and cognitive structure as structures of social action are from the outset of the same nature. However, with regard to the theoretical consequences this is not very helpful, since the assumption of the social constitution of cognitive structures, applied to the logico-mathematical structures of the knowledge of nature, implies that the mathematical form of the laws of nature is socially constituted as well—not a very convincing thought!

Closer to the revision of Piaget's theory suggested here are the efforts of developing a "dialectical theory of the development of cognitive structures" evident in the American discussions over the last few years based partly on a response to Hegel's philosophy and dialectical materialism, partly on a debate prompted by the results of Soviet psychology (see [41]; for the Soviet discussion of the works of Piaget see [10]). However, up to now, the contributions presented hardly amount to more than a theoretical program. The only comprehensive approach is Riegel's theory which expands Piaget's stages of cognitive development beyond the stage of formal operations adding a stage of "dialectical operations."

[4] See [4], p. 7ff.
[5] Piaget has frequently been criticized for the "impreciseness" of his empirical investigations. In the wake of the "cognitive reversal" many of his investigations were repeated as a result of this criticism claiming to confirm or refute Piaget's results by means of "exact" empirical methods. For the seriation experiment alone the following investigations conducted over the last few years and referring more or less directly to the experiment can be cited: [3]; [12]; [13]; [16]; [18]; [24]; [28]; [32]; [45]; [46]; [49]; [50]; [51]; [54]; [57]. What all these investigations have in common is that they transform Piaget's experiments into psychometric test arrays with which the existence or nonexistence of a cognitive structure is to be examined by means of the achievements generated whatever their definition may be. Only in one of these investigations formative data are collected at all ([57]). Consequently, the criterion for evaluating formative data which Piaget uses as an indication for the development of a cognitive structure is completely disregarded: the transition from empirical certainty to logical necessity. At best this criterion is used for the classification of answers which serve to explain certain problem solutions or it is operationalized as the "surprise" with which a child reacts to the unexpected result of an experiment.

The methodological discussion on the execution of the experiments is just as narrow. By way of example, the controversy between Braine and Smedslund on the question whether the test performance achieved with respect to the psychometrically operationalized seriation experiment permitted the conclusion that children think transitively was carried on without recourse to Piaget's methodology for deciding such a question (see [30], p. 412 ff.). The methodological discussion is essentially concerned with two questions: How should the problems be presented? How should the achievements be classified? (see [27]; [42]; [5]) The problem-solving process is largely ignored, and the process of interaction between the test person and the experimenter employed by Piaget for initiating problem-solving processes and directing them in such a way as to be able to generate theoretically relevant information is hardly ever seriously discussed. Under these circumstances the psychometrically produced data hardly allow any conclusions whatsoever with regard to the existence of cognitive structures. The interpretations of the results of such investigations rarely amount to more than unrestrained speculations on the basis of agreeing or disagreeing with Piaget's theory. Such investigations are nevertheless often presented with the claim of having demonstrated the untenability of Piagetian theorems. The recently published compilation by Siegel and Brainerd under the title *Alternatives to Piaget* may serve as an example ([47]).

[6] See [34], p. 4f.
[7] See [8], p. 171ff.
[8] [34], p. 16, protocol of the boy Wir (code name).
[9] [34], p. 25ff.
[10] See [34], p. 161ff., as well as the study expressly dedicated to this cognitive structure: [39].
[11] In Russell's version this antinomy has the following form: "Let w be the class of all those classes which are not members of themselves. Then, whatever class x may be, 'x is a w' is equivalent to 'x is not an x.' Hence, giving to x the value w, 'w is a w' is equivalent to 'w is not a w.'" ([55], p. 60)

If this argumentation is translated into the terminology of modern axiomatic theories, it can be represented as follows:

The universal set formation axiom states that for any statement $P(x)$ which can be constructed within a given formalized theory and which contains x as a free variable the following statement is true:

$$\exists w \forall x \, (x \in w) \Leftrightarrow P(x)$$

If the proposition $P(x)$ is substituted by the proposition $x \notin x$, the result is

$$\exists w \forall x \, (x \in w) \Leftrightarrow (x \notin x)$$

and if x denotes the set w, we get the following:

$$\exists w \, (w \in w) \Leftrightarrow (w \notin w) \, .$$

In his paper *Prädikationstheorie und Widerspruchsproblem* (Predicate Theory and Problems of Contradiction) Ruben called attention to the connection between this antinomy and the unrestricted extensional interpretation of predicates (see [43], p. 121).

[12] The fact that Russell's antinomy is a result of the reflective interpretation of a statement is not at all as obvious as it is with the so-called semantic antinomy which was already well known to the philosophers of ancient Greece: A Cretan said that all Cretans are liars and that all other statements made by Cretans were certainly lies. In this semantic antinomy the implicit statement "I am lying"—which can neither be true nor be false—is obviously interpreted reflectively as a statement concerning the relationship of this statement to its object (namely, the statement itself).

The so-called logical antinomies discovered towards the end of the 19th century, of which Russell's antinomy is an example, differ from such semantic antinomies in that they do not contain any metalinguistic concepts ("truth," "lie," etc.). According to the view presented here they are nevertheless the result of a reflective activity of thought. However, in this instance, the difference consists in the fact that the reflective activity of thought capable of being represented in metalinguistic terms is projected into an expanded system of object language that emerged from a reflective abstraction and is not free of contradictions.

This can be illustrated by means of Russell's antinomy. In preaxiomatic ("naive") set theory there was the tacit assumption that every proposition $P(x)$ with a variable x was fully defined by its extension—that is, it could be interpreted as the class of those objects for which it is true. This is a metalinguistic assumption which did not cause any problems as long as it was not used to turn the extension of a statement into its own object. If such reflective interpretations were tacitly avoided (because, strictly speaking, they were semantic considerations), no contradictions arose.

If, on the other hand, the metalinguistic consideration that a statement P (x) is fully defined by its extension, is added to set theory as a rule of inference from object language—that is, if

$$\exists w \forall x (x \in w) \Leftrightarrow P(x)$$

is accepted as an axiom, then the set theoretical formal system, tacitly restricted previously, is expanded by metalinguistic considerations in the guise of object language. Reflective inferences can be performed as formal deductions; their reflective nature only emerges with reference to its naive origin.

The argument presented here is based on such a reference. The proposition $x \notin x$ with the free variable x is a one-place predicate that can be stated about x the extension of which is postulated by means of the universal axiom of set formation presented above as the object of possible statements:

$$\exists w \forall x (x \in w) \Leftrightarrow (x \notin x) .$$

If this object w is introduced into the proposition universally valid for all objects x, then this is really the application of a predicate to its own extension. It is precisely from this that Russell's antinomy arises

$$\exists w (w \in w) \Leftrightarrow (w \notin w) .$$

In order to eliminate this contradiction, the validity of the set formation axiom must be restricted. The most common form of restriction is the use of Zermelo's "axiom of separation." According to this axiom only the following proposition is true of every proposition P (x) with the free variable x:

$$\forall a \exists w \forall x (x \in w) \Leftrightarrow P(x) \wedge (x \in a) .$$

Let us consider the effects of this restricted axiom! It states that a proposition P (x) can only select a subset from an already existing set a. Presupposing this axiom Russell's antinomy takes the following form:

$$\exists w (w \in w) \Leftrightarrow (w \notin w) \wedge (w \in a) .$$

This does not result in a contradiction, but in the proposition $w \notin a$. Since the set a from which w was selected was chosen completely at random, this conclusion means: However the set a was chosen, w could not have been contained as a member. Thus, Russell's antinomy turns into an inference that expresses the constructive nature of the self-application of the predicate $w \notin w$. Without giving proof I might add that Zermelo's "axiom of separation" divides propositions P (x) with a free variable x into those with an extant extension and those without it. The distinguishing criterion is the deductibility of the proposition

$$\exists w \forall x (x \in w) \Leftrightarrow P(x) .$$

The propositions $x \notin x$ and $x = x$ which can be interpreted reflectively are not among the propositions that meet this criterion. Which is to say, that there is neither the set of all sets that do not contain themselves as a member, nor the set of all sets per se (see [4], Chap. II, para. 1).

[13] If the concept of cognitive structure is used in accordance with the purported explanatory force and the general scope claimed in Piaget's theory of genetic epistemology, two questions initially arise that cannot be dealt with in this context. However, the problems associated with them should at least be pointed out.

The first question concerns the horizontal differentiation of the concept. Take a certain stage of cognitive development: Which cognitive structures are typical of it and what is their systematic relationship to each other? This question can initially be answered completely externally by asking which cognitive structures Piaget and his successors investigated and which principles they used in defining their questions. One thing becomes immediately apparent: At the center of the investigation were mainly those categories and concepts in the tradition of idealistic philosophy which were considered the universally valid categories of empirical experience par excellence. In particular, Piaget's most important investigations can be grouped around the categories and concepts of the Kantian system (to which there are numerous explicit references as well). All the important concepts of the system of "pure reason" the validity of which Kant assumes as *a priori* given in the pure science, are to be found as objects of Piaget's investigations: space, time, number, the logic of concepts, deductive analysis, causality, heaviness (Kant considers the proposition "all physical bodies are heavy" an *a priori* synthetic judgment); furthermore: the distinction of intuition and reason. Just as Kant stipulated the contrast between "pure reason" and "practical reason" with the "categorical imperative" as the universal principle, thus treating moral questions as cognitive questions requiring an understanding of this principle, Piaget examined the development of "moral judgment" as a process of the development of cognitive structures (see [33]). Only the concepts of the Kantian "judgment" and, consequently, aesthetic and teleological concepts were not considered.

This affinity of Piaget's investigations for the concept of reason of German idealism and their closeness to Kantian philosophy imply the existence of tacit assumptions with regard to answering the question of a systematic arrangement of cognitive structures. Piaget realizes this which is one of the reasons why he refuses to separate the psychological from the epistemological question of the development of cognitive structures. "Genetic epistemology" is meant to provide the context in which the cognitive structures form a unified system from which particular questions theoretically relevant to psychological research can be derived.

However, the attention which Piaget's work has by now received in the field of psychological research, hardly extends to this dimension of the problem. As cognitive structures, the objects of investigation introduced into psychological research by Piaget, appear to be worthy of attention in and of themselves. In the investigations on seriation mentioned above, the question whether these investigations are meaningful or not, for example, does not arise at all or arises only indirectly by viewing the investigation of seriation in terms of the verification or falsification of Piaget's theory.

The investigations on the socio-cognitive development are an exception (see [23]; [22]). The overarching theories providing the criteria of the relevance of particular questions are in this case the sociological (or socio-psychological) theories of human interaction. It is in this context that Piaget's question about the system of cognitive structures is asked and answered anew. This theoretical framework, however, is too narrow to provide new answers to the question concerning the categories of the understanding of nature. As a rule, the investigation of socio-cognitive structures is considered to complement Piaget's theory, and the questions concerning the theory of knowledge only arise with regard to the primacy of cognitive or socio-cognitive structures and their specific relationship to each other.

As far as Piaget's solution is concerned, it consists in the construction of a "logic" of its own in a system of "logical" and "infralogical" operations the genetic stages of which are to be characterized by certain comprehensive structures such as the "groupings" or the "INCR" group. This solution will not be discussed here in detail (see, e.g., [36]), since it is obvious that it is completely determined by the assumption of two cognitive modes (one empirical and the other logico-mathematical). Piaget's system is the attempt of demonstrating the independence of the system of cognitive structures of the cognitive contents in every detail by reconstructing their formal genesis. If the assumption is questioned, the problem of systematizing cognitive structures presents itself in a completely different way.

This becomes apparent in the revision of Piaget's theory by Bruner. Because of its universalism, Piaget's theory appeared to be a fairly unsuitable instrument for interpreting the results of intercultural comparisons. Nothing was more obvious than to assume that the development of cognitive structures was (to a limited extent) dependent on cultural contents, thus toppling Piaget's entire construction. Bruner calls his conception of cognitive structures as collective instruments of action "instrumental conceptualism." He characterizes it as follows: "In brief, it is a view that is organized around two central tenets concerning the nature of knowing. The first is that our knowledge of the world is based on a constructed model of reality, a model that can only partially and intermittently be tested against input. Much of the structure of our cognitive models is quite remote from any direct test, and that rests on what might be called an axiomatic base—our ideas of cause and effect, of the continuity of space and time, of invariances in experience, and so on. It seems not unlikely (no stronger phrase than that is justified) that some of this axiomatic structure informing our models of reality is already given in the innate nature of our three techniques for representing or 'modeling' reality: action, imagery, and symbolism. ... If the first central tenet of instrumental conceptualism is our idea of the model or representation and its constraints (the conceptualist side of the matter), the second is that our models develop as a function of the uses to which they have been put first by the culture and then by any of its members who must bend knowledge to their own uses." ([8], p. 319f.) Thus, Bruner constructs the dependence of cognitive structures (the enumeration of which reveals their origin in Piaget's theory and consequently refers to the principle of an *a priori*) in two stages. First, by means of a theory of representation he relates them to mediators between individual thought and cultural heritage. Second, he conceives of cognitive structures with regard to their cultural function as collective means of the representation of reality determined by the goals of their uses.

However, this construction is subject to specific assumptions that are questioned here. First, there are the assumptions of his theory of representation: According to this theory, representations are determined by three modes of representation (enactive, iconic, and symbolic representation) and their specific "constraints"; they are, furthermore, independent of any cultural determination. Second, there are the assumptions that are the basis of his concept of instrumentality: There is no connection between the cognitive instruments and the instruments of action that can be theoretically explained.

The implications of Bruner's revision of Piaget's theory of cognitive structures for the questions concerning its systematic representation turn out to be pragmatic in accordance with the abstractness of his reference to culturally determined uses. It is easier to find them in his educational program than in his efforts concerning the development of a psychological theory: Cognitive structures can be conveyed in school; their systematic nature follows from the structures of scientific disciplines (see [6]).

Piaget's "genetic epistemology" and Bruner's "instrumental conceptualism" are the only comprehensive answers I know of which cognitive psychology has provided concerning the systematic representation of cognitive structures. (Both emerged out of a concern which goes beyond narrowly defined psychological problems. The fact that specialized disciplines dominate scientific inquiry may serve to explain their uniqueness.) On the background of the considerations presented here both answers are unsatisfactory. Simply stated: the first, because it is an attempt to provide only a formal explanation for cognitive structures, the second, because it cannot in fact adequately conceive of the correlation between form and content and thus represents any form of thought developed by the institution of science as part of the culture and thus as valid.

Apart from the question concerning the horizontal differentiation of the concept of cognitive structure there is the second question concerning the vertical differentiation. How are the stages of cognitive development to be distinguished? How are the stages of the development to be determined?

The answer to this second question is of course dependent on the answer to the first question inasmuch as the characterization of each stage is affected by it. However, leaving aside the differentiations following from this, it can be stated that the periodization given by Piaget has largely been accepted in cognitive psychology (provided that the existence of cognitive structures is not denied on principle). Brief mention of this rough structure of individual courses of development should at least be made at this point.

Piaget distinguishes four developmental stages between birth and adolescence. In the first stage of "sensori-motor intelligence," which comprises the time up to about age two, the child acquires the ability of coordinating perception and action and thus a "practical" intelligence which is expressed in the coordination of the spatial, temporal, and causal relations of its actions. An important cognitive structure already acquired during this stage is the "schema of the permanent object," the ability of identifying certain objects as identical with themselves and of recognizing them. The second stage is the stage of "preoperational thought," which continues up to ages seven or eight. This stage is characterized by the acquisition of the "symbolic function" (something can be thought of as a substitute for something else), in particular by the acquisition of language. At this stage thought is characterized by its irreversibility, already sustained, however, by the ability of dealing with "representations." Using the terminology to be developed later, it can be said that the cognitive structure of linguistic competence is developed at this stage and, furthermore, a cognitive structure of semantics as well, which consists, however, initially only in references to sense perception; it is in particular not yet logico-mathematically structured. The third developmental stage is the stage of "concrete operations" which continues through ages eleven and twelve. At this stage the thought operations become reversible and, consequently, are organized in systems which are no longer confined to intuition. At this stage those cognitive structures develop which are defined as logico-mathematical structures; however, their formal representation is initially not yet independent of the concrete objects and actions or their conception. Language is not yet the representative of this thinking. This changes at the fourth stage, the stage of "formal operations." At this stage the ability to reverse the thought operations by means of the linguistic representatives of reality develops, notably on the basis of the cognitive structure of a semantics which is now logico-mathematically structured as well. Thought as "abstract hypothetical thought," conveyed by means of language and thus independent of sense perception and imagination now refers to its object.

[14] See [44], p. 254ff.

[15] The use of fingers to do arithmetic and the use of numeric symbols is already associated with groupings that prepare the ground for place-value systems. The invention of the reckoning board was able to make use of this precondition; however, in this case the grouping was of necessity expanded into a system (usually without an immediate change of the notation of the numbers as well). Thus, the reckoning board initiates the construction of a place-value system. At the same time the reckoning board does contain the zero in the form of unoccupied cells independent of any explicit denotation or conceptual elaboration. The extensive use of the reckoning board leads to developments that correspond to an expansion of the number concept. This can be demonstrated especially well in terms of the highly developed arithmetical technique in ancient China (see [21], p. 9ff., as well as the translation of a treatise on arithmetic from the early Han period: [52]). The form of division represented by dividends and divisors to be transformed successively amounted implicitly to a representation of fractions by means of numerator and denominator. Chi-

nese arithmeticians used it to determine the entire system of all fractional rules although they did not possess a symbolic representation of fractions (with the exception of a few individual symbols for $\frac{1}{2}$, $\frac{1}{3}$, etc.) and although linguistically they were only able to identify them as results of a process of division. The denotation of systems of linear equations by matrices on the reckoning board was just as generative. It provided a system of rules for the solution of systems of equations before the concept of an equation even existed. Moreover, the notation by matrices required the denotation of negative coefficients. This denotation was transferred to the numeric symbols; the reckoning board furnished the negative numbers, as it were, for free.

A particularly convincing example of the importance of calculating devices is provided by the calculi and their use in ancient Greece. The placement of calculi into sequences of geometrical patterns is one of the main sources of knowledge the Pythagoreans had of number theory. They follow empirically from the application of the method (see [62], p. 81ff.).

Ancient Egyptian arithmetic, on the other hand, is an example of how calculating devices can also limit thinking and conceptualization. Our knowledge of Egyptian arithmetical procedures comes primarily from two papyri. We do not know the calculating devices that were used; however, there is every indication that Egyptian mathematics used symbols from the very beginning and that reckoning boards were not used. Whatever the case may be, arithmetical procedures were based solely on the operations of addition, division into halves, and doubling, operations that could be performed by using a certain tabular representation of numerical notations without the use of any other calculating devices. All arithmetical procedures were reduced to these basic operations. The procedures of multiplication and division resulting from this are often considered to be nothing more than an historical oddity. However, an analysis of the fractions shows that Egyptian arithmetics generated a highly complex system of numerical notations with its own structure which does not contain multiplication and division as basic operations and, for this reason, does not permit the obvious representation of fractions as a numerical ratio (numerator and denominator). This explains the peculiar representation of fractions in Egyptian mathematics as sums of different unit fractions which, in its clumsiness, borders on folly (see [31]).

Most of the available studies on the historical development of the number concept do not adequately take into account the dependence of this development on the material means of arithmetic. If, however, the objective possibilities inherent in the available means of arithmetic are disregarded, a difficulty emerges. It becomes impossible to decide when a mathematical concept was formed. It is quite obvious that the emergence of a particular expression or symbol is too narrow a criterion. Thus, if the historical procedures and texts are interpreted with regard to the underlying cognitive structures (without paying attention to the means used), then the boundary between reconstruction and speculation becomes fuzzy. The uncertainties as to who first used the zero or the negative numbers or a pure place-value system are by no means merely attributable to inadequate historical records; rather, they are, in part at least, due to a disregard of the relationship between the material means of arithmetic and the cognitive structures based on them. The historical reconstruction must take into account that the material means of arithmetic already comprise the potential of cognitive structures which emerge in the form of corresponding concepts and mental operations as an outcome of their usage in historically relevant contexts.

The available studies on the ontogenetic development of the number concept, especially Piaget's study on which these presentations are largely based ([34]), are equally inadequate. Although Piaget relates the emerging operations of the cognitive structure of the number concept to material actions, their organization into a structure is derived from an autonomous logic of development. The actual preconditions of these actions (especially the availability of the material means) are always either assumed or not discussed at all. In particular, an analysis of the function of the means qua means in the environment of the child and consequently of the objective learning potentials present in this environment is lacking. The limited value of the reconstruction of the development of the number concept in developmental psychology concerning problems in mathematics education is due to this omission.

[16] The core of these brief reflections on the relationship between cognitive structures and material action consists in understanding the relationship of thought to its object (and, *a fortiori,* the relationship of language to its objects implied in speech) as a relationship actually established within material action by means of material mediation. This generation of relationships by means of material mediation presupposes the reversal of the relationship between ends and means. Whereas in animal behavior the means are subordinated to the ends and, as means, do not survive the attainment of the goal, this relationship is reversed in the specifically human relationship of acquisition which is especially evident in the material survival of the means qua means—that is, the relative permanence of cultural achievements.

The relationship of thought to its object (and, accordingly, the relationship of the linguistic symbols to the corresponding objects presupposed in speech) is generally called "meaning." Meaning is the imagined and considered conception of the real object (or of the object presupposed by linguistic symbols); the actual object (or the material linguistic symbol, respectively) becomes the vehicle of meaning as a result of its mental assignment.

These reflections must find expression in the concept of meaning. Meaning must be understood as an objective relationship established in the context of material action by means of material mediation, a relationship that is merely reflected in the ideal relationship of the thought to the object of thought (or, conversely, in the ideal relationship of the conceived object to the conceived symbol of the object, known in cognitive psychology as "symbolic function").

This theoretical conclusion which emerges when the individual development of cognitive structures is seen in the socio-historical context defining the material means as persisting means, constitutes one of the central arguments of the German "Critical Psychology" in opposition to the predominant theoretical schools of thought in Western psychology, in particular, however, in opposition to cognitive psychology. Holzkamp gave programmatic expression to the understanding of the concept of meaning to be derived from the genetic dependence of thought (and of language) on the social process of labor: "If the contents of perception are merely viewed in terms of their figural-qualitative properties then it is concealed that the perceived states of affairs in the real world are *meaningful* as such. ... Certain types of objects of perception, in particular the linguistic facts in the narrow sense, are the vehicles of *the meaning of symbols* with regard to their essential characteristic. They are 'representative' in character, represent something else, point beyond themselves. Words like 'hammer' or 'child' represent, symbolize, refer to certain states of affairs that they 'denote.'—However, in phenographical terms it is *wrong* to assume that the facts represented by linguistic symbols *are merely affected by the meanings of symbols, but are meaningless in themselves.* ... For this reason the *meanings of symbols* must be contrasted with a different kind of meanings, namely that of the objective or personal *meanings of objects.* In the meaning of an object there is no reference to a third entity, to something referred to as is the case with the meaning of a symbol; rather, the meaning of an object signifies *meaning in the context of human activity.* A 'hammer,' for example is not just the expression of a specific form and a specific type of coloring; rather, it is a complex objective unity of meaning which includes that it was made by humans, that it is used for hitting, its effective use, that one must be careful with it, etc., *all of this being a unified and unambiguous total characteristic of the hammer as a real, perceptible object.*—In this context, the objects are not perceived as units of meaning, but *in association with structures of meaning* which are materially given as well. I notice that the hammer which I put on top of the typewriter does not belong there, notice at the same time the relationship of the hammer with the tool box in the corner of the room, etc. The perceptible references of meaning between the objects of perception are inexhaustible, become increasingly fuzzy towards the edge of the field of perception in which the object is located and change with blurred fringes from that which is present to that which is merely envisioned." ([20], p. 25f.).

Let us compare this concept of meaning to that of cognitive psychology! Piaget has reconstructed the symbolic function from the development of material action. This necessarily leads to basing the meaning of symbols on the meaning of objects (which Piaget calls "elementary meanings" or "meanings of a lower order"). Piaget writes: "To assimilate a sensorial image or an object, whether through simple assimilation, recognition, or generalizing extension, is to insert it in a system of schemata, in other words, to give it a 'meaning.' Regardless of whether these schemata are global and vague or, as in the recognition of an individual factor, they are circumscribed and precise, consciousness does not know any state except in reference to a more or less organized totality. Ever since then it is necessary to distinguish, in every mental element, two indissolubly bound aspects whose relationship constitutes meaning: the signifier and the signified. With regard to 'meanings' of a higher order, which are also collective meanings, the distinction is clear: the signifier is the verbal expression, that is, a certain articulated sound to which one has agreed to attribute a definite meaning, and the signified is the concept in which the meaning of the verbal sign consists. But with regard to elementary meanings (significations) such as that of the perceived object, or even ... that of the simply 'presented' sensory images, the same applies." ([35], p. 189f.)

Thus, Piaget considers the symbolic meanings to be based on objective meanings explored in material action. However, the assumption of an autonomous logic of development of the schemata of action prevents this discovery from being pursued any further. In this case the generalization of the concept of meaning merely signifies that the attribution of meaning to linguistic symbols is preceded by a more comprehensive attribution of meaning with regard to all contents of perception. Thus, cognition must necessarily become a gigantic enterprise of interpretation. At the point where cognition is at its most concrete, the object of action vanishes into an idea: "The 'signified' of objective perceptions such as that of the mountain I see from my window or the inkwell on my table is the objects themselves, definable not only by a system of sensorimotor and practical schemata (climbing a mountain, dipping pen in ink) or by a system of general concepts (an inkwell is a container which ..., etc.), but also by their individual characteristics: position in space, dimensions, solidity and resistance, color in different lights, etc. Now the latter characteristics, although perceived in the object itself, presuppose an extremely complex intellectual elaboration; for example, in order to attribute real dimensions to the little spots which I perceive to be a mountain or an inkwell, I must place them in a substantial and causal universe, in an organized space, etc., and accordingly construct them intellectually. The signified of a perception—that is to say, the object itself—is therefore essentially intellectual. No one has ever 'seen' a mountain or even an inkwell from all sides at once in a simultaneous view of their different aspects from above and below, from East and West, from within and without, etc. In order to perceive these individual realities as real objects it is essential to complete what one sees by what one knows. Concerning the 'signifier,' it is nothing other than the few perceptible qualities recorded simultaneously and at the present time by my sensory organs, qualities by which I recognize a mountain and an inkwell. Common sense, which prolongs in each of us the habits of infantile realism, certainly considers the signifier as being the object itself and as being more 'real' than any intellectual construction. But when one has understood that every concrete object is the product of geometric, kinematic, causal, etc., elaborations, in short, the product of a series of acts of intelligence, there no longer remains any doubt that the true signified of perception is the object in the capacity of intellectual reality and that the apprehensible elements considered at a fixed moment of perception serve only as signs, consequently as 'signifiers.'" ([35], p. 190) In this instance the world is truly upside down!

The world as intellectual construction, this idea is too philosophical to be accepted without further ado in the context of a particular scientific discipline that experiences its subject matter as an object of empirical investigation in its actual form. Thus, Bruner's revision of Piaget's theory can ultimately be understood as an attempt to avoid these idealistic implications. He writes: "... we shall be concerned with intellectual growth as it is affected by the way human beings gradually learn to represent the world in which they operate—through action, image, and symbol. We shall attempt to show that these representations or constructions of reality cannot be understood without reference to the enabling powers of a culture and the heritage of man's evolution as a primate." ([8], p. 6) To this end Bruner complements Piaget's theory by means of the assumption that the cognitive structures can be effective and develop only by means of their representation in a material medium: in the medium of action, the medium of the image, and the medium of the symbol. This, however, does not avoid the idealistic implications at all. To the contrary, Piaget's comprehensive concept of meaning is once more restricted. Only that which already represents an idea (in the medium of the symbol, the image, or the action) can have meaning. Actions are only associated with meaning inasmuch as they facilitate representation. "... the idea of enactive *representation* (rather than simply action) is premised on the conviction that, insofar as action is flexibly goal-oriented and capable of surpassing detours, it must be based on *some* form or representation that transcends a direct linking of stimuli and responses." ([8], p. 11)

Thus, Piaget's strictly idealistic position in fact proves to be more materialistic than Bruner's position. This can be demonstrated in terms of the role of the means of actions which Piaget is granting them in terms of the development of intelligence: "This coordination of schemata, which clearly differentiates 'means' from 'ends' and so characterizes the first acts of intelligence properly so-called, insures a new putting into relationship of objects among themselves and hence marks the beginning of the formation of real 'objects.'" ([35], p. 263) According to Piaget, the elementary meanings emerge from the dissociation of ends and means, the discovery of new means by active elaboration of possibilities, and the invention of new means by means of mental combination, meanings which are represented by symbols only at a higher level (see [35], p. 320ff.). Bruner, on the other hand, is forced to make the assumption that the means of material action become means only as "amplifiers" of representations, as extensions of representations within the medium of action or of symbolic operation. This leads him to the conclusion that language is the genetic origin of work and not the other way around: "I would argue that language itself is not what is 'imposed' on experience—as already suggested in my rejection of Vygotsky's idea that language is internalized and becomes inner speech, which is tantamount to thought. Rather, language comes from the same basic root out of which symbolically organized experience grows. I tend to think of symbolic activity of some basic or primitive type that finds its first and fullest expression in language, in tool-using, and finally in the organizing of experience." ([8], p. 43f.) Bruner later expanded this assumption. It explains his attempt to interpret the development of language in terms of its pragmatic function and to provide an interactionistic foundation (see [7]).

[17] See [9], p. 3ff.

[18] Today the most widely used grammar in mathematics is that on which Bourbaki based his system (see [4], para. 1 and 2). It is mentioned here by way of example.

[19] Chomsky's studies on linguistic competence have found entrance into psychological research in numerous ways and have, in the course of the previous decade, without a doubt contributed to the emergence of an extraordinarily broad area of research on the ontogenetic development of language (see the annotated bibliography by [1]). At the same time, these writings encouraged theoretical developments which narrowly focused on questions concerning syntax (see Bruner's criticism in [7], p. 275ff.). The opposition mainly relies on the influence which Mead's interactionism has gained in cognitive psychology (see [17], p. 101ff.; [22]). Linguistic development is no longer to be understood primarily as the development of linguistic competence, but in terms of the background of an interactionist theory of "communicative competence." The criticism takes issue with the fact that the semantics of language is solely defined as a "semantic component" of grammar and that linguistic performance is treated merely as a realization of linguistic competence—sometimes disturbed by the external circumstances of speech.

The consequences resulting from the recourse to interactionism are particularly evident in Bruner's research: "Neither the syntactic nor the semantic approach to language acquisition takes sufficiently into account what the child is trying to do by communicating. As linguistic philosophers remind us, utterances are used for different ends and use is a powerful determinant of rule structures. The brunt of my argument has been that one cannot understand the transition from prelinguistic to linguistic communication without taking into account the uses of communication as speech acts. I have, accordingly, placed greater emphasis on the importance of pragmatic in this transition—the directive function of speech through which speakers affect the behavior of others in trying to carry out their intentions." ([7], p. 283) If my reflections on the function of the material means of action with regard to the development of meanings are correct, then the attempt to explain meanings pragmatically in terms of linguistic use is as one-sided as the overemphasis on syntax in Chomsky's theory.

[20] See [2].

[21] The first postulate claims that it is possible "to draw a straight line from any point to any point," the second that it is possible "to produce a finite straight line continously in a straight line" ([14], p. 154). The introduction of these postulates into a formal deduction obviously generates a statement corresponding to the use of a straightedge in the actual context which is linguistically represented in the statements. The same is true with regard to the use of the compasses for the third postulate which states that it is possible "to describe a circle with any centre and distance." In both cases formal rules concerning the formation of linguistic statements are identified directly with the envisioned performance of material actions using the appropriate tools. In this case the strictness of the Euclidean method is based on directly reflecting the mechanical generation of geometric figures.

Beyond that, however, the strictness of Euclidean proofs is conveyed by grammatical structures and their semantic reference to cognitive structures which refer to the mechanics of geometrical constructions. With regard to the methodology of proof the postulates are relatively insignificant. They do nothing more than ensure the existence of the objects to which the theorems and their proofs refer. For this reason their importance was not recognized for a long time. Geometry stagnated, remained "Euclidean," because it did not go beyond the framework of construction posited by the postulates.

The first thorough expansion of geometry was Descartes' geometry which gave rise to analytical geometry. What was new was that construction did not only make use of straight lines and circles, but of any curves capable of being represented algebraically. Descartes justified his method by pointing out the arbitrariness of being limited to straightedge and compasses. In connection with the description of a mechanical device for the construction of higher-order curved lines capable of being represented algebraically he argues: "... the point B describes the curve AB, which is a circle; while the intersections of the other rulers, namely, the points D, F, H describe other curves, AD, AF, AH, of which the latter are more complex than the first and this more complex than the circle. Nevertheless I see no reason why the description of the first cannot be conceived as clearly and distinctly as that of the circle, or at least as that of the conic sections; or why that of the second, third, or any other that can be thus described, cannot be as clearly conceived of as the first: and therefore I see no reason why they should not be used in the same way in the solution of geometric problems. ... I could give here several other ways of tracing and conceiving a series of curved lines, each curve more complex than any preceding one, but I think the best way to group together all such curves and then classify them in order, is by recognizing the fact that all points of those curves which we may call 'geometric,' that is, those which admit of precise and exact measurement, must bear a definite relation to all points of a straight line, and that this relation must be expressed by means of a single equation." ([11], p. 47f.). This is the whole secret of Descartes' discovery! Instrument building of the 16th and 17th centuries had long since expanded the practical possibilities for the construction of geometrical figures in such a way that Euclid's postulates no longer appeared to be evident presuppositions, but rather arbitrary limitations of thinking. Only a small step was needed to admit the possibilities created by the new means as possibilities of thinking in order to attain the theoretical means of a new geometry. However, several centuries of theoretical work were needed to make full use of these means.

[22] In his study on the theory of predication and the problem of contradiction, Ruben presented a critical evaluation of different semantic interpretations of the subject-predicate relationship using examples from the philosophical tradition ([43], p. 117ff.). The arguments used there can be applied without difficulty to the present discussion in cognitive psychology. However, Ruben uses the concept "meaning" in a different sense than in the present discussion.

[23] [53], p. 254.

BIBLIOGRAPHY

[1] Abrahamsen, A. A.: *Child Language. An Interdisciplinary Guide to Theory and Research.* Baltimore, MD: University Park Press, 1977.
[2] Bernal, J. D.: *Science in History.* London: Watts, 1954.
[3] Blackstock, E. G. & King, W. L.: "Recognition and Reconstruction Memory for Seriation in Four- and Five-Year-Olds." In *Developmental Psychology, 9* (2), 1973, pp. 266–267.
[4] Bourbaki, N.: *Theory of Sets.* Reading, MA: Addison-Wesley, 1968.
[5] Brainerd, C. J.: "Judgments and Explanations as Criteria for the Presence of Cognitive Structures." In *Psychological Bulletin, 79* (3), 1973, pp. 172–179.
[6] Bruner, J. S.: *The Process of Education.* Cambridge, MA: Harvard University Press, 1965.
[7] Bruner, J. S.: "From Communication to Language—A Psychological Perspective." In *Cognition , 3* (3), 1974/75, pp. 255–287.
[8] Bruner, J. S. et al.: *Studies in Cognitive Growth: A Collaboration at the Center for Cognitive Studies.* New York: Wiley, 1966.
[9] Chomsky, N.: *Aspects of the Theory of Syntax.* Cambridge, MA: MIT Press, 1965.
[10] Dawydow, W.: *Arten der Verallgemeinerung im Unterricht. Logisch-psychologische Probleme des Aufbaus von Unterrichtsfächern.* Berlin: Volk und Wissen, 1977.

[11] Descartes, R.: *The Geometry of René Descartes*. Trans. from the French and Latin by D. E. Smith and M. L. Latham. New York: Dover, 1954.
[12] Donaldson, M. & Balfour, G.: "Less is More: A Study of Language Comprehension in Children." In *The British Journal of Psychology, 59* (4), 1968, pp. 461–471.
[13] Estes, K. W.: "Nonverbal Discrimination of More and Fewer Elements by Children." In *Journal of Experimental Child Psychology, 21,* 1976, pp. 393–405.
[14] Euclid: *The Thirteen Books of Euclid's Elements: Translated from the Text of Heiberg with Introduction and Commentary by Sir Thomas L. Heath.* 2nd ed., 3 vols., New York: Dover, 1956.
[15] Goldmann, L.: *Dialektische Untersuchungen.* Neuwied: Luchterhand, 1966.
[16] Gollin, E. S., Moody, M. & Schadler, M.: "Relational Learning of a Size Concept." In *Developmental Psychology, 10* (1), 1974, pp. 101–107.
[17] Habermas, J. & Luhmann, N.: *Theorie der Gesellschaft oder Sozialtechnologie—Was leistet die Systemforschung?* Frankfurt a.M.: Suhrkamp, 1971.
[18] Harasym, C. R., Boersma, F. J. & Maguire, T. O.: "Sematic Differential Analysis of Relational Terms Used in Conservation." In *Child Development, 42* (3), 1971, pp. 767–779.
[19] Heidtmann, B.: "Systemwissenschaftliche Reflexion und gesellschaftliches Sein. Zur dialektischen Bestimmung der Kategorie des objektiven Scheins." In Heidtmann, B., Richter, G., Schnauß, G. & Warnke, C.: *Marxistische Gesellschaftsdialektik oder "Systemtheorie der Gesellschaft"?* Frankfurt a.M.: Verlag Marxistische Blätter, 1977, pp. 69–102.
[20] Holzkamp, K.: *Sinnliche Erkenntnis. Historischer Ursprung und gesellschaftliche Funktion der Wahrnehmung.* Frankfurt a.M.: Athenaeum Fischer, 1973.
[21] Juschkewitsch, A. P.: *Geschichte der Mathematik im Mittelalter.* Basel: Pfalz, 1966.
[22] Keller, M.: *Kognitive Entwicklung und soziale Kompetenz. Zur Entstehung der Rollenübernahme in der Familie und ihre Bedeutung für den Schulerfolg.* Stuttgart: Klett, 1976.
[23] Kohlberg, L.: "Stage and Sequence: The Cognitive-Developmental Approach to Socialization." In Goslin, D. A. (Ed.): *Handbook of Socialization Theory and Research.* Chicago, IL: Rand McNally, 1969, pp. 347–480.
[24] Lawson, G., Baron, J. & Siegel, L.: "The Role of Number and Length Cues in Children's Quantitative Judgments." In *Child Development, 45,* 1974, pp. 731–736.
[25] Luhmann, N.: "Reflexive Mechanismen." In Luhmann, N.: *Soziologische Aufklärung. Aufsätze zur Theorie sozialer Systeme.* Köln: Westdeutscher Verlag, 1970, pp. 92–112.
[26] Luhmann, N.: "Selbst-Thematisierungen des Gesellschaftssystems. Über die Kategorie der Reflexion aus der Sicht der Systemtheorie." In *Zeitschrift für Soziologie, 2* (1), 1973, pp. 21–46.
[27] Lumsden, E. A. & Kling, J. K.: "The Relevance of an Adequate Concept of 'Bigger' for Investigations of Size Conservation: A Methodological Critique." In *Journal of Experimental Child Psychology, 8,* 1969, pp. 82–91.
[28] Maratsos, M. P.: "When is a High Thing the Big One? In *Developmental Psychology, 10* (3), 1974, pp. 367–375.
[29] Miller, G. A., Galanter, E. & Pribram, K. H.: *Plans and the Structure of Behavior.* New York: Holt, Rinehart & Winston, 1965.
[30] Miller, S. A.: "Nonverbal Assessment of Piagetian Concepts." In *Psychological Bulletin, 83* (3), 1976, pp. 405–430.
[31] Neugebauer, O.: *Die Grundlagen der ägyptischen Bruchrechnung.* Berlin: Springer, 1926.
[32] Palermo, D. S.: "Still More About the Comprehension of 'Less'." In *Developmental Psychology, 10* (6), 1974, pp. 827–829.

[33] Piaget, J.: *The Moral Judgement of the Child.* Trans. by M. Gabain. New York: Harcourt, 1932.
[34] Piaget, J.: *The Child's Conception of Number.* Trans. by C. Gattegno and F. M. Hodgson. London: Routledge & Kegan Paul, 1952a.
[35] Piaget, J.: *The Origins of Intelligence in Children.* Trans. by M. Cook. New York: International Universities Press, Inc., 1952b.
[36] Piaget, J.: *Structuralism.* New York: Basic Books, 1970.
[37] Piaget, J.: "The Theory of Stages in Cognitive Development." In Green, D. R., Ford, M. P. & Flamer, G. B. (Eds.): *Measurement and Piaget.* New York: McGraw-Hill, 1971a, pp. 1–11.
[38] Piaget, J.: *Biology and Knowledge: An Essay on the Relations Between Organic Regulations and Cognitive Processes.* Chicago, IL: University of Chicago Press, 1971b.
[39] Piaget, J. & Inhelder, B.: *The Early Growth of Logic in the Child (Classification and Seriation).* Trans. by G. A. Lunzer and D. Papert. London: Routledge & Kegan Paul, 1964.
[40] Piaget, J. & Inhelder, B.: *Memory and Intelligence.* Trans. by C. Gattegno and F. M. Hodgson. London: Routledge & Kegan Paul, 1973.
[41] Riegel, K. F.: "Toward a Dialectical Theory of Development." In *Human Development, 18,* 1975, pp. 50–64.
[42] Rothenberg, B. B.:"Conservation of Number Among Four- and Five-Year-Old Children: Some Methodological Considerations." In *Child Development, 40* (2), 1969, pp. 383–406.
[43] Ruben, P.: *Dialektik und Arbeit der Philosophie.* Köln: Pahl-Rugenstein, 1978.
[44] Schurig, V.: *Die Entstehung des Bewußtseins.* Frankfurt a.M.: Campus, 1976.
[45] Siegel, L. S.: "Development of the Concept of Seriation." In *Developmental Psychology, 6* (1), 1972, pp. 135–137.
[46] Siegel, L. S.: "Development of Number Concepts: Ordering and Correspondence Operations and the Role of Length Cues." In *Developmental Psychology, 10* (6), 1974, pp. 907–912.
[47] Siegel, L. S. & Brainerd, C. J. (Eds.): *Alternatives to Piaget: Critical Essays on the Theory.* New York: Academic Press, 1978.
[48] Steiner, G.: *Mathematik als Denkerziehung. Eine psychologische Untersuchung über die Rolle des Denkens in der mathematischen Früherziehung.* Stuttgart: Klett, 1973.
[49] Townsend, D. J.: "Children's Comprehension of Comparative Forms." In *Journal of Experimental Child Psychology, 18,* 1974, pp. 293–303.
[50] Trabasso, T. & Riley, C. A.: "The Construction and Use of Representation Involving Linear Order." In Solso, R. L. (Ed.): *Information Processing and Cognition. The Loyola Symposium.* Hillsdale, NJ: Erlbaum, 1975, pp. 381–410.
[51] Trabasso, T., Riley, C. A. & Wilson, E. G.: "The Representation of Linear Order and Spatial Strategies in Reasoning. A Development Study." In Falmagne, R. J. (Ed.): *Reasoning: Representation and Process in Children and Adults.* Hillsdale, NJ: Erlbaum, 1975, pp. 201–229.
[52] Vogel, K. (Ed.): *Chiu Chang Suan Shu: Neun Bücher Arithmetischer Technik. Ein chinesisches Rechenbuch für den praktischen Gebrauch aus der frühen Hanzeit (202 v. Chr. bis 9 n. Chr.).* Braunschweig: Vieweg, 1968.
[53] Vygotsky, L. S.: "Thinking and Speech." In Rieber, R. W. & Carton, A. S. (Eds.): *The Collected Works of L. S. Vygotsky, Vol. 1.* New York: Plenum, 1987, pp. 37–285.
[54] Weiner, S. L.: "On the Development of More and Less." In *Journal of Experimental Child Psychology, 17,* 1974, pp. 271–287.
[55] Whitehead, A. N. & Russell, B.: *Principia Mathematica.* 2nd ed., 3 vols., Cambridge: Cambridge University Press, 1927.

[56] Wussing, H.: *Mathematik in der Antike. Mathematik in der Periode der Sklavenhaltergesellschaft.* Leipzig: Teubner, 1962.
[57] Youniss, J.: "Inference as a Developmental Construction." In Falmagne, R. J. (Ed.): *Reasoning: Representation and Process in Children and Adults.* Hillsdale, NJ: Erlbaum, 1975, pp. 231–245.

CHAPTER 3

PHILOSOPHICAL AND PEDAGOGICAL
REMARKS ON THE CONCEPT "ABSTRACT"*
[1982]

THE COMMON USAGE OF THE CONCEPT IN THE
DIDACTICS OF MATHEMATICS

The nature of mathematical objects seems to be appropriately characterized by the concept "abstract." Many theorists of education make use of this concept without giving it much thought. When used in this way, the concept "abstract" functions primarily as a metatheoretical concept. There is no mathematical definition of the concept. It is used when talking about mathematics.

The opposite of "abstract" is "concrete." Whatever is not abstract is concrete and vice versa. Both concepts can refer to the same set of facts. If that is the case, they typically characterize different forms of representation. I would hardly elicit any protest from theorists of mathematics teaching if I were to say that the usual definition of the convergence of a sequence constituted an abstract definition—the story of Achilles who tries apparently in vain to catch up to the turtle, a concrete example of convergence.

It is generally considered a worthwhile goal for students to learn to think abstractly in the course of mathematical instruction. Only practical experience imposes a number of restrictions: To be successful, it is necessary to proceed from concrete examples, and it is not advisable to advance to abstract objects too early.

All this sounds perfectly innocent. It is nevertheless somewhat astonishing that, although the concepts "abstract" and "concrete" are constantly being used in the literature on mathematics teaching, they themselves are only rarely scrutinized. As a rule, clear definitions are wanting. Yet a number of difficulties are obvious.

---

* Lecture delivered at the Institut für Didaktik der Mathematik der Universität Bielefeld, October 31, 1981.

Thus, for example, the normative didactic value that abstract thinking is a worthwhile goal is ambivalent. The goal of mathematics teaching is not just to encourage abstracting and abstract thinking, but the abstract concepts and results are to be applied to concrete examples as well. Consequently, "concretizing" also appears to be a meaningful activity in the context of mathematical teaching and—one might add—"concrete" thinking appears to be a worthwhile educational goal.

It seems, however, that we are not really dealing here with a specific problem of the didactics of mathematics, but with a much more general epistemological problem. I would, therefore, like to introduce some basic theoretical models of the process of abstraction which have been of some importance in the philosophical discussion concerning the concept "abstract" before addressing in detail the question of the meaningful use of the concept "abstract" in the didactics of mathematics.

ABSTRACTION AS THE ISOLATION OF QUALITIES

The most common concept of abstraction which—rightfully or not—is often called the Aristotelian concept of abstraction defines abstraction as the omission of qualities.

And since, as the mathematician investigates abstractions (for in his investigation he eliminates all the sensible qualities, e.g., weight and lightness, hardness and its contrary, and also heat and cold and the other sensible contrarieties, and leaves only the quantitative and continuous, sometimes in one, sometimes in two, sometimes in three dimensions, and the attributes of things *qua* quantitative and continuous, and does not consider them in any other respect. ...)([1], 1061a)

This common concept of abstraction is based roughly on the following conception of the process of cognition: Experience based on our sense perceptions constitutes the starting point of our knowledge of the world. This starting point is denoted by the term "concrete." The mind proceeds from the concrete experiences in two ways.

1. It decomposes the concrete experience of an object into individual qualities.
2. It emphasizes particular qualities by omitting other qualities—that is, by abstracting from them. In this way, general concepts emerge under which the concrete experiences are subsumed—as concrete examples of an abstract concept.

This process of isolating and omitting qualities is called abstraction. It is obvious that the common unreflected use of the terms "abstract" and "concrete" is also based on this concept of abstraction.

Such a concept of abstraction does, however, have a number of difficulties. One of these difficulties is that it appears to be entirely arbitrary which qualities are omitted and which are not. Abstraction appears as a completely random aimless activity. General concepts, however, as for example the basic abstract concepts of Euclidean geometry (point, straight line, circle, etc.) are not at all chosen at random and cannot be arbitrarily replaced by other concepts.

This difficulty was already known in Greek antiquity, and attempts were made to eliminate it. One possibility, for instance, was offered by Plato's theory of ideas. According to this theory every human being is already endowed with schemata of all abstract concepts as innate ideas which guide the process of abstraction in the cognition process. By means of recollection *(anamnesis)* the mind knows which qualities to omit or not to omit when abstracting concepts. However, an argument along these lines would hardly be convincing today.

A further—and presumably more serious—difficulty of this Aristotelian concept of abstraction consists in the fact that the discontinuous transition between two qualitatively very dissimilar domains is to be explained by means of a continuous process of omitting qualities of the concrete object. Concrete sensible impressions constitute the starting point of abstraction. In the end this is supposed to result in concepts with strict operational rules linked by logical necessities. The simple omission of qualities, for example, is thought to lead from a world teeming with forms, colors, and smells to the deductions of Euclidean geometry.

There are attempts to eliminate the difficulty by modifying the concept of abstraction in this instance as well. In the following, two attempts that complement each other are briefly outlined.

THE CONCEPT OF ABSTRACTION OF FORMAL LOGIC

On the one hand the use of the concept of abstraction can be confined to the logical-linguistic domain as practiced—partially following classic and scholastic models—by the so-called "classical logic" developed by the Cartesians. According to this conception the process of abstraction begins with already existing ideas which are already defined with regard to content and extent (intension and extension).

I call the *comprehension* of an idea those attributes which it involves in itself, and which cannot be taken away from it without destroying it; as the comprehension of the idea triangle includes extension, figure, three lines, three angles, and the equality of these three angles to two right angles etc.

I call the *extension* of an idea those subjects to which that idea applies, which are also called the inferiors of a general term, which, in relation to them, is called superior; as the idea of triangle in general extends to all the different sorts of triangles. ([2], p. 49)

The abstraction takes place by omitting attributes. In the course of this process, the content of the concept becomes more and more depleted. Progressively fewer attributes define the concept. The extent of the concept, on the other hand, becomes larger and larger. Ever more objects are covered by the concept. "The limited extent of our mind renders us incapable of comprehending perfectly things which are a little complex in any other way than by considering them in their parts, and, as it were, through the phases which they are capable of receiving. This is what may be termed, generally, knowing by means of *abstraction*."([2], p. 45)

The ... way of conceiving things by abstraction is, *when, in relation to a single thing, having different attributes, we think of one without thinking of another, although there may exist between them only a discrimination of reason* [ratione non re]; and this is brought about as follows: ... In the same way, having drawn on paper an equilateral triangle, if I confine myself to the consideration of it in the place where it is, with all the accidents which determine it, I shall have the idea of *that triangle alone*; but if I detach my mind from the consideration of all these particular circumstances, and consider only that it is a figure bounded by three equal lines, the idea which I form of it will, on the one hand, represent to me more accurately that equality of lines; and, on the other, will be able to represent to me *all equilateral triangles*. And if, not restricting myself to that equality of lines, but proceeding further, I consider only that it is a figure bounded by three right lines, I shall form an idea which will represent *all kinds of triangles*. If, again, not confining myself to the number of lines, I simply consider that it is a plane surface, bounded by right lines, the idea which I form will represent all rectilineal figures.([2], p. 46f.)

Later this concept of abstraction found entrance into the logic of classes and set theory and thus became an integral part of formal logic. Today every student of mathematics is familiar with the idea that by gradual abstraction the concepts of field, ring, group, semigroup, and finally the concept of set as the most general concept can be generated from the concept of number.

THE EMPIRICIST CONCEPT OF ABSTRACTION

On the other hand, it is possible to restrict the concept of abstraction to the domain of sense perception on the basis of strict empiricism. At the beginning of the modern era, such a concept of abstraction resulted, for example, from attempts to explore the role of mathematical procedures in the natural sciences and to distinguish the new method of the acquisition of knowledge from the traditions of thought of the feudalist epoch. According to this conception, abstraction does not result in logically necessary relations between concepts at all.

David Hume, for instance, who perhaps provided the most precise theoretical formulation of this variant, considers natural laws to be nothing more than perceptions linked by association which give rise to the formation of habits of thought: "... all the materials of thinking are derived either from our outward or inward sentiment ..." ([7], p. 19) "All inferences from experience, therefore, are effects of custom, not of reasoning." ([7], p. 25) By comparison, mathematical theorems, as "associations of ideas," are relatively insignificant for cognition, because "propositions of this kind are discoverable by the mere operation of thought, without dependence on what is anywhere existent in the universe." ([7], p. 25)

In both variations of the Aristotelian concept of abstraction, abstraction is no longer concerned with the development of concepts from the contents of perception. To be sure, the problem has simply been sidestepped in this case. Perhaps this unresolved difficulty helps to clarify why, since the eighteenth century, a completely different possibility has been entertained; that is to say, the idea that the abstract form of concepts does not arise from experience by means of abstraction, but is constituted by the subject in the process of cognition.

MODELS OF ABSTRACTION
BASED ON CONSTITUTION THEORY

The basic idea of models of abstraction generated by constitution theory is as follows: If the transition from perception to concept is so difficult to explain as a process of abstraction, the reason may well be that form and content of concepts have totally different origins. Accordingly, only the contents are based on experience and are isolated by means of abstraction. The forms of concepts and, implicitly, formal logic as well as mathematics are based on the activity of the subject. They are constituted by the subject and imposed on empirical experiences. The abstraction of the form is the result of the reflection of the mind on the activity of the subject.

Different models of the process of abstraction based on constitution theory differ chiefly with regard to the details of how the activity of the subject is defined. In what is probably the most influential constitution theory historically—that is, the constitution theory of the *Critique of Pure Reason* by Immanuel Kant, it is solely the reflection on mental activity ("It must be possible for the 'I think' to accompany all my representations"; [8], B 132), on which the constitution of the logical forms is based. In modern constructivism as well as in Jean Piaget's structuralism, it is, however, the objective activities of the subject from which the logical forms have allegedly been abstracted.

It is agreed that logical and mathematical structures are abstract, whereas physical knowledge—the knowledge based on experience in general—is concrete. But let us ask what logical and mathematical knowledge is abstracted from. There are two possibilities. The first is that, when we act upon an object, our knowledge is derived from the object itself. This is the point of view of empiricism in general, and it is valid in the case of experimental or empirical knowledge for the most part. But there is a second possibility: when we are acting upon an object, we can also take into account the action itself, or operation if you will, since the transformation can be carried out mentally. In this hypothesis the abstraction is drawn not from the object that is acted upon, but from the action itself. It seems to me that this is the basis of logical and mathematical abstraction. ([9], p. 15f.)

At this point, I really do not want to go into the strengths and weaknesses of some of the models of the process of abstraction based on constitution theory; rather, I would prefer to get straight to the point most important to me. In his evaluation of Kant's critique of reason Hegel gave a dramatic twist to the discussion of the role of the mathematical-logical forms of thought in the cognition process. In this context I am limiting myself to outlining the consequences for the concept of abstraction.

THE DIALECTICAL CONCEPT OF ABSTRACTION

Hegel doubted the usefulness of characterizing cognition as a process leading from the concrete to the abstract. He proved the starting point of cognition in individual sense perception to be totally abstract, because, at the beginning, the object of cognition appears to be completely isolated, detached from any of its existing contexts and relationships. Only the image of the object which emerges at the end of the cognition process could reasonably be considered concrete.

Maybe the following quote from Hegel will serve to alleviate the dullness of my exposition which by necessity must be brief:

> Whose thinking is abstract? That of the uneducated rather than the educated person. ... I need only give examples in support of my proposition that everybody would agree contained it. ... My good woman, your eggs are rotten, the shopper says to the peddler. What, the latter replies, my eggs rotten? You should say that of my eggs? You? Was not your father eaten in the ditch by lice, did your mother not run off with the French and did your grandmother not die in the poorhouse—you would do better to get a whole shirt instead of that frivolous scarf; everybody knows where you got this scarf and your hats; if it were not for the officers, many a woman would not be so dolled up ... it would be better you darned your socks!—In short, she had nothing good to say about her. She thinks abstractly and subsumes the former according to the scarf, hat, shirt, etc., as well as fingers and other parts, also the father and the whole tribe solely under the crime of having found her eggs rotten; every aspect of her is colored through and through by these rotten eggs, whereas the officers mentioned by the peddler—if, despite the gravest doubt, there is something to it—may have had the chance to see quite different things regarding her. ([5], p. 577ff.)

If we accept Hegel's objections to the customary usage of the concepts "abstract" and "concrete," a revision of the concept of abstraction becomes necessary that can briefly be characterized by the following two assumptions:

1. Beginning by subsuming experiences under abstract concepts, the cognition process essentially proceeds to a concrete image of the object. Abstraction is nothing but a trivial precondition. Cognition is essentially a process of concretization. It consists in linking abstract concepts in cognitive structures and using them in order to interpret facts and problems. It is only in this process of concretization that the value of individual abstractions becomes evident.

2. The logical characterization of concepts in terms of intension and extension covers the abstract concept only at the outset of cognition. Any use of the concept as a means in the cognition process serves to enrich it with subsumed experiences and to relate it to other experiences in numerous ways. It is in this way that the concept becomes more and more concrete.

I would like to add emphatically: This characterization is in my opinion also true of mathematical cognition, despite the fact that the latter seems to be confined entirely to the abstract. The axiomatically implicit definition of a mathematical concept constitutes, just like the explicit definition in an axiomatic system, no more than an abstract definition of a concept that provides a starting point—a starting point, furthermore, that must already have proved its usefulness in concrete situations in order to serve as a truly useful point of departure; there is no arbitrariness of abstraction at this point. The concrete content of a concept, on the other hand, comprises all theorems that can be formulated or proved with it or that stand in any relation to it—that is, all results of mathematical cognition.

If mathematical thinking as a whole seems nevertheless abstract, it is simply due to the fact that certain basic abstractions are not dissolved within the context of the discipline, but rather in the interdisciplinary use as tools for the acquisition of knowledge in order to cope with real problems. This, by the way, is equally true of any particular scientific discipline.

THE "NORMAL" PATTERN OF TEACHING

In the second part of my discussion of the concept of abstraction, I would like to try to demonstrate the possible benefits to be gained from a more precise definition of this concept by making use of a typical problem in mathematics teaching. Let me take as an example the problem of how to interpret the organization of lessons in mathematics teaching with regard to its function for the learning process in mathematics and the development of abstractions and concretizations.

A recently published empirical study by Dicther Hopf about the didactics and teaching methods in mathematics for the 7th grade of the German *"Gymnasium"* ([6]) contains a remarkable result: While numerous alternative teaching models are discussed in the literature on didactics, the practice of mathematics teaching in the Federal Republic of Germany—as far as it was covered by the study—is largely dominated by an established pattern. While it is true that deviations can be detected at each particular stage, it is also true that these deviations from the norm are by and large distributed randomly and do not coalesce into alternative concepts of teaching.

This empirical result raises two questions that must be considered before further suggestions are made with the aim of realizing alternative teaching concepts. On the one hand, the question emerges as to the external institutional conditions that obstruct a change with regard to the "normal" pattern of teaching. On the other hand it must be asked—and it is this question that will be pursued here—whether it might not be the case that internal conditions resulting from the function of this pattern for the development of learning processes are responsible for its stability.

Let us imagine the normal sequence of a mathematics lesson.

The sequence of a normal lesson can roughly be imagined as follows: First the teacher makes random checks of the homework—homework designed to immediately reinforce the topics discussed and given to the whole class, not tailored to groups or individuals, and about which there was no choice. ... Before the teacher proceeds with the introduction of a new topic, he or she repeats familiar topics—that is to say, material treated during the previous lesson—which may serve as a point of departure. This method of introducing a new topic is characterized by a procedure that the teachers themselves understand as an "educational dialogue" which they control. Apart from that there is a wide range of methods. It is extremely rare that the teacher limits him or herself to a blatantly deductive strategy. ... The introduction of new material is followed by a practice session, with everybody participating in the educational dialogue, however not in groups. Typically, simple problems are treated first, before the level of difficulty is gradually increased. Most of the problems are found in the textbook which plays an important role in mathematics teaching in general. ... At all the stages the teachers make sure that every student participates in class and that those students who remained silent are called on as well. These measures to control events in the classroom are designed for the normal teaching situation and would be absurd in the completely uncharacteristic classroom situation where small groups work together. ([6], p. 157ff. The references contained in the text to the items on which the statements are based were omitted here without specific identification.)

This typical sequence of classroom teaching in mathematics is characterized by two basic assumptions:

1. First, it is important to *understand* a newly presented topic on the basis of mathematical knowledge acquired earlier. The primary task of the collective development of the topic in the classroom dialogue consists in conveying this understanding.
2. When the topic is basically understood, it is necessary to establish it by *practicing*. Establishing the newly acquired knowledge is best accomplished by using it to solve problems (first collectively, later independently).

An investigation of the function of the normal sequence of classroom teaching is, therefore, necessarily connected to a theoretical explanation first of the process of understanding in mathematics education and second to the effect of exercises.

ON THE INTERPRETATION OF LEARNING PROCESSES IN COGNITIVE PSYCHOLOGY

It is the cognitive psychology inspired by the work of Jean Piaget and the structuralist science of education based on this tradition that opened up new possibilities in this area (see Chap. 1, 2, and 4). The appeal of cognitive psychology for the methodology of teaching is based on its ability to understand the student not only as an object reacting to actions related to the organization of classroom teaching. It allows for the possibility of understanding the student as an active subject interacting with the object to be studied and with the social environment during the learning process.

Let me simply summarize the most important preconditions for this paradigm switch here. The subject is not conceived as being directly dependent on its environment. The dependence on the environment is rather mediated by subjective interpretations and the actions guided by these interpretations. Every piece of information received and every action initiated by it involves the assimilation of the data to cognitive structures. These cognitive structures do not simply exist in the subject, but are based on systems of generative rules, which allow them to be activated when applied to individual cases and spelled out to the extent necessary. In this sense, every generative system of rules of interpretation potentially represents a closed system of possible relationships of empirical contents. By being assimilated to such a system, the particular piece of information—the isolated fact—is therefore integrated into a network of dependencies and possibilities of action.

In any process involving learning and experience two aspects must be distinguished: first, the assimilation of new experiences to existing structures and second, the modification of these structures themselves.

Since cognitive structures (as demonstrated above) constitute closed systems, their modification is not a continuous process. Rather, every learning process is interrupted by (observable) phases of the comprehensive reorganization of cognitive structures. The reorganization is—explicitly or implicitly—associated with reinterpretations of extensive areas of experience. The development of such phases of the reorganization of cognitive structures is decisive for the long-term success of learning processes.

UNDERSTANDING THE FACTS

After this brief review, back to the starting point—the investigation of the function of the normal sequence of classroom teaching. The first issue dealt with the relationship of a newly introduced topic to the act of understanding.

How should we imagine this act of understanding? Understanding constitutes an active achievement by the student rather than the reaction to a behavioral demand. Two essentially different aspects of understanding need to be distinguished in this context.

1. The concept of understanding covers the situation where an existing problem of which the student is aware is solved by the adequate assimilation to an already available cognitive structure (aha effect). The educational dialogue is obviously aimed at influencing this aspect of understanding and eliciting certain ways of assimilating newly introduced topics to existing cognitive structures.
   a. Previous experiences are actualized and presented as a problem with the obvious expectation that the students all have the tendency to interpret this problem in the same way.
   b. The student is provided with criteria to judge for him or herself whether he or she has interpreted the newly presented topics correctly and integrated them into the structured system of his or her already existing knowledge by jointly working on solving a given problem, thereby eliciting acts of understanding.
2. However, the concept of understanding also includes those cases where the successful assimilation takes place in a phase of restructuring cognitive structures. In cases like these, understanding essentially guides the process of restructuring. By comparison, the solution of specifically definable problems completely recedes into the background, since the decisive precondition is missing: The cognitive structure according to which the solution appears to be a solution is no longer stable.

This second case—much more significant for the development of learning processes in the long run—is obviously quite different from the first. Phenomenologically, however, both aspects of understanding hardly differ from each other. The teacher rarely has the opportunity to make a judgment with regard to the nature of understanding based on the nature of the student's reaction. As a result the difference is not reflected in how teaching is organized.

THE FUNCTION OF EXERCISES

Let us now turn to the second aspect of importance in a study of the function of the conventional organization of classroom teaching—that is, to the effect of exercises.

The interpretation of the role of exercises within the framework of a cognitively based approach always presents certain difficulties. It might be assumed that in the case of exercises we are dealing with the active examination of the object as a prerequisite for understanding. But it would really have to precede understanding; the exercise phase, however, follows the development of the topic in class discussion. Furthermore, the examination of the object would have to be exploratory and avoid stereotypical repetitions. It is, however, precisely the latter that are widely regarded as necessary conditions for the success of exercises.

Is the exercise phase superfluous? Certainly not. Every practitioner can attest to the fact that, given the conventional organization of classes, the exercise phase is absolutely necessary. It is always just the small number of students considered mathematically gifted who, once a new topic has been introduced, are capable of solving related problems on their own without further exercises.

Thus a safety net must be provided during the exercise phase for those students who failed to gain sufficient understanding of the topic during the introductory stage. The function of the exercise phase could be seen as an effort to decompose the intended cognitive structure into behavioral objectives in such a way that even those who fail to build up the global cognitive structure will be able to solve at least part of the exercises—with the expectation, of course, that a complete understanding of the topic will eventually emerge.

There are in fact many indications that the latter interpretation is correct. The analysis of typical sequences of interaction between the teacher and individual students during class discussions shows ([3]) that, as a rule, the interpretive patterns of the students differ radically from the interpretation that the teacher employed as a basis for class discussion. Observations of individual students who actively participate in class discussions show another phenomenon which supports the assumption that the teaching method aims at decomposing cognitive structures. Frequently numerous interpretations are offered by the students in rapid succession to find out which answer the teacher might be looking for at a given moment. The majority of such interpretations includes, or even emphasizes, the social organization of knowledge. Answers are sought to questions like: "What is the answer the teacher is expecting when she puts it *this way*?" "How can I elicit a piece of information from the teacher about the reaction he might be expecting?" In such a situation the most appropriate response of the teacher to the active learning behavior of the students consists indeed in moving on to the exercise phase by jointly working on problems.

ABSTRACTION AND CONCRETIZATION IN MATHEMATICS TEACHING

As a result, a remarkable difference concerning the function of the structure of the normal course of classroom teaching with regard to different groups of students has to be noted: In terms of its organizational structure the educational dialogue is aimed mainly at the "mathematically gifted" student. The differentiation resulting from this is offset during the practical exercise phase in which the rest of the students are brought up to the desired level—albeit by means of a qualitatively different learning process.

At this point a key problem of the theoretical interpretation with regard to the function of the typical sequence of classroom teaching emerges: How does this sizable developmental differentiation in the process of building up cognitive structures in mathematics teaching arise in the first place? If the interpretation of the function of the organization of classroom teaching outlined above is correct, then the deep-seated conviction of the practitioner that failures in mathematics teaching are ultimately due to the fact that the abstractness of mathematical subjects requires a specific "mathematical talent" actually has a very real basis.

I would like to use the remainder of my remarks to show that the reflections on the conceptual clarification of processes of abstraction and concretization presented at the outset can contribute to solving this problem.

Let us first make the attempt to relate the conventional conception of abstraction and concretization to the organization of classroom teaching. It is difficult to gain a clear picture of this relationship. The most obvious assumption would be that processes of abstraction take place when a new topic is introduced in the educational dialogue; concretizations, on the other hand, take place during the exercise phase if the exercises deal with applications to extramathematical problems.

Given the fact that the concept of abstraction in many of its variations refers to the genesis of abstract concepts, this classification seems to make a good deal of sense. Seen in isolation, the cognitive structures developed in the course of the educational dialogue constitute abstract relationships. It would, therefore, only be natural to characterize their development as an abstraction despite their integrative character.

This interpretation is, however, called into question due to the fact that these cognitive structures come about because of a reorganization of already existing structures. This would, at any rate, contradict the fundamental precondition of the so-called "Aristotelian" concept of abstraction that the process of abstraction is characterized by isolating and omitting qualities. At best, this process of reorganization could be understood as abstraction in terms of constitution theory. In this case the abstraction of cognitive structures would have to be explained theoretically in terms of the reflection taking place during the phases of reorganization. Jean Piaget espoused this very solution to the problem with his concept of "reflective abstraction." (see Chap. 1)

The aspect whereby the concept of abstraction would include the genesis of abstract concepts could, however, be dropped and abstraction could simply be defined as the process of assimilation to a cognitive structure. The main characteristics of this process are, after all, the isolation and omission of qualities of the assimilated object that do not appear to be essential.

However, the real difficulty is not with the concept of abstraction, but with that of concretization—and in this case Piaget's solution also is completely unsatisfactory. In the context of the usual cognitive explanations, I do not see a proper place for the concept of concretization and, consequently, for an adequate interpretation of problems relating to the application of mathematical cognitive structures to extramathematical situations. For the only form of concretization of any importance in this case is the concretization of cognitive structures by assimilating experiences and information. "Concretization" would thus only be another designation for this process, and "assimilation" would simultaneously be abstraction and concretization.

This is where my initial reflections come in. On the basis of the dialectical concept of abstraction it becomes evident that the construction of cognitive structures adequate to the objects cannot signify the conclusion of the learning process. Rather, they constitute the starting point characterized by the formation of abstract concepts. In this view, the introductory stage of a new topic where these cognitive structures are to be constructed is much less important than the subsequent stage of application. Consequently, the understanding of a topic is a necessary, but by no means sufficient condition for the formation of concepts. The definition of a concept within a framework of a system of relationships then becomes just an abstract starting point for:
1. the consolidation of previous learning experiences and their interpretation—that is, the reinterpretation of previous interpretations, and
2. the constant enrichment of the concept in the course of its application in various contexts.

Once the subject matter of a concept has been introduced and defined specifically as a concept for solving a certain problem with the result that it has even been understood in its full, emphatic meaning, it does nevertheless not yet attain its rich web of relationships until experiences have been reexamined, statements have been deduced, the application to intramathematical and extramathematical examples has taken place, the differentiation with regard to alternative interpretations has been accomplished, etc.

This result can be expressed as follows: The interpretation of learning behavior as actively interpreting information and experiences by means of cognitive structures and as modifying such structures in these assimilative processes merely describes the microstructure of the creation of competence. The process of concretizing concept formations, by necessity initially abstract and critical for long-range results, on the other hand, is characterized by the global structure of the curriculum—by the sequence of the subjects and the manner in which learning experiences are time and again taken up in new ways determined by biographical factors and subjected to expanded interpretations.

If this view is correct, step-by-step concepts for the organization of classroom teaching—notably the prevailing cycle determined by an introductory stage during class discussion, joint practice of examples, and individual practice by means of homework—are basically flawed: The scope of planning is determined by the usual forms of the step-by-step control of achievement and not by the long-range processes of the concretization of concepts—that is, especially by their numerous linkages and their integration within the contexts of mathematical theory and extramathematical application. Concretization is left to chance with the result that the stable cognitive structures transcending individual teaching phases and subjects may not at all be adequate to the mathematical objects. For example, as strategies designed to cope with the demands of participation in class discussion or employed to prepare for tests, they may be guided mainly by the social organization of mathematical knowledge. Or, maybe tested everyday interpretations concretized in countless difficult situations prevail in competition with mathematical modes of thought, although the teacher never encouraged such interpretations in the organization of the learning process. Phenomenologically, such cases of inadequate, global interpretive patterns for mathematical subjects appear to be precisely what is persistently called a lack of mathematical talent.

BIBLIOGRAPHY

[1] Aristotle: *The Works of Aristotle, Vol. VIII, Metaphysica.* Ed. by J. A. Smith and W. D. Ross. Oxford: Clarendon Press, 1908.
[2] Arnauld, A.: *The Port-Royal Logic.* Trans. by T. S. Baynes. 8th ed., Edinburgh: William Blackwood and Sons, 1851.
[3] Bauersfeld, H.: "Kommunikationsmuster im Mathematikunterricht—Eine Analyse am Beispiel der Handlungsverengung durch Antworterwartung." In Bauersfeld, H. (Ed.): *Fallstudien und Analysen zum Mathematikunterricht.* Hannover: Schroedel, 1978, pp. 158–170.
[4] Bauersfeld, H.: "Hidden Dimensions in the So-Called Reality of a Mathematics Classroom." In *Educational Studies in Mathematics, 11* (1), 1980, pp. 23–41.
[5] Hegel, G. W. F.: *Werke 2, Jenaer Schriften (1801–1807).* Frankfurt a.M.: Suhrkamp, 1970.
[6] Hopf, D.: *Mathematikunterricht. Eine empirische Untersuchung zur Didaktik und Unterrichtsmethode in der 7. Klasse des Gymnasiums.* Stuttgart: Klett-Cotta, 1980.
[7] Hume, D.: *Enquiries Concerning Human Understanding and Concerning the Principles of Morals.* Ed. by L. A. Selby-Bigge, 3rd ed. by P. Nidditch. Oxford: Oxford University Press, 1975.
[8] Kant, I.: *Critique of Pure Reason.* Trans. by F. M. Müller. 2nd ed., New York: Macmillan, 1927.
[9] Piaget, J.: *Genetic Epistemology.* New York: Columbia University Press, 1970.

CHAPTER 4

WHAT IS MATHEMATICAL ABILITY
AND HOW DO ABILITY DIFFERENCES EMERGE
IN MATHEMATICS EDUCATION?*
[1980]

The problem I would like to address concerns the ability differences in mathematics education. My presentation is divided into two parts: First, I will analyze conventional ideas about ability in mathematics education. In this part of my remarks I would like to try and show why I consider it necessary to reexamine ability in school mathematics. In the second part of my lecture I will proceed to outline an alternative. I will present some of the basic assumptions on the nature of thought, especially on those thought processes on which mathematical ability is based. And I will try to demonstrate that the emergence of ability differences can be more fully explained and that the mathematics teacher will gain new perspectives on teaching, if these assumptions are taken into account.

THE APPEARANCE OF MATHEMATICAL COMPETENCE
IN MATHEMATICS EDUCATION

Common Sense with Regard to Ability Differences

I am beginning with an exploration of the appearance of mathematical competence in mathematics education. On starting school, the six-year-olds are beginning with the systematic exploration of mathematical subjects and problems. Whatever they learned at home consists at most of ele-

---

* The manuscript published here is the text of a lecture given in April, 1979, at the colloquium on education of the mathematics department of the *Technische Hochschule* (Institute of Technology) in Darmstadt. I hesitated to agree to its publication because the theoretical approach outlined in the sections "The Concept of Cognitive Structure" and "Learning and the Development of Cognitive Structures" has implications that go beyond those discussed in the context of the present topic. A more comprehensive exposition and explanation is actually needed than can be provided within the confines of a discussion of

mentary skills with regard to counting and arithmetic. At this point, it does not make sense to speak of mathematical ability, unless this term is already used in a theoretical context which would make it possible to uncover and interpret mathematical abilities implicit in everyday activities. Apart from that, on starting school mathematical competence should only be referred to as a potential not to be realized until later, as a capacity to learn.

A few years later, the situation will have completely changed. By now, relatively stable ability differences between students have emerged, differences demonstrated by regularly conducted ability assessments. Furthermore, the belief in the existence of these ability differences is so strong that they actually become the norm for assessing ability: A statistical grade distribution deviating from the normal distribution is considered to stem from a faulty execution of the assessment—a test turned out "too well," hence it was too easy; or it did not turn out "well enough," hence it was too difficult. This, however, only as an aside. For it would no doubt be a mistake if this inversion of the logic of the ability assessment were to lead to the conclusion that ability differences, which emerge as a result of the assessment, are nothing but an illusion, a methodological artifact of the procedures of the assessment. Rather, we must assume that, together with the mathematical learning processes taking place throughout the school years, relatively stable ability differences develop in mathematics education.

A theoretical explanation of these ability differences is, however, not at all obvious despite the fact that these ability differences are generally accepted without a trace of skepticism. In fact, such an explanation was dispensed with for a long time in the practical educational context and the ability differences—more precisely, the ability differences in relationship to average ability—were, inasmuch as they were subject to change, attributed to the enthusiasm and the hard work or, conversely, the lack of interest and the laziness of the individual student. Apart from that, they were attributed to his or her mathematical talent, taken to be an indefinable and, in particular, unchangeable prior condition of individual mathematical ability.

---

a partial aspect of the topic. On the other hand, I could not very well do without at least presenting an outline of this more general approach if I wanted to show that the radical nature of my criticism of the conventional patterns of assessing ability is not the result of a one-sided consideration of a single aspect of learning processes taking place in the school context, but, in my judgment, is the necessary consequence of an interpretation of cognitive processes guided by "cognitive" principles. I am aware of the shortcomings of this presentation and must ask the reader to examine the seriousness of the deliberations outlined in what follows making use of the references listed in note 9.

When we are looking for explanations for these ability differences, we are, therefore, compelled to abandon this common-sense view of day-to-day teaching. We must look to the results of scientific research on the learning processes in the school setting.

The Concept of Ability as a Descriptive Empirical Category

First it would, however, be appropriate to give a little more thought to what it is we are thinking of when we speak of ability in mathematics education. What is this ability that is assessed in school, and how do we explain its diverse development in the course of an individual's school experience? I have already given an example indicating that the very concept of ability is more than problematic. I have pointed out that it can be applied to the situation as it exists on entering school only if additional theoretical assumptions are made. But such theoretical assumptions would themselves be in need of empirical controls—controls, moreover, that cannot be provided by the ability assessment conducted in schools, since this assessment is based on a particular understanding of mathematical abilities, which already presupposes that the results of the assessment provide empirical evidence of corresponding ability differences.

We could, of course, easily bypass these difficulties by taking seriously the empiricism of the ability assessment and by identifying ability with its actual appearance. The concept of ability would then become a category describing the results of learning processes rather than interpreting them, a category for classifying and ranking certain types of behavior which evolve in the course of learning processes in school. According to this definition, the mathematical ability of the students on entering school is indeed without exception almost nil, and the teaching goals define a level of ability cumulatively increasing throughout the school years. The ability assessment defines—more or less accurately—the deviation from the average level of ability as a measure of the individual level of ability. In this definition of the concept of ability a familiar reference point of the scientific preoccupation with achievement in school and with learning success is easily detected.[1]

This concept is, however, not free of problems either. That becomes evident if it is strictly applied to the ability assessment with regard to mathematics education. In order to demonstrate this, I will have to go back a little further.

Tasks as Instruments of Ability Assessment

If, true to empiricist criteria, the question concerning mathematical ability is limited to its external appearance and all questions regarding its essence are excluded, it can be said:

Mathematical ability becomes apparent in the process of solving problems typical of the classroom setting. Depending on the difficulty of the problems an individual is capable of solving, his or her ability is rated higher or lower. In school education the difficulty is determined by the stati-stical average—in complete agreement, by the way, with the psychometric concept of difficulty used to calculate the difficulty of test items: The fewer students capable of solving a problem, the more difficult it is considered to be.

A teacher, however, who is familiar with the subject matter taught up to the point of the assessment as well as with the previous achievements of his or her students has no need of determining the difficulty of every single problem empirically. Rather, criteria of the problems themselves can be identified which—given the assumptions stated above—permit an estimate of the degree of difficulty of a given problem.

For, as a matter of fact, every task presented in class for the purpose of assessing achievement is associated with one or more mathematical symbolic systems which are the object of class instruction. These symbolic systems are characterized by three things:

1. by a limited number of symbols, signs, and terms for these signs, etc., constituting the elements of the symbolic system and representing certain basic components and relations of the mathematical object,

2. by explicit or implicit formal rules making operations with these symbols and signs possible, and

3. by interpretations of content making use of informal language which render the symbolic system meaningful as well as useful, allow it to be interpreted in a nonmathematical context or, conversely, allow it to appear as a solution to certain general mathematical problems within a mathematical context.

The problems require an application of the rules, as determined by their meaning, to the symbols and signs with the intention of achieving a certain result defined by the problem. The task illuminates these problems either directly or as being framed in a colloquial context which reflects the interpretive background of the rules and symbols of those symbolic systems relevant to the task. The following may serve as an illustration of the first type:

a. the problem of dividing $\frac{2}{3}$ by $\frac{5}{8}$, or

b. the problem of proving the Pythagorean theorem by means of analytical geometry.

All word problems are examples of the second type. However, the extent to which it becomes necessary to deal with the interpretive context of such problems in order to isolate their formal core can vary considerably.

The Degree of Difficulty of Problems

If this phenomenalistic description of the problems employed to conduct the ability assessment is related to the previous learning process, whatever its theoretical interpretation, a simple classification according to characteristics emerges in which an increasing degree of difficulty is expressed:

The simplest type of problem results if the rules to be employed in the problem-solving pattern and the sequence in which they are to be applied were discussed and practiced in class and if it is clear from the way the problem is presented that it is precisely the application of this particular problem-solving pattern that is required. In this case, the expected performance is merely reproductive. The exercise basically does not contain a mathematical problem, but is simply testing the acquired knowledge and skills.

This first type of problem becomes, however, more difficult to the degree that the conditions stated above are not fulfilled. The difficulty increases if, for example, the rules to be applied were not treated in depth, but only marginally; if they were not practiced; if, although the rules are clearly defined by the problem, the combination required for solving it was never discussed in class; or if the problem-solving pattern was discussed and applied in class, but the problem at hand does not quite make it clear that it is this very problem-solving pattern that is to be applied.

At the end of the scale of difficulty thus defined we encounter a type of problem characterized by the following criteria: Although the problem defines the formal system within which it is posed, the problem was selected so as to ensure that the problem-solving pattern was not treated in class. Thus the expected performance consists in developing the problem-solving pattern independently.

Finally, the difficulty of the problem is increased to the extent that the problem requires an independent interpretation of the formal system and its manifold implications.

Thus, with regard to the relationship between the problems employed in assessing ability and classroom instruction, four types can be described which, for the sake of simplicity, I propose to call "complete algorithms," "incomplete algorithms," "heuristic problems" and "interpretive problems." They indicate an increasing degree of independence necessary for solving problems, and the difficulty of a problem can be assessed in terms of this degree of independence.

The Paradox of the Empiricist Concept of Ability

This brings me to the problems of the empiricist concept of ability mentioned earlier. It leads to a paradox. The mark of a competent student is, after all, his or her achievement measured in terms of his or her ability to solve objectively difficult problems. According to the given typology these are precisely those problems which require the student to find a solution which was not at all, or at least not adequately, taught previously. In short, the paradox consists in the fact that the mathematically more competent students can only be distinguished from the less competent students by giving them test problems they were not taught at all.

It is obvious that this paradox is based on the identification of ability with its actual appearance. It is based precisely on the fact that abstractions have been derived from the latent processes. For while these processes determine learning, they do not themselves become evident. Only their end result is expressed in the measurements for assessing achievement. These processes, which in fact ought to be regarded as the achievement, appear in the context of the empiricist concept of ability only in a negative sense, as something independent, as that which does not find expression in the stated learning objectives and the teaching process guided by those objectives.[2]

Let me summarize my remarks on the appearance of mathematical ability and the forms of assessing and evaluating mathematical achievements in the classroom associated with this appearance: Mathematical ability manifests itself when difficult problems must be solved, and, in this sense, mathematics education is to make mathematical performance possible. Herein lies a contradiction, for difficult problems are characterized by the fact that what they test has actually not been taught in the classroom. They actually do not test the knowledge and skills that are systematically taught in class, but rather independent thinking that goes beyond the reproductive application of acquired knowledge and skills.

In my view, the following conclusion must be drawn: We need a theory of mathematical competence capable of explaining exactly the kind of ability that cannot be understood as the direct result of teaching and that, at the same time, shows why some students acquire this competence while others do not or only to a much lesser degree. In particular, such a theory will have to explain how the individual forms of coping with the contradiction between creditable performances in the learning process and the performance measured as the discernible global results of learning by the assessment of performance determine as a precondition the development of ability.

PSYCHOMETRIC CONSTRUCTS OF ABILITY

Performance as the Result of General Traits

I am returning now to the question postponed at the beginning—that is, the question concerning a theoretical explanation of ability differences, and I already pointed out that, in turning to this question, we are leaving the common-sense level of the reality of classroom instruction. Given the empiricist concept of ability, the most common explanation of the differences of mathematical competence is based on the following assumptions:

The performance that can be achieved in solving problems depends on a number of traits which the students possess; the latter may, however, be present to varying degrees. The probability of solving each individual problem is defined by a specific function which assigns the appropriate probability for solving the problem to each and every combination of traits. The traits are conceived of as measurable variables largely independent of the contents of the problems. With regard to the individual student, the specific value of this variable can no doubt be influenced by learning processes; it can, however, only be changed over the long haul and must be considered more or less constant in terms of the shorter time frames of individual learning processes. This much with regard to the basic assumptions of the model.[3]

Comparable ideas of mathematical ability can of course be found in the popular views on mathematics education as well. Thus mathematical ability is frequently attributed to general traits such as an understanding of numbers, imagination, mental flexibility, objectivity, the ability for self-criticism, accuracy, etc. These undifferentiated ideas were clarified and refined and, most importantly, rescued from the realm of mere speculation only by scientific research working with this theoretical model. In this context, the problem of finding methods for testing the assumptions of the theory empirically has been especially prominent, since these assumption can no longer be directly inferred from the actual appearance of mathematical ability.

The Operationalization of the Concept of Ability

The basic characteristics of this operationalization of the concept of ability are as follows: Test items devised according to a number of criteria, for example, the criteria of objectivity and reliability (i.e., the fact that the result is independent of the persons evaluating the performance and the outcome can be reproduced when the same test is repeated or administered with analogous test items) take the place of problems in the classroom. If there is a high correlation between the results of specific items, it is assumed that the same combination of traits is necessary for solving the problems in question.

Items with a high correlation are combined into test batteries and used as gauges for underlying traits. Such test batteries are characterized by much higher indices for objectivity and reliability than single test items.

These test batteries are then combined into homogeneous or nonhomogeneous tests and are standardized using selected populations, that is, the distribution of the test scores of a population is ascertained by means of random samples, and their parameters are attached to the test as defining data. This standardization makes it possible to relate the test results to this particular population when the test is administered. It is, therefore, no longer the average performance of the individual class and, respectively, the results of teaching that particular class which provide the standard of abilities, but the diffuse performance average of a previously defined, more inclusive population sample, a performance average which is independent of specific learning processes.

The Construction of Traits

The inference as to the underlying traits basically takes place along the following lines: Different tests, homogeneous in themselves, will correlate with each other in different ways. These correlations are considered to be evidence of the traits which are necessary for solving different tests. Thus the results of the greatest possible number of different mathematical ability tests are examined to find out how many separate traits must be assumed in order to be able to explain the test results in terms of linear combinations of these hypothetical traits. According to the criteria of simplicity and plausibility, particular assumptions concerning such traits are selected from the multitude of logically possible assumptions capable of explaining the test results. These assumptions are considered to be the concrete instances of the general theoretical assumption concerning the existence of such traits.

The traits obtained in this fashion from psychometric test results are, therefore, specific theoretical constructs within the framework of the general explanatory model that seeks to reduce correct solutions of tasks and problems to personality traits.

The explanatory power of this model is, however, already jeopardized by the confusing variety of disparate constructs of traits generated by such psychometric examinations. This variety is, on the one hand, due to the dependence of the results on the specific method used to analyze the correlations and to determine the explanatory constructs of traits. On the other hand, it is decisive that there are hardly any criteria for these procedures that allow for the falsification of the basic assumptions of the model. To put it simply: These investigations always yield traits that can "explain" the test scores because the capability of explaining the test scores in question by means of these very traits is presupposed by the investigation and is not examined empirically in light of the empirical data. At best the consistency of the results of various investigations can be interpreted as an indication of the empirical content of the assumptions of the model.

The Results of the Investigation
of School Achievement Based on Psychometry

Keeping this in mind, let us ask the question: What is the result of the investigations of ability in mathematics education using the psychometric method?

It should first be noted that it yielded at least the following result: If mathematical achievements in school can indeed be attributed to a few basic traits, it must be remembered that the situation is obviously much too complex for a teacher who attributes the results of the assessment of school achievement to the existence or the absence of this or that trait—be it to the attentiveness of a student, or to the discipline in the classroom, or to intelligence, or to diligence outside the school setting—to be likely to make the right choice. In fact, almost all common-sense explanations for the development of mathematical ability proved to be invalid.

Let me give just a few examples: The assumption concerning the existence of a specific mathematical talent has not been borne out. Such an assumption is negated by the existence of factors common to achievements in different school subjects as well as by the multidimensional nature of achievements in specific subjects.[4] It was assumed that traits of the students independent of the teacher are expressed in the achievement. The clearest proof to the contrary is provided by the evidence for the existence of the "self-fulfilling prophecy" of the incorrect assessment of student abilities by the teacher.[5] It was assumed that valid instruments for predicting success in school could be obtained by means of measuring constructs of traits. This hope has been dashed, because the predictions can always be falsified by an appropriate change of the conditions under which learning occurs. It was assumed that classroom teaching could be adapted to the diverse abilities by forming groups that were homogeneous in terms of ability, thus improving learning results. This turned out to be erroneous. To the contrary, given favorable conditions, the opposite proved to be true.[6] And in particular the diversity of the results of empirical investigations shows: There is hardly any result that could be applied to mathematical learning processes in general. For each result that could possibly be interpreted as a general principle, there are results that falsify it when conditions are modified.

In this context, I must refrain from reporting on specific results. Instead, I would like to quote the overall assessment of an outstanding representative of this research tradition which seeks to explain school achievement in terms of human characteristics—that is, Benjamin Bloom.

When I first entered the field of educational research and measurement, the prevailing construct was: 1. *There are good learners and there are poor learners.* ... During the early 1960s, some of us became interested in the Carroll Model of School Learning, which was built on the construct: 2. *There are faster learners and there are slower learners.* ... During the past decade, my students and I have done research which has led us to the view that: 3. *Most students become very similar with regard to learning ability, rate of learning, and motivation for further learning—when provided with favorable learning conditions.*[7]

The practical conclusion of this result is the conception of "mastery learning"—that is, the idea of applying a new standard to teaching and of reorganizing it in such a way as to allow 90 percent of the students to attain 90 percent of the learning goals which define the desired achievement. It is, however, true that the central variable of the model, the variable of "quality of instruction" on which the success of mastery learning is based, constitutes an external variable for Bloom's theory of the connection between "human characteristics" and "school learning."

The Limits of Explaining Ability in Mathematics Education in Terms of Traits

When the ultimate goal of a research strategy which attempts to explain differences of ability in terms of traits consists in demonstrating that the learning process can be organized in such a way that the success is independent of the assumed traits, it is, in my view, high time to question this explanatory model in a much more thoroughgoing way.

Let us, therefore, take another look at the general structure of this explanatory model. Let us ask ourselves without bias whether it might not be the case that the basic common-sense assumptions concerning the development of achievements are totally inadequate with regard to their subject.

We have seen that the familiar ideas about ability in mathematics education as well as the psychometric investigations are based on a particular model of the learner. He or she is defined in terms of his or her traits—by personality characteristics that can, if at all, only be changed in the long run. Together with those conditions that are not dependent on the student, they constitute the input variables of a teaching model designed to explain the output variables, especially the output variable of achievement, the values of which will be determined by means of the procedures for the ability assessment in the school context or by means of standardized ability tests.

If we look at the simple basic structure of this explanatory model, a most problematic basic feature fundamentally opposed to the actual conditions in the classroom must, in my view, be noted: The student is viewed solely as an object of the learning process. To reduce the result of the learning process (i.e., the mathematical achievement represented by the solution of problems) to the conditions from which the achievement necessarily follows (i.e., to the combination of the subject matter that is the same for the students of a particular class and their different individual abilities) means that learning processes are seen as causal processes in analogy to processes in inanimate nature. Teaching is seen as nothing but a technology. The teacher must know the result of the learning process in advance; furthermore, he or she must know the laws of the learning process as well as the individual conditions in terms of specific values of the student variables; he or she can then create the conditions with regard to the teaching process that will allow learning to take place according to its laws and lead to the desired goal. Teaching equals the creation of conditions in agreement with the laws of learning so that learning will lead to the desired performance by the student. The problem raised at the outset—that is, that it is precisely the productive achievements of the student which must be explained, cannot be solved in this fashion.[8]

Today, it is in my judgment no longer possible to disregard this fact, if the results of scientific research with respect to common-sense assumptions about the development of school achievements in mathematics teaching are examined without bias. It leads to the demand which the teacher as well as the scientist cannot ignore: It is necessary to understand school achievements as productive achievements, as achievements of the learner who is the subject of his or her learning process. Theories and common-sense explanations that basically assume goal-oriented action only on the part of the teacher cannot explain the development of productive achievements.

ABILITIES AS COGNITIVE STRUCTURES

I am now getting to the second part of my presentation. Here I will try to show how the common ideas about the development of learning achievements in mathematics education—the entrenched preconceptions about natural hierarchies of abilities—must be revised so that the independent, productive mathematical achievement does not simply appear as the borderline case of reproductive learning processes which cannot be expressed theoretically.

I will first state the basic assumptions of a theory on productive learning achievements in the form of general theses which are then applied to the problem of the development of ability differences in mathematics education.[9]

The Concept of Cognitive Structure

My *first thesis* reads as follows:

Thinking is a form of the movement of consciousness. This form of movement is the result of internalizing actual movements, mediated in particular by objective actions of the cognitively active individual. In the simplest of cases these actions are closely connected with perception (in Piaget's terminology: sensori-motor schemata; in behavioristic terminology: stimulus-response sequences). In particular, this includes following movements with our senses, which makes perception itself the instrument of our actions that it is. At the developed level all forms of the active reorganization of reality appear as mediating actions inasmuch as they are the result of purposeful activity.

This thesis is directed at the familiar idea that mental action can be assessed in terms of its results and understood statically as a manifestation of a combination of mental abilities. This thesis implies that even the most abstract mental activity, as for instance thought processes in the context of mathematics education, cannot be understood at all without reference to an object external to it which has been acquired by means of objective actions.

The *second thesis* reads as follows:

Thinking moves in fixed schemata resulting from previous experiences with the object of thought. These schemata, called cognitive structures, are closely related to objective experiences on which, mediated by the actions of the conscious subject, they are based. They reflect the laws of movement of the objects to which thinking refers to the extent to which the movements of thought can be realized and thus confirmed within the context of objective action that gave rise to them. Within these schemata, developed thinking can move freely. If certain means are presupposed, it explores, so to speak, the dimension of the possibility which objective action opens up for the subject as a domain for setting goals and their realization.

According to this thesis cognitive structures play a key role in any kind of mental activity. The productive achievements of problem-solving processes are not reduced to cognitive abilities independent of content (although such abilities might well be important), but are considered to be the result of a thought process mediated by latent cognitive structures in which accumulated experiences with the object of thought are present. According to this thesis, the probability of solving a problem productively depends primarily on how well the particular cognitive structures of the individual who is trying to solve the problem fit this specific problem. To put it simply: Whether a problem is solved depends on whether the available knowledge is adequate for solving it. The concept of knowledge used here is, however, employed in a more comprehensive sense than is usually the case. Knowledge of an object is conceived not only as accumulated information on this object, but as structured knowledge, structured by the latent cognitive structures connecting this information in thinking and reflecting the logic of the possible changes of the object. That is to say that knowledge in this sense does not just refer to the appearance of the object—its factual empirical existence and its sensible qualities—but also to its essence emerging in the course of objective action.

The *third thesis* reads as follows:

While it is true that the cognitive structures, in which thinking moves, are based on the objective actions to which they indirectly refer, they are nevertheless themselves a necessary component of these actions. They constitute the referential foundation without which goal-oriented action is impossible. The ability to set realizable goals and to systematically realize them presupposes thought processes in which the actual process of transformation to be realized by action is anticipated in its essential components by means of cognitive structures.

The dependence of thinking on action must, therefore, not be understood in causal-mechanistic terms. Thought and action presuppose each other. Their true relationship can only be demonstrated in terms of developmental theory. It is a mistake to assume that, by engaging in exercises that have not yet been understood, action-oriented experiences that would lead to an adequate cognitive processing of this activity can forcibly be achieved—as is attempted with learning problem-solving patterns by rote in certain task-oriented training programs. Conversely, it is an oft repeated illusion on the part of educational reformers to assume that practice can be replaced by insight and that rule-governed operations with a subject not yet understood can be banished from the learning process once and for all by means of a teaching method based on understanding.

The *fourth thesis* reads as follows:

In the thought process we make use of objective means; it is because of this that the cognitive structures gain stability and that their development becomes irreversible. These objective means are: pictorial representation, language, and writing as well as—for mathematical thinking—artificial symbolic systems derived from the former. In all these cases, we are dealing with systems of rule-governed actions with objective representatives of the contents of thought that make it possible to "mechanically" simulate movements of thought in cognitive structures as objective actions carried out with the representatives. This representation of movements of thought presents itself to thinking as the distinction between content and form of thought—as a process of abstraction. Only the form as the general component of a cognitive structure is represented. Relating this form to the particular contents of individual experiences is a reproductive and a productive achievement to be accomplished in thinking again and again with every problem.

The fourth thesis enables us to determine the specifics of mathematical thinking—that is, to define the specific ability to be developed in the course of mathematics education. Mathematical thinking separates form from content; its subject is the form as content. The object of this type of thought is the mediation of thought by objective representations which are manipulated formalistically—that is, abstracted from their meaning, their reference to content. Arithmetic, for instance, historically emerged as a theory of calculating aids abstracted from the contexts of their practical application. Euclidean geometry is the theory of construction using straightedge and compasses abstracted from their uses. In both cases the abstraction appears as an idealization, as a release from the concrete limitations of calculations that can actually be performed and constructions that can actually be performed, although arithmetic and geometry were in fact basically theories of the representation of cognitive operations in these objective means. And—last, but not least—structural mathematics of the twentieth century is based on the formalization of the language of proof. The formalization represented by the structural concept presupposed the total independence of the form of the mathematical proof from the content of the mathematical argumentation.

Thus, mathematical ability must be defined by means of those cognitive structures that are based on the actions carried out with the objective means of thought, and they run the gamut from the calculating and construction aids of antiquity to the symbolic systems of structural mathematics. The abstractness of the performance demanded in mathematics education is based on the merely indirect reference to the primary experiences concerning everyday actions.

With this theoretical definition of the concept of ability the first of the problems identified at the outset has been solved: the elimination of the paradox contained in the empiricist concept of ability. If the concept of ability is defined in terms of the problems to be solved, then the concept of ability is guided by the goals of the learning process on which teaching is, and must be, based. The independence of the student, however, is not based on these goals of the learning process, but on the means at his or her disposal in order to reach these goals. As a consequence of the theses presented above, the concept of ability is expressly guided by those means that establish independence. It is ultimately these very means that also determine the abilities which correspond to the conventional concept of ability. It is, however, true that we are unlikely to gain insight into ability in the above sense by means of the procedures of the conventional ability assessment, since there are always many ways to solve a problem or to fail to do so. We must do without a considerable part of the usual control, if we take the development of independence as our standard. However, what we stand to gain in this process are criteria for organizing teaching in such a way that we are no longer hemmed in by the seemingly natural, insurmountable limitations of the individual mathematical learning process, the naturalness and insurmountability of which are suggested by the conventional concept of ability.

The definition of what constitutes mathematical ability is, however, only the very first step. The theses presented here have not yet said anything about the development of cognitive structures. Before I turn to this development, I would like to sum up the four theses that define the concept of cognitive structure:

Thinking moves in fixed patterns. They are based on action-oriented experiences. The movements of thought are organized in systems—that is, the cognitive structures. These structures reflect laws of movement of reality that determine the content of the action-oriented experiences. On the other hand, the cognitive structures are themselves components of goal-oriented action as its necessary basis of orientation. They achieve their stability by being represented in rule-governed operations using objective means of thought: image, language, writing, symbol. This representation appears in thought as a process of abstraction, as the distinction between form and content.

Learning and the Development of Cognitive Structures

We now come to the development of cognitive structures.

My *first thesis* reads as follows:

Cognition is an active process. It is simultaneously a process of learning in the sense of absorbing new information and a learning process in the sense of modifying existing cognitive structures. Learning in the first sense takes place by assimilating new information to already existing cognitive structures. However, this process of assimilation is, at the same time, also a process of the modification of the cognitive structure (in Piaget's terminology: an accommodation of the structure) in two ways: The structure is generalized; by assimilating new contents, it is extended. At the same time the structure is differentiated; the newly assimilated contents differ from each other, and with each new content the structure is, therefore, related to other structures in an increasing variety of ways.

We had seen at the beginning that the model which views the student merely as an object of the teaching process, cannot adequately explain the phenomena of independence and productivity evident in the processes of solving difficult problems in those cases where the problem-solving pattern had not already been dealt with in the course of classroom instruction. The thesis stipulating the active nature of each and every learning process completely changes the view. Even if the reproductive learning processes seem to be nothing more than the absorption and the unmodified reproduction of information or the repetition of patterns of action as presented, according to this thesis they actually constitute merely the manifestation of a latent productive learning process that changes the cognitive structures. Even the most rigid learning environment, where all the learning conditions are controlled by the teacher, cannot prevent the students from interpreting new information in the context of their individual cognitive structures, from processing this information thoughtfully and productively, and from developing their thinking in this process—albeit under conditions that, as a rule, are not conducive to learning.

The success of this active process will be determined by the relationship of the student's individual cognitive structures and past experiences incorporated in these structures to the newly acquired information and the tasks to be mastered in class. If the teaching process is, however, guided exclusively by the modification of the actual performance as it appears in problem solutions, and if the learning success is measured exclusively in terms of that performance, if, in other words, teaching does not include the latent processes of the productive processing of new experiences in thinking, then the development of the cognitive structures that determine learning ability and competence is ultimately left to chance. Under these circumstances it would indeed appear that mathematical competence as expressed by productive ideas to be employed in solving problems for which the problem-solving patterns had not been presented in class is not a product of teaching, but of a randomly distributed mathematical talent.

The *second thesis* reads as follows:

The modification of a cognitive structure is not a continuous process. The development takes place in successive stages, each separated by a phase of the restructuring of the cognitive structures. This organization in stages is a consequence of the formal closure of each cognitive structure. The structuring is caused and advanced by contradictions in the thought process (called, as a rule, intrinsic motivation). It is a process of advancement in the sense that the new structure contains the old one as one of the components; it has been integrated as a partial structure. However, in this process a profound transformation of the meaning of this structure takes place.

For a mathematician the necessity of perceiving the developmental process of cognitive structures as a process of stages is especially easy to understand. Let us, for instance, imagine that we wanted to proceed in a continuous manner from deliberations concerning natural numbers to those concerning whole numbers by, for example, continuously adding one negative number after the other to the natural numbers. It is obvious that the first number added would immediately violate the relative homogeneity of the system of natural numbers and be incompatible with the basic laws of natural numbers.

But what follows from this thesis of discontinuity (concerning the developmental process of cognitive structures) in terms of explaining the development of mathematical competence? For some of the very basic cognitive structures of human thinking we have very accurate analyses of the process of their development, for example, for the structures of spatial thought and for the most general logical structures. These investigations show that cognitive structures, even if they are, objectively, not compatible with certain pieces of information, are remarkably stable due to their relative efficiency. Nobody throws out the familiar schemata of his or her thought simply because occasionally new experiences cannot be assimilated without contradiction. And a schema that works is indeed vastly superior to a correct new idea that has, however, not yet developed into a schema due to the very quantity of experiences assimilated to the former schema. This is at least true for most problems which present themselves to thinking. This goes to show that teaching guided solely by the conventional assessment of ability and its criteria of success not only leaves the development of cognitive structures to chance, but that expecting the student to convert learning results immediately into performance is actually counterproductive.

The *third and last thesis* concerning the development of cognitive structures is intended to answer the question as to what it is that is guiding the restructuring of cognitive structures of thinking in the transitional phase. It reads as follows:

The restructuring of cognitive structures is guided by reflection. Reflection is that kind of processing of thought which is based on the fact that thinking is simultaneously component and reflection of objectively mediated, goal-oriented activity. As long as a cognitive structure remains stable, it is reflectively expanded by reflection—that is, experiences are assimilated to it at a metalevel. The movement of thought within this structure, therefore, becomes more and more effective by means of reflection. In the transitional phase of restructuring a cognitive structure, on the other hand, reflection determines the direction of the process. Reflection concerning the objective mediation of actions in which the failure of the old structure becomes evident, guides the movement of thought with regard to the dimension of the expanded possibilities offered by the newly discovered means.

What follows from this concept of reflection for the organization of learning processes aiming at the development of cognitive structures is that not only is it necessary to provide the opportunity for self-directed, objective thought and action in the classroom (the transition to "active methods" in Piaget's words), but that content has an important function as well, for it is the usefulness of the available objective means of this thought and action for acquiring the content that is decisive for the success of learning.[10]

I am thus finally returning to the problem raised at the beginning—that is, the problem of the emergence of ability differences. Let me, however, first briefly summarize the theses concerning the development of cognitive structures as well:

Cognition is an active process of the modification of cognitive structures. In the learning process new contents are assimilated to existing structures, while the existing structures are generalized and differentiated by means of this assimilation at the same time. The cognitive structures are relatively stable, and their modification is not a continuous process. Their development takes place in stages separated by phases of restructuring. Each transitional phase is characterized, first, by the construction of a more comprehensive structure into which the previous structure is integrated and, second, by a transformation of the meaning of the latter in the course of its integration. The restructuring is guided by the reflection on the objective mediation of actions which constitute the basis of the cognitive structure.

Mathematical Competence as the
Result of the Construction of Cognitive Structures

I am now returning to the question posed at the beginning of how ability differences in mathematics education are to be explained and how they arise. If the thesis is correct that learning is accompanied by a latent process of the development of cognitive structures, then there is a simple explanation for the emergence of achievements attained in the process of solving difficult mathematical problems which—according to the terminology introduced at the beginning—are characterized by heuristic problems and problems of interpretation; such achievements cannot be explained solely in terms of a causal-mechanistic model of the learning process. A problem-solving pattern creates a connection between the initial conditions of a task or a problem and the solution in the form of mental operations by means of which the problem is solved. This is not an arbitrary correlation. Rather, it is strictly determined by the laws of the object to which the problem refers. During the learning process preceding the presentation of the problem, information about this object was assimilated to the cognitive structures of the learner. Consequently, to the extent that these structures are consistent with the laws of the object, the student not only absorbs this information and is able to reproduce it, but by integrating this information into his or her thinking, he or she at the same time productively acquires the ability of independently developing problem-solving patterns for tasks which are completely new to him or her.

The acquired contents are in any case latently interpreted within the framework of existing cognitive structures. This specific achievement of the student does not normally become visible in the solutions of problems used to measure performance. It only appears quantitatively as the ability to solve more types of problems—including those the average student is unable to solve—in less time and with greater certainty in precisely the kind of situation where the assimilation to cognitive structures is commensurate with the acquired contents.

At this point, I do not want to explore what an independent problem-solving process would look like in detail. However, in order to prevent a mechanistic interpretation, I would like to point out that in all cases calling for solving more difficult problems, the cognitive structure to which the problem is assimilated can of course be relatively far removed from the real nature of the object to which the problem refers. It is, furthermore, sufficient that the problem-solving procedure is anticipated correctly merely in its essential features by the assimilation to a cognitive structure, since this is sufficient to divide the problem into components that are less complex and can be dealt with separately. In this way, the problem is gradually reduced, until it can be based entirely on familiar problem-solving patterns.

Thus it is not only the illuminating flash of inspiration that is necessary for productively solving a heuristic problem—a flash of inspiration that, in the context of the theory presented here, must be interpreted as the correct assimilation of the problem to a cognitive structure which accurately reflects the object of the problem in its essential aspects. The solution of such a problem also requires systematic, detailed work guided by the cognitive structure. Through such work the particular ideas which represent guideposts in the heuristic process of finding the solution, are developed into the construction of a complete problem-solving pattern.

How is this explanation of productive mathematical achievements to be related to the conventional view of teaching and learning processes and the resulting organization of learning?

The evaluation of the success of learning processes is guided by the short-term achievements which the student demonstrates in the process of solving problems. The entire system of ability assessment in our school system is essentially governed by this principle, and—in some way or other—students as well as teachers equally conform to this assessment of the organization of their learning and teaching processes respectively.

If the theory presented here is accurate, there is, however, obviously a blatant discrepancy between the kind of performance that emerges in these solutions and the student's actual overall achievement in the learning process. The essential result of the learning process, the latent development of the cognitive structures, is not at all adequately reflected by the evaluation of the achievements. For the solution appears either as a reproductive performance and, in this case, does not give any indication concerning the productive aspect of the learning process which is crucial for mathematical competence; this is the case where the problem objectively simply demands the reproduction of knowledge and skills taught and practiced in class. Even if, in such a case, productive components figured in the problem-solving process, they are devalued as components of a reproductive performance in the evaluation. Or, on the other hand, the solutions merely indicate that individual students were capable of an independent performance above and beyond what was taught, whereas others were not; this would be the case if the problem presented heuristic problems or problems of interpretation.

What follows from reflection on these experiences?

The feedback of the system of evaluation, referring the successes of the learning process back to this learning process itself and its conscious control by the student, therefore, consists in the fact that, as a rule, he or she receives the information that a much more limited performance is expected of him or her—that is, only a reproductive performance—than he or she is in fact regularly accomplishing in the learning process—that is, the productive achievement of integrating what has been learned into the structures of his or her thinking. Conversely, however, this feedback continually demonstrates to him or to her that, at any given time, he or she is capable of accomplishing this reproductive performance only to a limited extent. Thus, the more the learning process is guided by the misinformation from the feedback of the evaluative system and not from a feedback that consists in the primary information on the object of learning, the more the productive process of the development of cognitive structures will be obstructed.

As a rule, more importance is attributed to the feedback of the evaluative system in the consciousness of teachers and students alike when the achievements are more modest. With the decline of the performance as it appears in the solutions to problems, the prospects of improving competence in the learning process in the school context also decline. Given the learning conditions described above, which presumably quite accurately reflect the reality of mathematics education, and provided that the explanation is accurate that the ability differences are due to the development of cognitive structures, then the development of achievement in mathematics education resembles a divergent system. Ironically, the negative effect of this system on the performance of the majority of the students is attributed to its victims as a personal failing by means of the ability assessment.

NOTES

[1] The identification of ability in terms of its actual appearance is the principle of behavioristic psychometry as well as—in normative terms—of the behavioristic concept of educational objectives. The basis for this identification is the classification of behavior and the description of ability by means of the concepts generated in this way (see [5], [1]). An especially extreme form of this identification can be found in the guidelines for mathematics education at the secondary level in Hesse *(Hessische Rahmenrichtlinien für den Mathematikunterricht in der Sekundarstufe I)*. In these guidelines the expected performance with regard to the subject of "negative numbers," for example, is defined by a list of 120 educational objectives. This seemingly absurd example is, as is the compilation of detailed lists of educational objectives in general, based on the idea that it is necessary to

provide the teacher with a full list of the behavior patterns that define the performance to be elicited in order to improve classroom teaching and to adjust the ability assessment accordingly. Ability is the cumulative outcome of acquired behavior patterns. Learning is a specific form of behavior modification which, as a matter of principle, is understood as a cumulative process. Thus Benjamin S. Bloom writes in the introduction to his book ([2]), "The book could also be summarized in a single mathematical formula $I_2 = I_1 + f(E_{2-1})$, where 'I' represents quantitative measures of a characteristic at two points in time and 'E' represents the relevant environmental characteristics during the intervening period."

2. This paradox becomes serious because the identification of ability with its appearance is the principle of ability assessment in schools. Grades do not refer to credible achievements in the process of learning, but to its global result. Furthermore, they do not refer to this result in such a way that they measure the specific learning success, that is to say, the difference of the level of ability before and after a learning sequence, but to the global level of ability, to the accumulated competence compared to the average as it is expressed in the actual performance when solving more or less difficult problems. The paradox has its reality in the classroom. It is not possible to understand the development of achievements in mathematics education and the emergence of relatively stable ability differences—whatever the definition of the concept of ability—without a theory which takes into account the effects of this paradox on the development of ability (see [4]).

3. The survey [7] provides a comprehensive documentation of the results of research regarding school achievement based on these basic assumptions. The entire second volume in particular is dedicated to school achievement in mathematics education.

4. See [7], part 1, p. 31ff., as well as the overview of dimensional analyses of school achievement in mathematics education [7], part 2, p. 334ff.

5. See [7], part 1, p. 19f.

6. See [7], part 1, p. 15ff. International comparisons have shown that comprehensive school systems, where teaching takes place to a larger extent in groups that are heterogeneous with regard to ability, are equal to selective school systems with regard to the top abilities in mathematics education, but are superior with regard to the levels of mathematical achievement of an entire class (see [6]).

7. Quote from the preface of [3].

8. If the paradox of the empiricist concept of ability presented at the beginning is placed in the context of this causal-mechanistic explanatory model, one more aspect becomes evident: If learning processes are in fact to be viewed as goal-directed only, if they are, as indicated, the result of a goal-directed teaching process, then the progress of knowledge from one generation to the next can only be viewed as a contingent coincidence. For learning is viewed as a function of teaching, and teaching stops where the knowledge of the teacher has its limits. Reproductive learning processes do not lead beyond reproduction. Complementing the abilities which can be acquired reproductively by assuming the existence of productive abilities, for example, a "problem-solving ability," is obviously not sufficient for eliminating this paradox.

9. Since I am at this point only concerned with the adequate definition of a concept of mathematical ability, I am refraining from referring to the theoretical traditions which the following assumptions on productive mental performance are indebted to and from explicating the many explanatory accomplishments and empirical confirmations of these assumptions. Instead let me refer to [8], Chapter 1, Chapter 2.

10. At this point I would like to add a few brief remarks for those who are familiar with Piaget's theory of the development of cognitive structures. While it is true that Piaget explains the development of cognitive structures in terms of the reflection on the actions of the learners, he nevertheless limits reflection to the form of actions. He thus calls this process reflective abstraction and consequently assumes an autonomous logic of the development of cognitive structures independent of the contents of learning. For this reason it

is difficult to draw conclusions from Piaget's theory with regard to the organization of teaching. The process of the development of cognitive structures is, however, not merely a process of the transition of one formal structure to another, more general one, but at the same time also a process of the transformation of meaning, of the reinterpretation of all the contents assimilated to the structure. In my opinion it is exactly at this point where the real problems arise which the learner must overcome in the development of his or her cognitive structures. For that reason, I am giving a more general definition of the concept of reflection in the thesis on the restructuring of cognitive structures by means of reflection, and I believe that I have thereby achieved a better explanation of the process of the development of cognitive structures. More on this in Chapter 1.

BIBLIOGRAPHY

[1] Bloom, B. S. (Ed.): *Taxonomy of Educational Objectives; the Classification of Educational Goals, by a Committee of College and University Examiners.* New York: D. McKay, 1956.
[2] Bloom, B. S.: *Stability and Change in Human Characteristics.* New York: Wiley, 1964.
[3] Bloom, B. S.: *Human Characteristics and School Learning.* New York: McGraw-Hill, 1976.
[4] Bölts, H. et al.: "Leistungsbewertung im Mathematikunterricht." In *Westermanns Pädagogische Beiträge, 30,* 1978, pp. 264–267.
[5] Mager, R. F.: *Preparing Objectives for Programmed Instruction.* San Francisco: Fearon, 1962.
[6] Postlethwaite, N.: *School Organization and Student Achievement. A Study Based on Achievement in Mathematics in Twelve Countries.* Stockholm: Almqvist & Wiksell, 1967.
[7] Roeder, P. M. & Treumann, K.: *Dimensionen der Schulleistung,* 2 parts. Deutscher Bildungsrat: Gutachten und Studien der Bildungskommission, 21. Stuttgart: Klett, 1974.
[8] Skemp, R. R.: *The Psychology of Learning Mathematics.* Harmondsworth: Penguin, 1971.

CHAPTER 5

MATHEMATICS EDUCATION AND SOCIETY
[1984]

INTRODUCTION

It must be stated at the very outset that the topic of mathematics education and society is not an essential part of the prevailing discourse in education.[1] Those who design and plan mathematics education usually do not at the same time give serious thought to the circumstances that determine the domain of the educational efforts. Therefore I do not feel obliged to discuss primarily the "state of research," to delineate "approaches," and to cite "investigations" when presenting the facts relevant to the topic. This would presuppose a consistent or, at the very least, a coherent, albeit possibly controversial, educational discourse on this subject.

Nor do I intend to treat the topic of mathematics education and society in the usual fashion as merely another topic on which to articulate a personal opinion without implications for educational efforts. The following presentation takes this aspect into account: The facts that I consider essential to pedagogical theory will be outlined broadly rather than in detail, although they are normally not, and in part for good reason, at the center of theoretical considerations.

Public discussion of mathematics education has for some time been dominated by the basic assumption that mathematics education in public schools constitutes a social requirement; above and beyond this, it is often even assumed that society does and must exert constant pressure on mathematics education in order to expand it and make it more effective, because, despite considerable effort, it does not yet meet the stated expectations. The much quoted resolution of the Standing Conference of Ministers of Education and Cultural Affairs of the Länder in the Federal Republic of Germany *(Kultusministerkonferenz)* of October 1968 on the modernization of mathematics education begins, for example, with these words:

Progress in mathematics and the fact that mathematical approaches are penetrating the sciences which are of significance for the economy, the society, and the state, necessitate a modernization of mathematics education at all levels. ... A person will be in a position to solve the problems facing him or her in the modern, technical world only, if he or she has gained insights into scientific methods and an understanding of mathematical structures." ([48], resolution No. 611, p. 1)

In 1976, the German Association of Mathematicians *(Deutsche Mathematiker-Vereinigung)* expresses the same view in its memorandum on mathematics education:

> In a world more and more subjected to mathematical methods the helplessness with regard to mathematical ideas and methods ... has disastrous consequences. More and more young people are in danger of failing in their education or occupation because of insufficient mathematical training. ([17], p. 1)

The challenge for the educator seems to be first and foremost to devise better and more effective methods, to give more and more students an ever more profound understanding of mathematical problems, facts, concepts, and theories.

In the climate created by this kind of basic assumption not much interest is to be expected on the part of educators with respect to the question concerning the relationship of mathematics education and society. The predetermination of the educational goals seems to be all too obvious. The exploration of new possibilities seems in and of itself meaningful, be it in terms of an effort to deal with the four color problem in class or with the methodological planning of how to teach about congruent transformations in finite Euclidean planes. As long as there are difficulties in doing fractions, any suggestion to overcome them is valuable. If skepticism thrives, because unreasonable goals cannot be achieved, because insufficient understanding of their meaningfulness has a demotivating effect or, at times, leads to noncooperation or maybe just because the familiar school routine is threatened, this appears to be an invitation to increase the educational endeavors once again, while scaling down the demands a little. In short, in the context of the planning of mathematics education, there is, after a brief acknowledgement, a tendency to move on to the matter of everyday educational planning in the narrow sense. Relevant educational concerns are understood to be the domain of the practitioner. Inclusive theories of mathematics education and comprehensive investigations of its social function are not much in demand.

This attitude of many educators is reinforced by the unique nature of the subject. The question of the social relevance of mathematics education seems to be inappropriate or at least extraneous to the subject. Mathematics, understood as the discipline of formal systems, as the science of universal structures capable of being described axiomatically, seems to be concerned with objects of ahistoric consistency. At best the question whether innate ideas, constructions, or conventions determine the nature of mathematical objects seems to be worthy of consideration. It is this ahistoric subject and not its historically changing usefulness that is of interest to the educator. It is this that must be taught in school. Those who would discuss the relationship of mathematics education and society in other than declamatory terms, are easily suspected of not being sufficiently conversant with this matter or even of belonging to the camp of the numerous secret foes of mathematics education. Anyone who would embark on asking questions about mathematics education in a context other than that of the predetermined political goals of mathematics education must remember this.

THE VICINITIES OF MATHEMATICS— MATHEMATICS OF THE VICINITIES

The efforts to attain ever more precisely defined objectives for mathematics education by means of structural analyses are generally aimed at defining the learning steps in each case in such a way as to create optimal conditions for the subsequent learning steps. Such a procedure of determining the objectives contains as an unstated premise a circular answer to the question about the meaning and usefulness of mathematics education: We are learning mathematics in order to keep on learning mathematics—in elementary school for middle school, in middle school for job training programs or for high school, in high school for college. But for every student the day will come when this sequence of reasons breaks. The number of those who will continue, professionally or as an avocation, to perceive each learning stage as the precondition for further mathematical knowledge until the day they die is minute. For the rest, the day will come when the question as to the meaning of mathematics education presents itself in a different way: Is it alright just to forget what has been learned or does it have a place outside mathematics as well?

To begin with, the obvious question might be raised whether the subject matter is applicable outside of mathematics. Applicable to what? Surely not to any given problem in any given area. It does not make sense to try to "mathematize" any problem whatsoever; the problem must be inherently mathematical for the application of mathematical methods to make sense. Mathematics, so to speak, has vicinities which, while they can be observed from the vantage point of mathematics, can be adequately conceived with the help of the methods of mathematics alone neither in terms of their inherent nature nor in their differentiation from areas that cannot be understood at all in terms of mathematics. Those who want to find out why mathematics education is worthwhile will first of all have to explore the vicinities of mathematics.

What is it that makes the vicinities of mathematics different from mathematics itself? The following example might help to focus on some initial reflections on this question!

In 1590, the Nuremberg mechanic, Johannes Richter (1537–1616), known as Praetorius, who later became a professor of mathematics, invented an instrument that is in use to this day: the plane table (see [8], p. 589). A detailed description of the instrument which is used to produce maps true to scale in the field was published and thus made accessible to the public only two years after his death by his student Daniel Schwenter (1585–1636) (see [8], p. 666ff.). It consists of a portable drawing board with an alidade, a rule equipped with simple or telescopic sights used for the determination of directions, that makes it possible to plot the bearings directly onto paper.

The use of the instrument is simple. First, a baseline is marked off, measured, and transferred to the paper true to scale. The plane table is positioned at one end of the baseline in such a way that the plotted and the actual baselines point in the same direction. Then, by aligning the sights of the alidade the bearings of all objects to be shown in the plan are taken and mapped onto the paper at the one end of the drawing of the baseline. The same procedure is repeated at the other end of the baseline. Each of the objects is located at the intersection of the two corresponding bearings drawn on the paper.

Can that be considered mathematics? There can be no doubt that the "useful little geometric table, invented by the estimable and world-renowned mathematician, the late M. Johannes Praetorius,"[2] would not have been invented without the latter's knowledge of mathematics, exceptional for his time. There can further be no doubt that the method of using the plane table gives rise to a host of divers mathematical considerations. Thus, in modern school books the use of the plane table frequently serves to demonstrate the practical relevance of the concept of geometric similarity, for instance, when the objective is to prove that the image of an area taking shape on the drawing board is indeed true to scale.

Such mathematical considerations do not, as in this example, necessarily remain without practical consequences for the use of the method. Thus, in

1617 the Leyden scientist Willebrord Snellius published in his treatise *Eratosthenes Batavus* a solution to the following problem: The location of a place is to be determined by means of the measurement of two angles obtained by taking the bearings of three known objects (see [8], p. 705). It is true that his answer to this problem was at first disregarded. But soon the problem became so urgent for surveying that it became the subject of several independent mathematical investigations. With the aid of the solution to this problem, for example, the exact location of the plane table could be determined using nothing but a map of the surroundings.

However, even if there is no doubt about the close association between the use of the plane table and mathematical considerations, the application of this method does not by itself constitute mathematics. To the contrary: The invention of the plane table helped to solve precisely the problem of drawing a map of an area without the mathematical skills necessary for converting angles and lines from field measurements to be shown in a map of the area by means of construction and calculation. Now a method was available that delivered this map automatically by using a mechanical rule. Other methods of surveying known at the time were by no means superseded; rather, they were augmented by a version that differed only slightly with respect to the necessary mathematical skills and had its own advantages and disadvantages.

This process of increasing mechanization also characterizes the subsequent development of the method of surveying by means of the plane table. Today the method is no longer used as described above. Since the last century, the alidade has been equipped with a telescope that has a device for optically measuring distance. Thus, it was no longer necessary to determine any longer a baseline and to move the plane table. Bearings of the objects to be located are taken and drawn to the paper. Then the distances measured are transferred directly on these viewing lines. This procedure constitutes progress for the quick and uncomplicated production of maps of a given area. At the same time, however, the splendid mathematical idea on which the original method was based and which to this day gained the method using the plane table entrance into the school books has become practically irrelevant.

What can be learned from this example? First, in discussing the question as to what kind of mathematical knowledge is necessary and how it should be acquired, it has to be observed that a careful distinction must be made between the method of using a plane table and the mathematics implied by it. This is the distinction between mathematics and its area of application, that is to say, its vicinities. The distinction is fundamental. It concerns the objectives that are pursued and the means used in the process. With regard to the example above: In the vicinities of geometry the main motivation was to develop a method for registering property as an unquestionable basis for determining the tax rate and for settling disputes. Mathematics, on the other hand, was concerned with generating ideas for this purpose to be secured by means of constructions and deductions in the context of the conceptual system and with the aid of the methods of construction of Euclidean geometry that had been handed down and refined.

As important as this distinction may be, it is nevertheless equally important to emphasize the substantial identity between mathematics and its vicinities. Both fields are concerned with the same entities; the example cited above deals both in mathematics and in surveying practice with the relationships of position and size of similar geometric figures. The common view that the vicinities of mathematics are nothing but an external precondition for the development of mathematics needs to be revised in view of the substantial connections between the conceptual foundations of mathematics and the social problems and historical developments in its vicinities. Mathematics is profoundly influenced by the problems and the solutions of these vicinities, whereas the fundamental differences between mathematics and its vicinities have their origin in external conditions of social development, in particular in the specific forms of the division of labor between practice and mathematical theory that are useful in the mathematical vicinities. These conditions are always in need of discussion, and the separation of mathematics from its vicinities that mathematics takes for granted, is called into question, if the objectives of mathematics education are to be defined in terms of their social function.

Excursus: The Timelessness of Mathematical Truth

Nothing fosters the perception that mathematics is unaffected by social developments and that mathematics education is above the ties of educational institutions to their social function more than the timelessness of mathematical truths. What is the origin of this assurance of the validity of mathematical propositions?

The certainty does not arise from the vicinities of mathematics. There is almost nothing in these vicinities that remains untouched by social change. None of the real problems to which mathematics can be applied resembles, for instance, the theoretical problem of the squaring of the circle—a problem, furthermore, that, for all practical purposes, had been solved with sufficient accuracy long before it became a theoretical problem. None of the problems of the vicinities of mathematics retains its character throughout history. Such a problem can always be solved in a completely different way than is the case, more or less satisfactorily, in a given situation with the available means. What is necessary for this is ingenuity, often a little luck—sometimes even some knowledge of mathematics—and usually time, lots of time, as much time as is required for the changes of the social conditions and of the forms of the appropriation of nature which make possible, suggests or unequivocally demand particular solutions.

As an example the vicinities of arithmetic—that is, calculating—might be called to mind. Numbers are timeless, but the practical methods of dealing with them are largely determined by the context in which they are used. Nobody could have provided Ptolemy in his time with a calculating device of the quality of an electronic calculator. Had he had such an instrument, he would have known how to use it, but between the Babylonian calculating tables which he presumably used and the electronic calculating devices of our time there are 2,000 years of an eventful history of the most diverse methods and instruments with which calculations could be performed or circumvented. Their multiplicity reflects the diverse purposes and social objectives for which, based on the available means, they were invented. Today, most of them have become useless, and many have been forgotten altogether.

Given this background of the historical variability of calculating devices and methods, the arithmetical laws appear as the paragon of timeless universality due to their permanence and their absolutist claim. And even if historical developments within the conceptual system of mathematics similar to those of other sciences cannot be overlooked, mathematics is on the whole surrounded by the very aura of consistency that, in contrast to the changing calculating methods, is characteristic of the laws of arithmetic.

Only a comparison of mathematical methods with the problem-solving methods of the vicinities of mathematics exposes the source of this apparent timelessness. It is characteristic of mathematical methods that they are limited to a small number of admissible means. In contrast to the vicinities of mathematics where any means contributing to the solution of a problem is acceptable and none appears to be justified if it cannot properly compete with other procedures, the investigation of mathematical problems is carried out with the help of an exceedingly limited number of explicitly stated methods of thinking. In the course of history, these methods have in turn been differentiated to such a degree that only academically trained specialists are in a position to make full use of their possibilities. Whatever looks as if it cannot be addressed by means of these limited methods, is by definition eliminated from mathematics at the very outset; a "mathematical problem" is defined as a problem that, as a matter of principle, is within the scope of objectives that can be realized using these methods. It would be fair to say that there is no other science in which the methods of investigation constitute the subject to an equal extent.

In particular, the actual purpose that gives rise to a problem in the vicinities of mathematics is disregarded in mathematics. It does not influence the thought process. The possibilities of the arsenal of mathematics are employed to the fullest extent whether the effort can be justified in a practical context or not. The result is sometimes much more fruitful than the original problem, and this abundance may be more than can be used in the vicinities of mathematics. Frequently, the initial problem even remains unresolved in this process. Sometimes hundreds of years pass before a mathematical problem that arose out of the vicinities of mathematics is solved; the cause may long since have faded from view and the practical problem may have been resolved in a different way.

It is precisely this abstraction from the actual context of the vicinities of mathematical problems that appears as the timelessness of mathematical truth. As long as the methods of mathematical thinking being used are kept constant, the problems that can be addressed with their help and that, hence, emerge as "mathematical problems," can be clearly isolated, and no historical event can make the solution to a problem which can be achieved by applying these methods seem obsolete. Thus, the question with regard to the timelessness of mathematical truth refers to a totally different type of question not at all surrounded by the aura of timelessness, that is to say to the question of the socio-historical conditions that gave rise to mathematics as a science defined by its methods in the context of the problems arising from its vicinities.

TRANSMISSION OF KNOWLEDGE PREDETERMINED BY SPECIFIC ENDS

The procedures employed in the vicinities of mathematics as well as the knowledge gained in the process of their application and the problem solutions achieved with their help must be transmitted, if they are not to be lost. To become part of the cultural heritage, the mathematical and quasi-mathematical methods of architecture, surveying, astronomy, bookkeeping, and similar areas of the vicinities of mathematics not only had to be invented, successfully and properly applied to the problems at hand, and to retain lasting usefulness thus ensuring continued interest in them, but most of all they needed effective organizational and institutional forms of being passed on from one generation to the next.

The usual form of this transmission of knowledge has little or nothing to do with general education. For the cultural survival of a method it is usually sufficient that a small number of professionalized experts be trained in the proper use of the method and acquires the necessary skills. Indeed, for this transmission mathematical training in the proper sense is not even necessarily required. As a rule, the necessary quasi-mathematical skills are taught as an integral part of the specific professional training guided by the particular social purpose. For an understanding of the relationship between mathematics education and society, it is therefore necessary to realize that, to a large extent, the socially necessary techniques in the vicinities of mathematics do not demand a general mathematical education at all, because they are handed down in the context of direct instruction in the use of procedures adequate for the particular purposes. In view of this way of transmitting knowledge about mathematical and quasi-mathematical procedures of the vicinities of mathematics, mathematics education in the narrow sense appears to be a relatively insignificant marginal activity, and its remoteness from practical problems and goals seems to be a consequence of its rather limited significance for the broader application of mathematical skills and ideas in practice.

In view of this situation the basic structure of this process of ends-determined transmission of knowledge deserves our attention, even if, or precisely because, it is generally deemed not to contribute very much to genuine mathematical education. Three passages from textbooks used for such ends-determined transmission of knowledge may serve as a point of departure:

First example
Example of a round field of diameter 9 *khet*. What is its area?

Take away $\frac{1}{9}$ of the diameter, namely 1; the remainder is 8. Multiply 8 times 8; it makes 64. Therefore it contains 64 *setat* of land.

Second example
Someone has a room which is 50 ells long and 40 ells wide which he wants to have paved with valuable stones. Four of these stones are 1 ell long and 1 ell wide. The question is: How many of these stones are needed to pave the room? Answer: 8,000 stones.

Solution: Multiply 50 by 40 ells which is 2,000 ells, and then multiply by 4 stones which are needed for one square ell.

Third example
What was the interest rate at which 3,000 dollars that were loaned out had grown to 5,619 dollars after 16 years?

Solution: 3,000 dollars increased to 5,619 dollars; one dollar would have increased to 5,619 : 3,000 = 1,873. It is known that this sum was reached after 16 years. The table of compound interest indicates that the amount of 1,873 dollars is reached after 16 years if the interest rate was 4 percent.

The texts from which these examples were taken could not have been chosen more arbitrarily, and with respect to their mathematical content these examples have almost nothing in common. The first example has been chosen from the *Rhind Mathematical Papyrus* (see [9], problem 50), a handbook that was used in Egypt by officials of the middle kingdom roughly 4,000 years ago. It presents a problem designed to teach a method for calculating the area of a circle which, while not exactly accurate mathematically, was sufficient for practical purposes. The second example comes from *The Improved and Complete Book of Arithmetic* ([28], p. 109), a volume on arithmetic by Johann Hemeling published in 1705. Hemeling was among a number of experts on arithmetic who taught the merchants and craftsmen of the emerging bourgeoisie the arithmetical skills that were becoming more and more important. Finally, the third example is gleaned from a contemporary textbook entitled *Arithmetic for Hairdressers* ([33], p. 39) and serves to practice the use of a table of compound interest which makes knowledge of exponential functions unnecessary.

As different as these examples may be, they have one thing in common: They represent the basic methodological model of transmitting knowledge determined by specific ends—that is, passing on those narrowly defined practical mathematical skills that are needed in the vicinities of mathematics. The basic component of this type of teaching is the "problem." The problems which serve to initiate the learning process reveal the relationship between the process of transmitting knowledge and given purposes with regard to the structure as well as, in part, even the content. Inasmuch as learning is determined by and does not go beyond the problems, it is not open like scientific research, but embedded in the fundamental process of transmitting knowledge about available problem-solving patterns. It is the problem-solving procedures that are learned, and they are taught in the form of instructions for reproducing the sequence of steps necessary for executing the operations of a given procedure. All that is usually necessary is the reproduction of the solution in the concrete case; the mathematical background need not even be discussed. Often the simplicity of the problem-solving operations even disguises the mathematical background as, for instance, in the above examples the mathematical content of the problems of squaring the circle, of the area concept, and of exponential growth.

This basic educational model of transmitting knowledge determined by specific ends does, however, has an inner dynamic which reverses the preconditions of the model. While the goal initially determines the type of problem and thus the choice of the problem-solving procedures to be learned, in the course of teaching it recedes completely into the background precisely because it is presupposed and only the teaching of the procedures is considered to be of any significance. The problems no longer determine the means to be used for solving them, but the procedures to be learned determine the kind of problem to be posed. While the latter still refer to the application of the procedures to be learned, they no longer represent the complexity of real applications; rather, the reference of the problems to the applications merely serves to "dress up" the problem-solving procedures which are abstracted from solutions of real problems. This development toward stereotyped problems follows the logic of the means employed to solve them and not the logic of the objects to which these means are to be applied.

This characterization applies not only to the mathematical training of practitioners in the vicinities of mathematics. It is, to a certain extent, also true of mathematics education proper, of the kind of instruction solely devoted to the development of mathematical thinking. Mathematical thinking begins where the formalism of the problem-solving procedure that is merely reproduced reaches its limit, but mathematical thinking in this sense is rare in mathematics education. Who is not familiar with the complaint of those literate in mathematics about the misuse of the formal treatment of mathematical methods and the unenlightened reproduction of acquired problem-solving procedures for routine problems. However, in everyday mathematics education the unreflected use of available problem-solving procedures is of far greater importance in working on mathematical problems than is admitted by the educational and pedagogic objectives in curricula and handbooks on pedagogy.

There is good reason to condemn the mechanical use of formulas and symbols in mathematics education. If pure mathematics were to be the judge, everyday mathematics education would not stand much of a chance to find approval. However, in the present context we are exploring the relationship of mathematics education to society and its requirements, and in this case the objectionable fact merits attention mostly as an indication that in day-to-day mathematics education the principles of transmitting knowledge determined by specific ends is of greater importance than many pedagogues care to admit.[3] We are reminded that mathematics instruction which is limited to pure mathematics is merely a special form of transmitting mathematical knowledge. This instruction, mathematics education in the proper sense, obviously has much in common with the way knowledge in general is commonly transmitted in the vicinities of mathematics—maybe because it satisfies certain social requirements in this way and renders a service which is not adequately acknowledged or maybe because attempts to free it from the petrified forms of instruction that are determined by specific ends but have become useless have failed.

THE TRAINING OF THE MIND

Despite its similarity to the transmission of knowledge determined by specific ends, it cannot be completely overlooked that mathematics education in the proper sense cannot be reduced to the function of transmitting useful mathematical knowledge. Mathematics education is associated with a peculiar notion according to which it is of intrinsic value to understand the abstract and often arcane thinking of mathematicians. On this view, those who remember the mathematics instruction they had to endure only with loathing and bitterness, are considered nothing but pathetic creatures lacking direction, good will, or ability. Such a view can hardly be interpreted as an adequate expression of the usefulness of mathematical skills as a means of mastering the problems in the vicinities of mathematics.

The very origin of mathematics and, consequently, of any kind of instruction in mathematics gives pause because of its obscure detachment from the practical purposes in the vicinities of mathematics.[4] The term "mathematics" is of Greek origin, and close scrutiny of the available sources for the earliest signs of abstract mathematics leads to the conclusion that, considering its very nature, mathematics itself indeed has its origin in Greek antiquity. At any rate, despite the insights Egyptians and Babylonians gained with the help of their arithmetic methods, they remained in the vicinities of mathematics even where their calculations no longer primarily served practical purposes. However ingeniously Egyptians and Babylonians operated with numbers, they did not yet have a number concept in the true sense. Thus, neither of these two developed civilizations possessed, for example, a word that could, without reservation, be regarded a term for the abstract concept of "number."

Sometime in the 6th century B.C., when the art of the dialog acquired social significance in the life of the early citizens of Greece and when early methods of this "dialectical" art, the art of verbal contest, were expressed in rules and used intentionally, someone applied the newly acquired method of "definition" to the concept of number for the first time and made fundamental inferences. Tradition ascribes this achievement to a conservative political thinker of the landed gentry; his name was Pythagoras. There is no text that could illuminate the background, but this much is certain: Speculation with numbers was a part of the esoteric doctrine of a secret political society founded by him. The doctrine of this society was called "liberal art"; for it was unencumbered by the practical purposes of the calculation techniques to which number speculation was opposed.

Thus, the "doctrine" (Greek: "mathema") of the Pythagoreans was completely different than instruction aimed at transmitting ends-determined mathematical knowledge that is necessary in a practical context. Rather, it was a doctrine of a conspiratorial brotherhood rife with divers religious elements, passed on within the context of strict rituals and prescriptions its members were subjected to.

We are familiar with the contours of the mathematical elements of this doctrine from later accounts. It comprised the subjects of "arithmetic," "geometry," "astronomy," and "harmonics." Pythagorean mathematics, structured in this fashion, not only became the basis for mathematics education in Greek antiquity, but also the blueprint for the "quadrivium" of the upper division of the seven liberal arts in medieval universities. There is every indication that the theorems about even and odd which Euclid included in book IX of his *Elements* belong to the oldest part of Pythagorean arithmetic, theorems such as for example proposition 21: "If as many even numbers as we please be added together, the whole is even." Such theorems have outlasted Pythagorean doctrine and remain valid to this day. Their obscure origin seems insignificant in view of the fact that, to this very day, we are relating them to the same numbers Pythagoras did, and their qualities have not been changed by any social event, revolution, and progress. There can be no doubt: The politically motivated detachment from practical purposes of the secret Pythagorean doctrine did not only have obscure consequences. In a certain sense the "mathemata" of Pythagoras were mathematics in our contemporary sense; Pythagoras had left the vicinities of mathematics and entered its core territory.

However, the doctrine of the Pythagoreans most certainly would not have survived their activities had it not been incorporated early on into a different program with goals just as opposed to practical concerns. The idea that the preparation for participation in the political life of the polis by the members of a nascent bourgeoisie demanded a special training of the mind to be achieved by a public education system, initially propagated by the Sophists, quickly gained ground. The teachings of the Pythagoreans seemed to be predestined for such an education system, for the goal of their kind of instruction was not to learn "for professional purposes, to become a practitioner, but in the way of liberal education, as a layman and a gentleman should." ([60], 312b) The detachment from practical purposes became a basic characteristic of the Greek educational ideal aimed at shaping thinking and feeling. And in this situation the "mathemata" of the Pythagoreans became a central part of the Greek system of higher education during the Hellenistic period (see [21]; [52]; with regard to mathematics education in the Middle Ages: [24]).

In his treatise on the state Plato made a detailed case for this. As the starting point of a concept of mathematical education which has persisted up to the present time, his arguments are still worthy of consideration. On arithmetic he wrote for instance:

...we should induce those who are to share the highest functions of state to enter upon that study of calculation and take hold of it, not as amateurs, but to follow it up until they attain to the contemplation of the nature of number, by pure thought, not for the purpose of buying and selling, as if they were preparing to be merchants or hucksters, but for the uses of war and for facilitating the conversion of the soul itself from the world of generation to essence and truth. ... You see, then, my friend, ... that this branch of study really seems to be indispensable for us, since it plainly compels the soul to employ pure thought with a view to truth itself. ... Again, have you ever noticed this, that natural reckoners are by nature quick in virtually all their studies? And the slow, if they are trained and drilled in this, even if no other benefit results, all improve and become quicker than they were? ... And, further, as I believe, studies that demand more toil in the learning and practice than this we shall not discover easily nor find many of them. ... Then, for all these reasons, we must not neglect this study, but must use it in the education of the best-endowed natures. ([60], 525c–526c)

The Greek conception of mathematics education no longer determined by specific ends is so closely intertwined with the emergence of mathematics itself that it is difficult to distinguish the growing independence of mathematical contents in the educational canon from the growing independence of these contents in the pursuit of mathematical problems in science (see [31]). According to the Greek conception, the occupation with mathematical problems already abstracted from their actual context was to be pursued in education for its educational effect on the mental attitude of the student. Consequently, the process of abstracting actual problems from the vicinities of mathematics, essential to the emergence of mathematics as a "liberal art," was reproduced in the educational context. Thus, from its inception, the idea of mathematics education no longer determined by specific ends is ambivalent as far as its social function is concerned. On the one hand, this idea provided the framework for the possibly unique character of mathematics to provide scientific knowledge that anticipates future practical problems; 1,500 years later, modern science could take up Greek mathematics with hardly a change. On the other hand, this idea simultaneously established a tradition to consider education for its own sake a status symbol.

Even to some contemporaries of antiquity the value of such an education seemed questionable. Isocrates, for instance, a contemporary of Plato's and, like Plato, an initiator of the Greek educational ideal, took exception to the detachment from practical purposes, and one of the most famous mathematical problems, the problem of the duplication of the cube, appeared to him to be a prime example for the useless hair splitting which he sought to keep out of education (see [52], p. 176ff.).

Criticism of the idea of mathematics education transmitting knowledge that is no longer determined by specific ends becomes more poignant on the part of those who were politically opposed to aristocratic-conservative Platonism and its idea of the creation of elites. The story goes that Diogenes who was nicknamed "the dog" because of his brazen protest against the social organization of the *polis* believed that "we should neglect music, geometry, astronomy, and the like studies, as useless and unnecessary." ([20], Vol. 2, p. 75) He would wonder "the mathematicians should gaze at the sun and the moon, but overlook matters close at hand." ([20], Vol. 2, p. 29f.; on the philosophy of Diogenes, see [27]) By way of example Epicurus, whose criticism resulted from his theoretical approach emphasizing the withdrawal into the private sphere, might also be mentioned. Even in Cicero's disparaging account of the Epicureans the essence of their criticism of the idea of a mathematics education no longer determined by specific ends becomes evident. Cicero had one of Epicurus's disciples defend him against the charge that he was ignorant as follows:

You are pleased to think him uneducated. The reason is that he refused to consider any education worth the name that did not help to school us in happiness. Was he to spend his time, as you encourage Triarius and me to do, in perusing poets, who give us nothing solid and useful, but merely childish amusement? Was he to occupy himself like Plato with music and geometry, arithmetic and astronomy, which starting from false premises cannot be true, and which moreover if they were true would contribute nothing to make our lives pleasanter and therefore better? Was he, I say, to study arts like these, and neglect the master art, so difficult and correspondingly so fruitful, the art of living? No! Epicurus was not uneducated: the real philistines are those who ask us to go on studying till old age the subjects that we ought to be ashamed not to have learnt in boyhood. ([10], p. 75).

Whatever one's judgment of the idea, developed in ancient Greece, of an intellectual education by means of an occupation with a mathematics released from practical goals, it should be kept in mind that this idea was developed in a political context in which education was expected to bestow the proper consciousness upon an aristocratic elite. For it becomes evident that the struggles during the 19th century between technical school education and grammar school elite education from which our present-day curriculum for mathematics education in secondary schools essentially emerged, by no means represented merely a temporary eclipse of the discussion about pedagogical objectives by controversies motivated by social policy, a discussion only superficially concerned with the subject matter of mathematics education and its objectives (see [66], p. 116; [49], p. 39ff.). Rather, at issue was the very essence of the idea of a mathematics education unencumbered by specific ends and purposes. And who, in view of the fact that to the present day educational concepts are being developed for a selective educational system that differs only in degree from the schools of the past reserved for members of the upper classes, would want to declare without further ado that the Greek idea and its political implications have become totally obsolete?

MATHEMATICS AS A PROFESSION

In the preceding section, connections between the esoteric origin of educational concepts of antiquity and their detachment from considerations of usefulness were shown. It would, however, be extremely lopsided, if the reasons and consequences of doing mathematics for its own sake were discussed solely under this aspect. Abstraction from practical objectives is not unique to the articulation of mathematical facts and problems motivated by pedagogical considerations; to a far greater degree it characterizes the activities of the professional mathematician. Let us, therefore, once more examine history, but with a different perspective, and let us this time not inquire into the origin of mathematics, but into that of the mathematician (see [73]).

It is true that some of the Pythagoreans already called themselves "mathematikoi," but "mathema" initially meant nothing more than "doctrine." The first "mathematicians" simply made use of this term to indicate that they were followers of the Pythagorean doctrine.

The peripatetics, the followers of the philosophy of Aristotle, subsequently narrowed the meaning of the concept of "mathema" to include the four subjects of the quadrivium, "arithmetic," "geometry," "astronomy," and "harmonics" and, given this restriction, defined certain philosophers as mathematicians. They thus gave the concept of "mathematics" its specific orientation, but it goes without saying that, even after this shift of meaning, mathematicians were not the specialists of a particular scientific discipline that we designate as mathematicians today. Evidently, the question must be defined as follows: How long has it been that mathematics existed not only as a separate subject detached from practical objectives, but that there were also specialists in this subject—that is, the "mathematicians"—who devote themselves exclusively to this subject?

Were there any mathematicians as early as the Middle Ages? While medieval universities had chairs for mathematics, the duties of such a chair tended to be discharged as an external obligation rather than a rewarding calling. Let us, for instance, examine the life of perhaps the most distinguished medieval mathematician—Nicholas Oresme (1323?–1382).

At the age of 25, Oresme became a student at the Collège de Navarre at the University of Paris. Mathematics education there left much to be desired. Oresme quickly advanced from being a student to becoming a teacher, and in 1356, eight years after he got there, he became head of the Collège—an indication that, as a prerequisite for this office, he earned a doctorate in theology. In 1361, he left the Collège and rose through the ranks of the church hierarchy. As a secretary to Charles V he was entrusted with diplomatic assignments. In 1377, five years before he died, he was elected Bishop of Lisieux. That certainly does not look like the career of a mathematician as we understand it.

Was the situation different with regard to the mathematicians of the Renaissance? Did change come with the "new science" of early bourgeois development?

Girolamo Cardano (1501–1576), often considered a mathematician because of his "Ars magna sive de regulis algebraicis" was—if he is to be classified according to his professional position as a scientist—a physician and later a professor of medicine. However, he also lectured on geography and architecture in order to earn a living and was probably as partial to playing dice as to mathematics. His rival, Niccolò Tartaglia (1500?–1557) was too poor to earn academic degrees. He earned his living primarily as a teacher of arithmetic. That seems to come somewhat closer to mathematics, but Tartaglia also was certainly not a mathematician in our sense.

Here is a brief summary of the situation! François Vieta (1540–1605) and Pierre de Fermat (1601–1665) could be described as lawyers. René Descartes (1596–1650), Johannes Kepler (1571–1630), Galileo Galilei (1564–1642), and Isaac Newton (1643–1727) are usually not even considered mathematicians, although the designations of philosopher, astronomer, and physicist are no less misleading. For Gottfried Wilhelm Leibniz (1646–1716) no adequate professional description can be found at all. In any case, he was more of an historian than a mathematician, because, in his function as councillor to the Duke of Hanover he was charged with writing the history of the house of Brunswick. In exchange for his position he dedicated 40 years of his life to this task. Leonhard Euler (1707–1783) first applied for a teaching position in physics in Basel, then he tried to get a teaching position in physiology and anatomy in Petersburg, was finally appointed to a position in physics that became available in Petersburg and only after that shifted to a position in mathematics that subsequently became available. His publications also indicate that he cannot be categorized as a mathematician without reservations. This applies more or less to the other mathematicians of the late 18th century, to Lagrange, Monge, and Laplace.

Let us summarize: Up to the end of the 18th century it does not make much sense to classify the scholars and scientists in terms of particular disciplines. Scientific training usually comprised several disciplines. The educated scholar was qualified for a variety of professions that could at best be attributed to particular disciplines with regard to the teaching professions and even then did not necessarily lead to a corresponding specialization of scientific accomplishment. Furthermore, many scientists were entrusted with practical assignments that did not allow for a narrow specialization. Mathematics as a discipline did exist, the mathematician specialized in mathematics did not. There was mathematics education, but it was always situated within the context of a lively exchange with related disciplines and practical applications, and the overall education of individual scientists attested to a unity of science beyond the boundaries of individual disciplines.

The situation did not change until the 19th century (see [50], especially p. 59ff.; [45], especially p. 108ff.; [4]; [36]; [53]). Let us take a look at a typical biography of a mathematician that illustrates the change that took place at that time. Initially Carl Gustav Jacob Jacobi (1804–1851) studied the classics and mathematics in Berlin. However, while still at the university, he decided to dedicate himself fully to mathematics. In a letter to his uncle he gives the following explanation for giving up his studies of the classics:

For the time being, I must drop them altogether. The colossal hunk presented by the writings of an Euler, Lagrange, Laplace requires the utmost energy and exertion of thought, if it is one's goal to penetrate to the core rather than just scratch the surface. (quoted according to [73], p. 376)

This judgment was unquestionably accurate. The ideas and achievements accumulated during the 18th century could no longer be fathomed without intensive concentration on mathematics, and without such understanding successes, equal to the existing level, could no longer be attained despite the most brilliant ideas. In 1824, Jacobi writes in a letter:

What I have done was tough work, and what I am doing now is tough as well. Neither hard work nor memory lead to success in this endeavor; they are the subordinate servants of the movements of pure thought. But tenacious, brain-racking thinking elicits more vigor than the most persistent hard work. If, therefore, by dint of this continual habit of thinking I have gained some vigor, let nobody think it was easy for me, perhaps because of some fortunate talent. It is hard work, very hard work that I have to engage in, and often the anguish of thinking mightily shook my health. The awareness of the vigor gained, however, is the most wonderful reward for this work and provides the encouragement to continue rather than giving in. (quoted according to [73], p. 376f.)

Soon after, not only the awareness of the personal energy gained, but also a brisk career as a university mathematician was the reward for focussing all available energy on mathematics. After quickly passing the state examination for secondary school teachers, the standard conclusion of a course of studies in mathematics at the time, his first scientific paper was accepted in 1825 both as the qualification for the doctorate and for lecturing at the university level *(Habilitation)*. Jacobi gave his first courses at the University of Berlin, moved to Königsberg as a lecturer in 1826, became an assistant professor in 1827, and an associate professor in 1829; he became a member of the London Royal Society in 1833 and a corresponding member of the Berlin Academy in 1829; in 1836 and 1844 he advanced to become first an external member and then a regular member in addition to becoming a corresponding member and an external member of the Paris Academy in 1830 and 1844 respectively. He became the head of a school of mathematics that produced many well-known mathematicians, for example O. Hesse, J. G. Rosenhain, F. Joachimsthal, and L. Seidel.

In total harmony with the nature of his work Jacobi was convinced that the honor of science consisted in not being of any practical use. In a letter to his brother, a physicist and engineer, he exclaimed: "The highest achievement in science as in art is always impractical!" (quoted according to [73], p. 385) In the 17th century, scientific questions were at one and the same time ideological questions; in the 18th century, the ideological program was in part realized—as outlined in a scientific world view. Jacobi's scientific career represents a new type of science towards which the 19th century is headed: the specialized scientific discipline without interdisciplinary approaches to scientific questions. The ideological conflicts that once applied to science as an institution, were relegated to the private sphere. Galileo was forced by the inquisition to renounce his scientific beliefs. Such a fate could no longer befall Jacobi; his conflict with the powers that he had had nothing to do with his scientific activities: During the unrest of March 1848 in Berlin, Jacobi gave a speech in a liberal club in which he appeared to sympathize with republican ideas. As a consequence his salary was withheld which, for a while, led to great financial hardship for his family. A friend from Gotha, the astronomer P. A. Hansen, took his family in and, until his death in 1851, Jacobi was unable to move his family back to Berlin.

A contemporary historian of mathematics has commented on this incident:

> He was now permitted to stay on in Berlin until his health should be completely restored but, owing to jealousies, was not given a professorship in the University, although as a member of the Academy he was permitted to lecture on anything he chose. Further, out of his own pocket, practically, the King granted Jacobi a substantial allowance.
>
> After all this generosity on the part of the King one might think that Jacobi would have stuck to his mathematics. But on the utterly imbecilic advice of his physician he began meddling in politics "in benefit his nervous system." If ever a more idiotic prescription was handed out by a doctor to a patient whose complaint he could not diagnose it has yet to be exhumed. Jacobi swallowed the dose. When the democratic upheaval of 1848 began to erupt Jacobi was ripe for office. On the advice of a friend—who, by the way, happened to be one of the men over whose head Jacobi had been promoted some twenty years before—the guileless mathematician stepped into the arena of politics with all the innocence of an enticingly plump missionary setting foot on a cannibal island. They got him.
>
> ... who can wonder that his allowance from the King was stopped a few days later? After all even a King may be permitted a show of petulance when the mouth he tries to feed bites him. ([6], p. 332f.)

Jacobi's biography clearly demonstrates that the emergence of the "mathematician" in the 19th century was a consequence of a fundamental transformation of mathematics as a science. In retrospect the mathematician Felix

Klein who, half a century later, became an ardent proponent of a reform of mathematics education designed to take the changes in the status of science into account, described the transformation of mathematics thus:

> The ... specialization of the sciences begins ... with the pressure of the manyfold, considerably expanded tasks. Mathematics splits off from astronomy, geodesy, physics, statistics, etc. The number of specialized mathematicians grows immensely and is spread out over the most remote nations. In this prodigal development of specialized research it is no longer possible even for the most universal of minds to achieve a synthesis of the whole internally and to bring it to fruition externally. ... There can be no doubt that the life of science lost many valuable characteristics in this hectic modern development. Just think of the admiration the small group of distinguished men evokes in us who represented our science in the 18th century! As academics, without national boundaries, continually exchanging their thoughts by corresponding with each other, they combined the most fruitful scientific work with an ideal and in every respect well-rounded development of their own personalities. In this picture it is only one feature that the scholar of that time was well-versed even outside his own field and always knew himself to be in close touch with the development of science as a whole. ... Although we left behind ... and were forced to leave behind the university ideal of the 18th century for many reasons, it seems fitting to remember its advantages from time to time when we look at how science is usually conducted today. There are hundreds of working mathematicians in every civilized country each of whom toils only in a tiny corner of his science. Understandably, his little corner seems to surpass everything else in importance. He publishes the fruits of his labor in diverse papers in several far-flung journals maybe even printed in different languages. The presentation, addressed only to a few colleagues in the same specialty, refrains from any indication of a connection with larger, more general questions. It is, thus, possibly even hard to understand for a colleague with slightly different interests. For a wider circle it is unpalatable. ([45], Part 1, p. 3ff.)

Was this still the kind of mathematics to be learned because of its general usefulness? Was that still the kind of mathematics accorded special educational value because of the general intellectual training it afforded? The transformation of mathematics raised the question with regard to the purpose of mathematics education in a novel way.

THE EMERGENCE OF SCHOOL MATHEMATICS

Can the mathematician serve as a model for mathematics education? Should mathematics education in school be organized as preparation for the scientific discipline of mathematics? Obviously, the justification for such a trend decreases as the specialization and the professionalization of mathematicians increases. The prominence of mathematics in secondary education cannot be explained by the need of having to insure the continuance of the relatively small group of professional mathematicians. And the traditional educational ideal of training the mind regardless of specific ends by means of mathematics is becoming questionable in view of the fact that mathematics has changed from an integrative science, considered by modern science as an interdisciplinary ideal of deductive rigor and lucidity, to an esoteric and inaccessible discipline specializing in the abstract, namely that "colossal hunk" which requires the student of mathematics to forego a well-rounded education. To the extent that mathematics education turned into an institutionalized specialty and the mathematical thinker became a professionalized specialist, to that extent the idea of mathematics education was headed for a crisis.

This did not make itself felt very much during the 18th century. A high school graduate was not yet expected to have any, or at most scant, mathematical knowledge; mathematics education began at the level of the noble academies and universities and was integrated into the overall goal of the educational program of these institutions. Not until the 19th century did the conditions develop that gave mathematics education an independent character (see [42]; [58], p. 268ff.; [66], p. 112ff.; [63], p. IXff.). In Prussia, for instance, it was the development of a secondary school system aimed at preparing students for the university which created the preconditions for a possible confrontation between mathematics education and scientific discipline in the first place. In 1810, the first regulations for the examination of prospective teachers were enacted, in 1812, requirements for the *Abitur* (at that time a prerequisite for entering the university) were first issued, and in 1816, a curriculum was developed, the so-called Süvern curriculum which, however, was not immediately publicized. Finally, a set of rules relating to the final high school examination issued in 1834 excluding non-graduates from access to the university was the most decisive turning point.

With regard to the relationship of the contents of the curriculum to mathematics as a special scientific discipline two features of the Süvern curriculum are noteworthy. First, mathematics education—in comparison to, for example, science education—was given extraordinary prominence. Mathematics, together with Latin, Greek, and German, was one of the four core subjects and, next to Latin, was accorded the largest number of hours per week. Second, the objectives of mathematics education articulated in this initial curriculum were—compared with later curricula—extremely ambitious.

The strong position of mathematics education was due to the influence of neoclassicism. Following educational ideas of antiquity, mathematics education was believed to have the same importance in the development of the mind as Latin and Greek. This strong position was not determined by the content, but by the ability attributed to it of encouraging a particular mental attitude. The concept of "formal education" served to establish this idea as a pillar of the educational objectives of this century.

These demands resulted from the fact that the lower level of the education at the universities and the noble academies was incorporated into the secondary school system (approximately up to the analysis of the infinite series and Taylor's series). Thus, the mathematician initially did indeed become the model for mathematics education, and this tendency was reinforced by the fact that the qualification to teach at the secondary school level became the standard degree for mathematics students. Many outstanding mathematicians of the 19th century like Grassmann, Kummer, Plücker, Steiner, and Weierstrass taught, at least for a time, at secondary schools. The "Rules for the Examination of Candidates for the Teaching Profession at Secondary Schools of December 12, 1866" contained the following regulation:

> Only those candidates are to be regarded as having the qualification for teaching the upper grades who, in the course of the examination, prove themselves to be trained mathematicians and have mastered higher geometry, higher analysis, and analytical mechanics to the extent that they are able to successfully conduct their own investigations in these fields. ([70], 2nd div., p. 76)

Heinrich Emil Timerding, Professor of Applied Mathematics in Strassburg and subsequently of Descriptive Geometry in Brunswick, one of the most dedicated reformers of the turn of the century, declared with regard to the intended contents of teaching during the 19th century:

> All this is appropriate and justified for the purpose of preparing for theoretical mathematics. However, are we to believe that the same considerations were transferred without a thought to the secondary schools as well where future mathematicians are most certainly not trained? As improbable and puzzling as it may seem, for almost a hundred years mathematics instruction in schools has been conducted as if all students were about to study mathematics later on. ([66], p. 121)

The response was, however, not long in coming. The Süvern curriculum was never put into effect. Johannes Schulze who, from 1818 to 1840, served under the Minister of State Count von Altenstein in the newly created "Department for Educational and Medical Affairs" was entrusted with supervising the secondary school system and who allegedly said: "One line by Cornelis Nepos contains more educational value than all of mathematics" (according to [66], p. 122), reduced the number of hours per week at the upper level by one third and significantly lowered the requirements with regard to the subject matter. In addition, the Department deemed it necessary to issue the following memorandum in 1826:

> The ministry has had accasion to notice that several secondary schools *(Gymnasien)* have been neglecting to help students attain the altogether indispensable skill of ordinary arith-

metic. Contrary to the intention of the ministry some secondary schools introduce mathematics proper already in the lowest grade. Thus, instruction in ordinary arithmetic is, on the one hand, dropped altogether. On the other hand, in other secondary schools, where instruction in ordinary arithmetic does take place, it is not conducted with the required emphasis on practical exercise, or it is not separated accurately and carefully enough from instruction in mathematics. Since proficiency in arithmetic is required in any occupation whatsoever and since we know from experience that a deficiency in this skill is not easily eliminated later in life, but is, at the same time, experienced as a burden, the ministry feels compelled to require that in all secondary schools mathematical instruction proper not commence until the third year. In the two lowest grades skills in arithmetic are to be practiced without any admixture of mathematics .... ([70], Section 1, p. 98)

This memorandum did not remain the only one of its kind. Ministerial decree of December 13, 1834:

The ministry cannot accept the petition to continue the past practice regarding mathematics instruction at the Joachimsthal *Gymnasium* of including spherical trigonometry and the topic of conic sections in the upper grades as well. ... In its regulations the ministry intentionally did not require graduates to be familiar with spherical trigonometry and the topic of conic sections, because ... there were always only very few graduates who could meet the requirements issued in the regulation of June 4 of this year. ([70], Section 1, p. 99)

Circular decree of December 1, 1854:

... I, therefore, request of the Royal Provincial School Faculties that they pay attention to mathematics instruction in particular. First and foremost it must be strictly observed not to extend it beyond the limits set by the examination regulations as has been the case in some schools .... ([70], Section 1, p. 101)

These decrees illustrate the conflict between the social requirements of secondary education and the specific mathematical requirements of the mathematicians which grew more intense during the 19th century. The growing opposition of neoclassicist theoreticians of education to an overly dominating position of mathematics and the sciences in the educational canon did not just represent a conservative attitude directed against the educational innovations of the French Revolution; for the increasing difficulties of bringing the changing requirements with regard to the subject in line with the main ideas of higher education were all too obvious in the classroom (see [35]).

Ludwig Wiese, between 1852 and 1882 head of the Prussian Department for Higher Education, describes his personal experiences with mathematics education in his memoirs:

The mathematical zest which we ... had brought to it was not nurtured in the *Gymnasium* and was soon entirely extinguished. The teacher of the last several grades always came to class late. He was dreaming, gave an academic lecture and finally called on those students, whom he considered capable, to repeat it. Later I found out that he was an engaging man with a philosophical bent. However, he did not know how to teach and engage an entire class. (quoted according to [58], p. 278)

Wiese, unlike his predecessor Eilers, was not among those who laid the blame for the revolutionary events of 1848 on the schools and who claimed that mathematics and science education in particular conveyed a utilitarian and materialistic attitude. To the contrary, he put his trust in the reasonableness of scientific judgment and adopted, for example, a suggestion of the Königsberg mathematician and student of Jacobi's, Richelot, as it appeared in the regulation from the set of rules for aspiring secondary school teachers of 1866 quoted above (see [50], p. 99). In his memoirs, however, he feels compelled to state:

> It is regrettable that the trend of the mathematical sciences at the universities is relinquishing more and more the connection with what is the task of secondary schools. There is a gap between the university sciences and the schools which makes things difficult for the young person at the outset of his studies. Later many an aspiring teacher coming from the university does not learn to bridge that gap for some time. (quoted according to [50], p. 208)

Among the critics of the tendency of mathematics education to emulate mathematics training at the universities, were in particular the professors of the institutes of technology, founded for the most part in the middle of the 19th century. Oskar Schlömilch, Professor at the Dresden Institute of Science and subsequently an official with the Ministry of Education of Saxony, provides a vivid description of the situation as it developed:

> In many towns with 8,000 to 12,000 inhabitants the mathematician of the *Gymnasium* or technical high school is the only person expected by the public to have any knowledge about a steam engine, caloric engine, electric telegraph, centrifugal extractor, etc. Maybe there is a chamber of commerce, the tailor and the glove maker read the papers, and the mathematician is called upon to explain; however, as a rule the latter does not know anything about such things (certainly not, if he studied mathematics at the university to become a secondary school teacher), and in that case the illustrated magazine is his source. (quoted according to [50], p. 105)

Given the background of this situation, it becomes understandable why, towards the end of the century, school mathematics was almost inseparably linked to an attitude opposing the requirements of professional mathematics. Within the movement of engineers who pushed secondary school reform, for example, the view prevailed that mathematics education at the university level was far too abstract to serve as a model for mathematics education in secondary schools. In secondary schools mathematics should function merely as an auxiliary to other sciences, especially physics. This opinion was also accepted at a conference held in Jena in 1890 which laid the groundwork for establishing the "Association for the Advancement of Mathematics and Science Education" *(Verein zur Förderung des mathematisch-naturwissenschaftlichen Unterrichts).* At the actual founding meeting of the Association in Brunswick in 1891, a statement of principles was unanimously adopted which read:

> In general the students at the secondary school level are not yet capable of recognizing mathematical qualities in the phenomena presenting themselves in their own lives. The cause of this must be sought primarily in the circumstance that the applications of mathematical theories frequently consist in artificial examples instead of referring to situations which present themselves in reality. For this reason the system of mathematics instruction in secondary education, its complete independence as a subject notwithstanding, must from the outset be planned to take into account obvious applications (physics, chemistry, astronomy, and business). ([51], p. 15)

Such were the conditions that, towards the end of the century, gave rise to a reform movement which defined the nature of school mathematics, the subject of mathematics education at the secondary level, up to the present time: Professional mathematics set the norms; yet it was unable to provide the necessary guidance for defining the mathematical components of an educational canon responsive to the pressing social requirements. These conditions explain the success of the man who, in his role of champion of the reform, appears to have shaped it according to his own understanding of what constitutes mathematical knowledge to such an extent that many people named this reform for him: Felix Klein.

According to the prevailing opinion, the activities of the mathematician Klein (full professor in Erlangen, Munich, Leipzig, and, from 1886 until his retirement, in Göttingen; see [50], p. 211ff.; [51], p. 17ff.; [34], p. 83ff.) in the fields of science policy and educational policy took place at a time when he had long since passed the apex of his productivity as a mathematician. It is true that, in his inaugural lecture (1872) at Erlangen, he already addressed programmatic aspects of the organization of secondary school and university education; at the same time his remarks were, however, entirely couched in terms of the neoclassical concept of formal education. His true commitment did not begin until the 1890s. In 1892, he founded the Göttingen Mathematical Society *(Göttinger Mathematische Gesellschaft)* together with Heinrich Weber. In 1895, the Association for the Advancement of Mathematics and Science Education which had been established a few years earlier and in which he became actively involved from then on, met at his request in Göttingen. At this meeting, he initiated a change in the direction of the Association with regard to the relationship of school and university mathematics with his speech "On Mathematics Education at Göttingen University with Special Emphasis on the Needs of Prospective Teachers at the Secondary Level." Furthermore, in 1898, the "Göttingen Association for the Advancement of Applied Physics and Mathematics"

(Göttinger Vereinigung zur Förderung der Angewandten Physik und Mathematik) was established at Klein's initiative with the participation of prominent industrialists; its objective was a more pragmatic university policy. Klein was appointed to the commission on schools that met in June of 1900 and negotiated questions of secondary education (see Negotiations on Questions of Secondary Education [67], especially p. 153ff.). In 1904 the educational commission of the Association of German Scientists and Physicians *(Gesellschaft Deutscher Naturforscher und Ärzte)* was established at Klein's suggestion in which he took an active part as he continued to do in the organization which succeeded it—that is, the German Commission for Mathematics and Science Education *(Deutscher Ausschuß für den mathematischen und naturwissenschaftlichen Unterricht,* see [25]). In 1908, he became chairman of the International Commission on Mathematical Education *(Internationale mathematische Unterrichtskommission)* established in Rome in connection with the International Conference of Mathematicians. There can be no doubt that Klein occupied a key position with regard to the educational reform resulting from the conflict between university mathematics and school mathematics.

What accounted for his success in this role? It is probably due to the fact that he found himself between the camps and could not be identified with any of the conflicting parties. He was guided by the following principle:

Mathematical thinking must be fostered in school in terms of its full independence; however, its content must be related in as lively a manner as possible to the other responsibilities of the school, i.e., to the different components of the general education characteristic of each particular type of school. ([41], p. 14f.; see also [67], especially p. 153ff.).

This guiding principle displays all the characteristics of a pragmatic compromise, and in their practical implementation Klein's suggestions never got beyond such a compromise. These suggestions did, however, have the power to persuade others, because Klein firmly believed in them and was able to create the illusion that a solution of the conflict between university mathematics and school mathematics was in principle possible and was close at hand.

Klein confronted the opponents of university mathematics among educators, the engineers and the professors at the institutes of technology, who sought to reduce mathematics education to the minimal program of an "elementary mathematics," with his program of an "elementary mathematics as viewed from a higher standpoint" ([44], Vol. 1). He, the prominent scientist, was given credence when he claimed that mathematics was on the verge of generating an internal fusion principle that could reverse the effects of specialization. His determined defense of university mathematics seemed sincere, because he had accepted the consequences resulting from his position between the camps as far as science policy was concerned: He had long since become an outsider in mathematics who used all means available to him to restore the link between mathematics and its applications on a new basis. His diagnosis of the situation:

> We would be farther along, if the past century had not been rather unfavorable to a more wide-spread dissemination of general mathematical knowledge. ... If neohumanism takes an interest in mathematics at all, it is only interested in its formal values, not its substance. By contrast, our science was greatly appreciated during the era of rationalism. However, it also proved to be a particular hindrance that most of the mathematics scholars, especially at the universities, were governed exclusively by the interests of their specialty. Should the hope be in vain that, at the beginning of the 20th century, there is a change in the offing in favor of the tendency presented here? ([43], p. 49f. For a contemporary evaluation of Klein's reform see [3])

Today we know that this hope was futile indeed. Mathematics has become far more abstract, specialization has progressed further, and contact with the applications has weakened. The situation is still critical, but nobody expects a basic solution any more: School mathematics remains suspended between the pretensions of professional mathematics and the social demands aimed at integrating the knowledge generated by the particular sciences.

MATHEMATICS EDUCATION IN THE CONTEXT OF MODERN INSTITUTIONAL CONDITIONS

In the spring of 1959, the French mathematician Jean Dieudonné, charter member of the legendary group "Nicolas Bourbaki" which deeply influenced the conception of mathematics in the 20th century (see [26]), addressed nearly 50 scientists, educators, and government experts in education from 18 countries on the status of mathematics and the inevitable implications for school mathematics. The speech and the meeting became world-famous as the manifestation of "New Math," arguably the most powerful reform movement in the history of mathematics education. An excerpt from Dieudonné's speech may serve to give an impression of the sweeping goals and expectations that were associated with this reform movement at the outset:

> If the whole program I have in mind had to be summarised in one slogan it would be: *Euclid must go!*
> This statement may perhaps shock some of you, but I would like to show you in some detail the strong arguments in its favor. Let me first say that I have the deepest admiration for

the achievements of the Greeks in mathematics: I consider their creation of geometry perhaps the most extraordinary intellectual accomplishment ever realised by mankind. It is thanks to the Greeks that we have been able to erect the towering structure of modern science.

But in so doing, the basic notions of geometry itself have been deeply scrutinised, especially since the middle of the 19th century. This had made it possible to reorganise the Euclidean *corpus,* putting it on simple and sound foundations, and to re-evaluate its importance with regard to modern mathematics—separating what is fundamental from a chaotic heap of results with no significance except as scattered relics of clumsy methods or an obsolete approach.

The result may perhaps be a bit startling. Let us assume for the sake of argument that one had to teach plane Euclidean geometry to mature minds from another world who had never heard of it, or having only in view its possible applications to modern research. Then the whole course might, I think, be tackled in two or three hours—one of them being occupied by the description of the axiom system, one by its useful consequences and possibly a third one by a few mildly interesting exercises.

Everything else which now fills volumes of "elementary geometry"—and by that I mean, for instance, *everything* about triangles (it is perfectly feasible and desirable to describe the whole theory without even *defining* a triangle!), almost everything about inversion, system of circles, conics, etc.—has just as much relevance to what mathematicians (pure and applied) are doing today as magic squares or chess problems! ([57], p. 35f.)

Twenty-one years later, in the summer of 1980, the same Jean Dieudonné addressed the 4th International Congress on Mathematics Education attended by 2,000 participants from many different countries and, again, spoke of the performance and competence of modern geometrical thinking ([19]). He spoke, for instance, about the astonishing interpretations such complicated problems as the problem of the classification of the simple Lie-groups in n-dimensional Euclidean spaces have found, about the contributions of differential geometry to the general theory of relativity, about the development of functional analysis on the basis of topological concept formations and ways of thinking. But this time the mathematician Dieudonné refrained from any conclusion regarding mathematics education. In his lecture there was not one word about school mathematics, not one word on its current state, not one word about worthwhile changes of the contents or methods of teaching.

This event is symptomatic for the development that took place in the meantime (on this point see [12], p. 40ff.). At the beginning of the 1960s many shared Dieudonné's opinion that it was obvious that discrepancies between school mathematics and the status of mathematical science as well as the changing social requirements regarding mathematics education demanded a reform of mathematics education and that the real problem simply consisted in bringing this fact to the attention of policy makers and those affected. There can be no doubt that this eventually happened, although it is probably true that external events such as, for instance, the "sputnik shock" in the US or, some years later, the "educational catastrophe" prognosticated for the Federal Republic of Germany, were of greater influence than the power of persuasion of the reformers' arguments. In any case, decisions were made in many countries—and within each of these countries on very different levels and in completely different areas—aimed at reforming school mathematics. In the US, for instance, the enactment of the National Defense Education Act (1958) led to funding a host of projects for the development of new educational materials, the establishment of Research and Development Centers and Educational Laboratories (see [38]); in Great Britain the government supported centers for teachers and school-based development projects (see [30], especially p. 171ff.); in France a central commission for the reform of mathematics education headed by the mathematician André Lichnerowicz was set up (1966; see [29], p. 104ff.); in the Federal Republic of Germany the resolution of the *Kultusministerkonferenz* "Recommendations and Guidelines Concerning the Modernization of Mathematics Education in Public Schools" *(Empfehlungen und Richtlinien zur Modernisierung des Mathematikunterrichts an den allgemeinbildenden Schulen)* resulted, for example, in the preparation of new curricula in all federal states (see [12], p. 221ff.); in socialist countries as, for example, in the USSR, a commission for the revision of mathematics education headed by the mathematician Andrey Nikolaevich Kolmogorov was established (1966; see [16]; [39]); in Hungary a project headed by Tamas Varga was started (1962), and it was decided to introduce the program developed by this project into the schools (1972; see [65]), etc.

But by the middle of the 1970s the general mood had changed. In the US, for instance, Morris Kline's pamphlet *Why Johnny Can't Add* became a bestseller, and a back-to-basics movement advocated the return to traditional educational contents (see [47]). In France the chairman of the Academy of Science disassociated himself from the reform in a policy statement, and the association of mathematics teachers, the *Association des Professeurs des Mathématiques de l'Enseignement Public,* adopted a counter program, the "Charter of Caen" (see [29], p. 126ff.). In the USSR the mathematician Lev Semenovich Pontriagin assumed the role of spokesman for the opposition to the reform in the pages of the leading party publication "Communist" (see [39], p. 111). In the Federal Republic of Germany the Tübingen mathematician Walter Felscher publicly supported the popular slogan "new math makes you sick" summoning scientific arguments![5]

It is fair to assume that this change of attitude was essentially due to the recurring economic crises which began in the mid-1970s. As a result, in many countries reforms already introduced could no longer be financed, new plans hardly found any support, and the results of the discussions on educational policy of the 1960s were suppressed or repudiated. This change of attitude has been justified by various arguments. There is the claim that the reform of mathematics education was the cause of an appalling decline of performance with the result that the basic mathematical requirements of professional or vocational training, vocational schools, and the universities are no longer met. There is the opposite claim that all arguments along these lines are merely unsubstantiated attempts to justify the curtailment of reform programs which took place for entirely different reasons. And finally, there is the compromise position that the reform was essentially reasonable and successful, but that certain excesses led to problems, and the time has now come to reinstate proven traditions.

These three lines of argument are so shallow and usually betray such a lack of real information about the reform process and its effects that they do not, in my opinion, deserve serious attention. In specific instances they may well have some validity; their general approach in terms of evaluating the situation can, however, be easily refuted. The claim to be taken quite seriously, however, is the following: The change of attitude could hardly have taken place, if proponents as well as opponents did not share the experience that almost everywhere a wide chasm is apparent between the expectations initially associated with the aims of the reform and the situation actually existing in the classroom today.

This chasm takes on many different forms, and—this probably being the most important observation holding any promise—the causes for their development are as numerous as the social conditions that determined the different directions of the reform. The circumstances under which a similar change of attitude took place in the most diverse countries serve as a reminder that, in view of the complexity of the institutional conditions determining contemporary education in general and mathematics education in particular, the means available to educational policy have shown themselves to be deficient in their ability to influence specifically what is taught and what is learned. The basic model of educational thought prevalent in mathematics teaching that is primarily preoccupied with designing teaching models and constructing learning sequences for obvious or, possibly, externally determined objectives based on educational and psychological information, proves to be essentially inadequate in this situation because of the detachment from the diverse social and historical contexts in which mathematics education takes place.

Those critical of the "New Math" reform movement have, at least in part, created an awareness of this situation which subsequently led to a number of investigations in the area where mathematics, sociology of education, educational economics, and curriculum research intersect; these studies treat the relationship of mathematics education and society differently than generally was the case before.[6] The contents and the organizational forms of mathematics education are analyzed in terms of the characteristics of partial functions of the educational system and its institutions, that is to say, in terms of their dependence on the dynamics of this system which develops in a variety of interdependent ways ranging from functional relationships and dependencies on resources to channels of communication with other subsystems within the social context and, in this process, exhibits particular forms of stability and change that appear as an obstacle to attempts of reform and/or as a specific impetus even under constant external conditions.

In such studies the basic theoretical and metatheoretical concepts of mathematics teaching, the validity of which was hardly ever questioned before, are of necessity defined and discussed in terms of their social context: educational goals in terms of the strengths of institutions to determine educational activities, educational resources such as school books and materials produced by development projects as being determined by investment decisions and marketing strategies and/or financial policy and implementation processes respectively, achievement tests in terms of instruments for directing the flow of students and for distributing status, etc. However, overall such studies have so far raised more questions than they have answered. The deficits of the conventional theory of mathematics teaching, clearly demonstrated by the fate of the "New Math" movement, have not yet been balanced.

Thus, Robert B. Davis, in his evaluation of three studies on the successes of the "New Math" movement in the US, comes to the following conclusion:

A number of worthwhile innovative curriculum improvements were developed in the United States. Their impact has been somewhere between slight and insignificant. Not only has this phenomenon not been explained, it has not really been studied, although the three studies ... have made an important beginning. ([16])

Ian Westbury, whose analytical distinction between intended, implemented, and realized curriculum gave a new interpretation to the function of evaluative studies, summarized the case studies presented at an international conference on the stability and change of mathematics curricula in 1980 as follows:

... the case studies all lead, it would seem, to the conclusion that changes in an *implemented* and *realized* curriculum that is outside the trajectory of an existing order is not easily achieved, at least in some societies. With this kind of conclusion in hand two different sets of considerations would seem to emerge: First, what is it about the schools and their curriculum systems that lead to this kind of comparative stability in the face of efforts at change and reform? Second, what can we learn about the process of *systemic* change from those places in which change was achieved? ([69], p. 517)

The "New Math" movement and everything that went with it was a complex phenomenon with various goals of differing scope. At issue were not only the coordination of the canon of school mathematics with new developments in mathematics at the university level, but also the consequences of the changed role of mathematical methods in various areas resulting from the development of microelectronics. The change in the status of mathematics education from being a privilege limited to institutions of higher learning to becoming the organized center of a basic general mathematics education, a "mathematics for the majority"[7] was also of significance in this context. It is difficult to find historical parallels for this worldwide movement. The first aspect was already at issue in connection with Klein's reform, although the conditions were different and the extent was not comparable. With regard to the other two aspects it would be necessary to take a look at historical processes such as the alphabetization or the introduction of arithmetic as a general cultural skill to find comparable examples for turning specific abilities into basic qualifications of the population as a whole.

This comparison clearly shows the uniqueness of the present situation. In the past, even the general educational institutions largely had particular functions. If a shift in these functions necessitated changes in the curricula, the limited scale of the institutional framework allowed for the transparency of the conditions for the success of reforms and for the anticipation of the consequences. Or, conversely, if general functions of educational institutions actually were at issue, as was the case with the introduction of mandatory education or with the creation of school mathematics, the results were so neutral in terms of their functions that the unfolding of the developmental processes could largely be left to the inner dynamics of educational institutions without any guidance. Thus, the failures of the "New Math" movement seem comparatively harmless given the fact that, towards the end of the last century, that is to say, roughly 150 years after the introduction of mandatory education, the average illiteracy rate in Prussia probably still amounted to about 10 percent and that, in some areas in the provinces east of the Elbe river, 20 percent of the men and 25 percent of the women still signed their marriage certificate with an "X" at this time, which means that they could not even write their names (see [13], p. 514f.).

On the other hand, the future problem is unmistakable: We are no more in a position today to deliberately modify and improve general education by changing educational contents and learning patterns than 100 years ago. This is due to the fact that the transformation of social systems still takes place mainly in quasi-elemental processes involving competition and struggle among special interest groups and institutions.[8] Even though educational systems are in principle subject to state control and, hence, open to democratic decisions, it is, for the time being, completely unthinkable that the development of the general educational level could in any way be adjusted to the rapid development with which the potential for the exploitation of natural resources and of technology is expanding. However, how much longer can we afford to be daunted with regard to in the field of social development in the face of such relatively minor goals as the improvement of basic mathematics education and put our trust into the efficacy of traditional institutions and the stability of established problem-solving patterns, while, at the same time, creating a reality we are unable to control by means of traditional skills and behavior patterns, a reality including global environmental changes, risky technologies, and a system of mutual deterrence that, once its destructive power calculated in advance is set in motion, is designed to escalate quickly?

NOTES

[1] Cf. the studies on the didactics of mathematics by Lenné ([49]) and Bölts ([7]). Aspects of this topic are treated in Münzinger ([54]) and Neander ([56]).

[2] Described in these words in the title of the Third Treatise of "Geometria practica nova et aucta" by Schwenter; see [8], p. 669.

[3] The unreflected use of available problem-solving patterns has often been the subject of radical pedagogical criticism, for instance, from the vantage point of traditional secondary education in [71], p. 45ff. or in terms of the genetic principle in [68], p. 41ff. However, this aspect of teaching also tends to be viewed negatively as far as conventional objectives are concerned; see, for example, [72], p. 21f.; [74], p. 47ff.

[4] With regard to the social conditions of the emergence of mathematics cf. [14].

[5] See the lead story of the news magazine *Der Spiegel*, No. 13, 1974, 62ff. The criticism had its effect. The resolution of the *Kultusministerkonferenz* on the Modernization of Mathematics Education was revised for the lower grades in 1976 ([48], resolution No. 136), and several of the federal states partially or completely rescinded their decision of teaching new math at the elementary level (especially Baden-Württemberg, Rheinland-Pfalz, and Schleswig-Holstein).

MATHEMATICS EDUCATION AND SOCIETY   145

⁶   Of particular interest are evaluations by experts, commission reports, etc. produced by
    political advisers engaged in preparing policy decisions which treat problems of mathe-
    matics education in the context of comprehensive developments of educational policy, for
    example, the report of the education commission of the German Education Council issued
    in 1975 ([18], especially p. 117ff.) for the Federal Republic of Germany, the report of the
    National Science Foundation ([55]) for the US, the Cockcroft Report of which only a few
    papers have been published to date ([1];[2]; [62]) for Great Britain. In addition, analyses
    of particular countries or comparative analyses dealing with the reform process are of in-
    terest, especially [12]; [16]; [23]; [29]; [30]; [37]; [38]; [39]). Finally, individual studies
    on particular questions have been undertaken, for example, on the problem of defining
    mathematical qualification in terms of social requirements ([15]), on the problem of
    mathematics education in the German "Hauptschule" (German school type comprising
    grades one through nine or ten; [13]), on the social organization of information in math-
    ematics education ([59]), on the role of the textbook ([40]), etc. Furthermore, the confer-
    ence report ([64]) deserves special mention as well as the publications of the Second
    International Mathematics Study of the International Association for the Evaluation of
    Educational Achievement to be issued in the near future which, to a greater extent than
    the first study, takes the external conditions of mathematics education into account as
    well.
⁷   "Mathematics for the majority" is the name of a British project that is representative of
    this aspect of the reform of mathematics education. See [22], p. 674f.
⁸   Regarding this point, the systematic overview of international experiences with reforms
    of educational systems by Dalin ([11]) deserves to be mentioned.

BIBLIOGRAPHY

[1]   Advisory Council for Adult and Continuing Education (ACACE): *Adults' Mathemati-
      cal Ability and Performance.* Leicester: ACACE, 1982.
[2]   Bailey, D.: *Mathematics in Employment (16–18).* Report, February 1981, University of
      Bath, 1981.
[3]   Behnke, H.: "Felix Klein und die heutige Mathematik." In Behnke, H. et al. (Eds.): *Ma-
      thematische-Physikalische Semesterberichte zur Pflege des Zusammenhangs von
      Schule und Universität,* Vol. 7. Göttingen: Vandenhoeck & Ruprecht, 1961, pp. 129–
      144.
[4]   Bekemeier, B.: "Zum Zusammenhang von Wissenschaft und Bildung am Beispiel des
      Mathematikers und Lehrbuchautors Martin Ohm." In Høyrup, J. & Bekemeier, B.: *Stu-
      dien zum Zusammenhang von Wissenschaft und Bildung.* Materialien und Studien,
      Vol. 20. Bielefeld: IDM, 1980, pp. 140–228.
[5]   Bekemeier, B. et al. (Eds.): *Wissenschaft und Bildung im frühen 19. Jahrhundert.* Ma-
      terialien und Studien, Vol. 27 and Vol. 30. Bielefeld: IDM, 1982.
[6]   Bell, E. T.: *Men of Mathematics.* New York: Simon & Schuster, 1986.
[7]   Bölts, H.: *Kritik einer Fachdidaktik. Eine ideologiekritische Analyse der gegenwärtigen
      Mathematikdidaktik in der BRD.* Weinheim: Beltz, 1978.
[8]   Cantor, M.: *Vorlesungen über Geschichte der Mathematik,* Vol. 2. 2nd ed., Leipzig:
      Teubner, 1913.
[9]   Chace, A. B. (Ed.): *The Rhind Mathematical Papyrus.* 2 vols., Oberlin, OH: Mathemat-
      ical Associaton of America, 1927/1929. (Reprint Reston, 1979.)
[10] Cicero: *De Finibus Bonorum et Malorum. With an English Translation by H. Rackham.*
      New York: Putnam's Sons, 1931.

[11] Dalin, P.: *Limits to Education Change.* London: Macmillan, 1978.
[12] Damerow, P.: *Die Reform des Mathematikunterrichts in der Sekundarstufe I. Vol. 1: Reformziele, Reform der Lehrpläne.* Stuttgart: Klett, 1977.
[13] Damerow, P.: "Wieviel Mathematik braucht ein Hauptschüler?" In *Neue Sammlung, 20,* 1980, pp. 513–529.
[14] Damerow, P. & Lefèvre, W. (Eds.): *Rechenstein, Experiment, Sprache: Historische Fallstudien zur Entstehung der exakten Wissenschaften.* Stuttgart: Klett-Cotta, 1981.
[15] Damerow, P., Elwitz, U., Keitel, C. & Zimmer, J.: *Elementarmathematik: Lernen für die Praxis? Ein exemplarischer Versuch zur Bestimmung fachüberschreitender Curriculumziele.* Stuttgart: Klett, 1974.
[16] Davis, R. B. et al.: *An Analysis of Mathematics Education in the Union of the Soviet Socialist Republics.* Columbus, OH: Ohio State University, 1979 (ERIC Clearinghouse for Science, Mathematics, and Education).
[17] Deutsche Mathematiker-Vereinigung e.V. (DMV): *Denkschrift: Zum Mathematikunterricht an Gymnasien.* Hektogr., Freiburg: 1976.
[18] Deutscher Bildungsrat: *Die Bildungskommission, Bericht '75. Entwicklungen im Bildungswesen.* Verabschiedet auf der 48. Sitzung der Bildungskommission am 13. Juni 1975. Bonn: Bundesdruckerei, 1975.
[19] Dieudonné, J.: "The Universal Domination of Geometry." In *Zentralblatt für Didaktik der Mathematik, 13,* 1981, pp. 5–7.
[20] Diogenes Laertius: *Lives of Eminent Pilosophers. With an Englisch Translation by R. D. Hicks.* 2 vols., Cambridge, MA: Harvard University Press, 1970.
[21] Dolch, J.: *Lehrplan des Abendlandes. Zweieinhalb Jahrtausende seiner Geschichte.* 2nd ed., Ratingen: Henn, 1965.
[22] Fletcher, T. J.: "Applications of Mathematics in English Secondary Schools." In Steiner, H.-G. (Ed.): *Comparative Studies of Mathematics Curricula. Change and Stability 1960–1980.* Materialien und Studien, Vol. 19. Bielefeld: IDM, 1980, pp. 661–706.
[23] Griffiths, H. B. & Howson, A. G.: *Mathematics Society and Curricula.* London: Cambridge University Press, 1974.
[24] Günther, S.: *Geschichte des mathematischen Unterrichts im deutschen Mittelalter bis zum Jahre 1525.* Berlin: Hofmann, 1887.
[25] Gutzmer, A. (Ed.): *Die Tätigkeit der Unterrichtskommission der Gesellschaft Deutscher Naturforscher und Ärzte.* Gesamtbericht. Leipzig: Teubner, 1908.
[26] Halmos, P. R.: "'Nicolas Bourbaki.'" In Kline, M.: *Mathematics in the Modern World. Readings from Scientific American.* San Francisco: Freeman, 1968, pp. 77–81.
[27] Heinrich, K.: "Antike Kyniker und Zynismus in der Gegenwart." In Heinrich, K.: *Parmenides und Jona. Vier Studien über das Verhältnis von Philosophie und Mythologie.* Frankfurt a.M.: Suhrkamp, 1966, pp. 131–209.
[28] Hemeling, J.: *Neuvermehrter und vollkommener Rechenmeister.* Hannover: 1705.
[29] Hörner, W.: *Curriculumentwicklung in Frankreich. Probleme und Lösungsversuche einer Inhaltsreform der Sekundarschule (1959–1976).* Weinheim: Beltz, 1979.
[30] Howson, G., Keitel, C. & Kilpatrick, J.: *Curriculum Development in Mathematics.* Cambridge: Cambridge University Press, 1981.
[31] Høyrup, J.: "Influences of Institutionalized Mathematics Teaching on the Development and Organization of Mathematical Thought in the Pre-Modern Period. Investigations in an Aspect of the Anthropology of Mathematics." In Høyrup, J. & Bekemeier, B.: *Studien zum Zusammenhang von Wissenschaft und Bildung.* Materialien und Studien, Vol. 20. Bielefeld: IDM, 1980, pp. 7–137.
[32] Høyrup, J. & Bekemeier, B.: *Studien zum Zusammenhang von Wissenschaft und Bildung.* Materialien und Studien, Vol. 20. Bielefeld: IDM, 1980.
[33] Ihnow, G. & Knöss, C.: *Fachrechnen für das Friseurhandwerk.* Stuttgart: Klett, 1965.

[34] Inhetveen, H.: *Die Reform des gymnasialen Mathematikunterrichts zwischen 1890 und 1914. Eine sozioökonomische Analyse.* Bad Heilbrunn: Klinkhardt, 1976.
[35] Jahnke, H. N.: "Zum Verhältnis von Bildung und wissenschaftlichem Denken am Beispiel der Mathematik. Eine Kontroverse um den mathematischen Lehrplan der preußischen Gymnasien 1829–30 und ihr methodologischer Kontext." In Bekemeier, B. et al. (Eds.): *Wissenschaft und Bildung im frühen 19. Jahrhundert.* Materialien und Studien, Vol. 27 and Vol. 30. Bielefeld: IDM, 1982, pp. 1–225.
[36] Jahnke, H. N. & Otte, M. (Eds.): *Epistemological and Social Problems of the Sciences in the Early Nineteenth Century.* (Papers from a workshop held at the University of Bielefeld 1979.) Dordrecht: Reidel, 1981.
[37] Keitel, C.: "Entwicklungen im Mathematikunterricht." In Projektgruppe Bildungsbericht (Ed.): *Bildung in der Bundesrepublik Deutschland. Daten und Analysen, Vol. 1.* Reinbek: Rowohlt, 1980, pp. 447–499.
[38] Keitel, C.: *Reformen des Mathematikunterrichts in den USA. Geschichte, Reformkonzeptionen und Curriculumentwicklung.* Dissertation, Bielefeld: 1981.
[39] Keitel, C.: "Mathematics Education and Educational Research in the USA and USSR: Two Comparisons Compared." In *Journal of Curriculum Studies, 14,* 1982, pp. 109–126.
[40] Keitel, C., Otte, M. & Seeger, F.: *Text, Wissen, Tätigkeit. Das Schulbuch im Mathematikunterricht.* Königstein: Scriptor, 1980.
[41] Klein, F.: "Über eine zeitgemäße Umgestaltung des mathematischen Unterrichts an den höheren Schulen." In Klein, F. & Rieke, E. (Eds.): *Neue Beiträge zur Frage des mathematischen und physikalischen Unterrichts an den höheren Schulen.* Leipzig: Teubner, 1904a, pp. 1–32.
[42] Klein, F.: "Hundert Jahre mathematischer Unterricht an den höheren Preußischen Schulen." In Klein, F. & Rieke, E. (Eds.): *Neue Beiträge zur Frage des mathematischen und physikalischen Unterrichts an den höheren Schulen.* Leipzig: Teubner, 1904b, pp. 63–77.
[43] Klein, F.: "Bemerkungen zum mathematischen und physikalischen Unterricht." In Gutzmer, A. (Ed.): *Die Tätigkeit der Unterrichtskommission der Gesellschaft Deutscher Naturforscher und Ärzte. Gesamtbericht.* Leipzig: Teubner, 1908, pp. 44–57.
[44] Klein, F.: *Elementarmathematik vom höheren Standpunkte aus.* 3 vols., Berlin: Springer, Vol. 1: 1924 (3rd ed.), Vol. 2: 1925 (3rd ed.), Vol. 3: 1928 (3rd ed.).
[45] Klein, F.: *Vorlesungen über die Entwicklung der Mathematik im 19. Jahrhundert.* 2 Parts. Berlin: Springer, Part 1: 1926, Part 2: 1927. (Reprint 1979.)
[46] Klein, F. & Rieke, E. (Eds.): *Neue Beiträge zur Frage des mathematischen und physikalischen Unterrichts an den höheren Schulen.* Leipzig: Teubner, 1904.
[47] Kline, M.: *Warum kann Hänschen nicht rechnen? Das Versagen der Neuen Mathematik.* Weinheim: Beltz, 1974.
[48] Ständige Konferenz der Kultusminister in der Bundesrepublik Deutschland (KMK) (Ed.): *Sammlung der Beschlüsse der Ständigen Konferenz der Kultusminister der Länder in der Bundesrepublik Deutschland.* Loseblattsammlung. Berlin: 1963ff.
[49] Lenné, H.: *Analyse der Mathematikdidaktik in Deutschland.* Stuttgart: Klett, 1969.
[50] Lorey, W.: *Das Studium der Mathematik an den Deutschen Universitäten seit Anfang des 19. Jahrhunderts.* Leipzig: Teubner, 1916.
[51] Lorey, W.: *Der Deutsche Verein zur Förderung des mathematischen und naturwissenschaftlichen Unterrichts E.V. 1891–1938. Ein Rückblick zugleich auch auf die mathematische und naturwissenschaftliche Erziehung und Bildung in den letzten fünfzig Jahren.* Frankfurt a.M.: Salle, 1938.
[52] Marrou, H. I.: *Geschichte der Erziehung im klassischen Altertum.* München: dtv, 1977.

[53] Mehrtens, H., Bos, H. & Schneider, I. (Eds.): *Social History of Nineteenth Century Mathematics: Papers from a Workshop, Technische Universität Berlin, 1979*. Boston, MA: Birkhäuser, 1981.
[54] Münzinger, W.: *Moderne Mathematik, Gesellschaft und Unterricht. Zur soziologischen Begründung der modernen Mathematik in der Schule.* Weinheim: Beltz, 1971.
[55] National Science Foundation, Curriculum Development and Implementation for Pre-College Science Education (NSF): *Report,* prepared for the Committee on Science and Technology. U.S. House of Representatives Ninety-Fourth Congress. First Session. Washington, DC: U.S. Government Printing Office, 1975.
[56] Neander, J.: *Mathematik und Ideologie. Zur politischen Ökonomie des Mathematikunterrichts.* Starnberg: Raith, 1974.
[57] Organisation for Economic Co-operation and Development (OECD): *New Thinking in School Mathematics.* Paris: OECD, 1964. (First published OEEC, 1961.)
[58] Pahl, F.: *Geschichte des naturwissenschaftlichen und mathematischen Unterrichts.* Leipzig: Quelle und Meyer, 1913.
[59] Pfeiffer, H.: *Zur sozialen Organisation vom Wissen im Mathematikunterricht.* Materialien und Studien, Vol. 21. Bielefeld: IDM, 1981.
[60] Plato: *The Collected Dialogues of Plato Including the Letters.* Edited by E. Hamilton & H. Cairns. Princeton, NJ: Princeton University Press, 1961.
[61] Projektgruppe Bildungsbericht (Ed.): *Bildung in der Bundesrepublik Deutschland. Daten und Analysen.* 2 vols., Reinbek: Rowohlt, 1980.
[62] Sewell, B.: *Use of Mathematics by Adults in Daily Life.* Leicester: ACACE, 1981.
[63] Steiner, H.-G. (Ed.): *Didaktik der Mathematik.* Darmstadt: Wissenschaftliche Buchgesellschaft, 1978.
[64] Steiner, H.-G. (Ed.): *Comparative Studies of Mathematics Curricula. Change and Stability 1960–1980.* Materialien und Studien, Vol. 19. Bielefeld: IDM, 1980.
[65] Szendrei, J.: "A Case Study in the Development of Geometry Teaching in Hungary." In Steiner, H.-G. (Ed.): *Comparative Studies of Mathematics Curricula. Change and Stability 1960–1980.* Materialien und Studien, Vol. 19. Bielefeld: IDM, 1980, pp. 347–353.
[66] Timerding, H. E.: "Die Verbreitung mathematischen Wissens und mathematischer Auffassung." In Hinneberg, P. (Ed.): *Die Kultur der Gegenwart. Part III: Die mathematischen Wissenschaften,* unter Leitung von F. Klein. Zweite Lieferung. Leipzig: Teubner, 1914, pp. 50–161.
[67] *Verhandlungen über Fragen des höheren Unterrichts, Berlin 1900.* Halle: Verlag der Buchhandlung des Waisenhauses, 1901.
[68] Wagenschein, M.: *Ursprüngliches Verstehen und exaktes Denken,* Vol. 1. 2nd ed., Stuttgart: Klett, 1970.
[69] Westbury, I.: "Conclusion." In Steiner, H.-G. (Ed.): *Comparative Studies of Mathematics Curricula. Change and Stability 1960–1980.* Materialien und Studien, Vol. 19. Bielefeld: IDM, 1980, 509–522.
[70] Wiese, L. (Ed.): *Verordnungen und Gesetze für die höheren Schulen in Preußen.* 2nd ed., Berlin: Wiegandt und Grieben, 1875.
[71] Wittenberg, A. I.: *Bildung und Mathematik. Mathematik als exemplarisches Gymnasialfach.* Stuttgart: Klett, 1963.
[72] Wittmann, E.: *Grundfragen des Mathematikunterrichts.* Braunschweig: Vieweg, 1974.
[73] Wussing, H. & Arnold, W. (Eds.): *Biographien bedeutender Mathematiker. Eine Sammlung von Biographien.* Berlin: Volk und Wissen, 1975.
[74] Zech, F.: *Grundkurs Mathematikdidaktik. Theoretische und praktische Anleitungen für das Lehren und Lernen im Fach Mathematik.* Weinheim: Beltz, 1977.

CHAPTER 6

PRELIMINARY REMARKS ON THE RELATIONSHIP OF THE PRINCIPLES OF TEACHING ARITHMETIC TO THE EARLY HISTORY OF MATHEMATICS
[1981]

IN RETROSPECT: REPETITION IN UNISON

We often take the ordinary for granted. It is, therefore, sometimes useful to take a look at historical models in order to gain a better understanding of one's own position. Heinrich Pestalozzi, the eminent educational reformer of the early 19th century, introduced his pupils to multiplication by means of a table which he called the table of units (cf. Fig. 1). With this table in mind his pupils had to repeat in unison the lines of the multiplication table up to ten. Because of his enthusiasm for Pestalozzi's educational ideas Baron von Türk, legal advisor to the Duke of Mecklenburg-Strelitz, gave up his position and his secure income in order to dedicate himself entirely to education.[1] In his writings we find a vivid description of how Pestalozzi's table of units could be used.

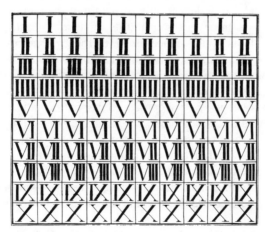

Figure 1: Pestalozzi's table of units as modified by the use of Roman numerals used by Baron von Türk. (Source: [29])

150   CHAPTER 6

This table may be used in the following way, especially where the goal is to make the children speak.
Teacher (pointing with a rod to the first and then the second square of the position of the I).
One times 2 is 2 times 1.
(The children repeat: one times 2 is 2 times 1.)
Teacher (first pointing to the first square, then to the second square of the 2nd position or the position of the IIs).
Two times 2 is 4 times 1.
(The children repeat in unison: two – times – two – is – four – times one.)
The teacher continues in this vein until he says (pointing to the first and the second square of the position of the Xs): 10 times 2 is 20 times one.[2]

Elsewhere von Türk describes his impressions when he observed Pestalozzi teaching arithmetic this way for the first time:

> When, from a distance, I first heard the monotonous: Four times one is one times 3 and one times the third part of 3. – Five times one is one times 3 and two times the third part of 3 etc., repeated at the same time by all the boys of a large class in the already somewhat harsh Swiss dialect,this first impression was, needless to say, not pleasant. However ... I could not help but admire the speed, the assurance, and the confidence with which the boys solved the most difficult problems and provided detailed explanations of the nature of the solution. ... This imposing result of the method, this assurance and confidence with which the youngsters performed calculations, their lively, frank demeanor; the fact that they never made a mistake while calculating or at least noticed and corrected their own mistakes at once, that external intrusions could not distract them proved to my satisfaction that the method by which they were *thus* trained deserved attention.[3]

What, from our own vantage point, is our reaction to this report? We are bound to doubt its veracity, or at the very least we must doubt that Pestalozzi's educational successes can be attributed to such a method of repetition. Pestalozzi understood the structure of the number concept first and foremost as a process of transmitting linguistic meaning by means of demonstration.[4] Our doubts concerning the efficacy of the method of repetition are based on the almost universally accepted fundamental assumption of the contemporary methodology of mathematics education that at best language has only a secondary function with regard to the development of mathematical concepts. We are convinced that mathematical and arithmetical thinking in particular is based on the active interaction with concrete objects and that the structure of this activity is internalized in the form of mental structures which in turn confer meaning upon the terms of the specialized language of mathematics. It is this basic assumption that arouses our distrust concerning the efficacy of the method of repetition. We assume that operating with logical blocks and Cusenaire rods, drawing, folding, pasting, and cutting—to name but a few of the diverse operations with structured materials deemed meaningful today—are more useful starting points for the development of mathematical concepts.

ACTION AS THE STARTING POINT FOR MATHEMATICAL THINKING?

On closer view, however, this basic assumption proves to be more of a rule of thumb for designing learning units than a strict theorem of a developed theory of the mathematical learning process. In particular it remains unclear how much significance for the learning process as a whole should be attached to action as the point of departure in concrete situations. For there can be no doubt that the verbal communication of mathematical information becomes more and more important as the learning process progresses.

In recent years the assumption that mathematical thinking is a result of the internalization of schemata of action has been advanced mainly by the work of Jean Piaget, and the interest aroused by his investigations can probably be considered the reason that this assumption has found such widespread acceptance among mathematics educators today.[5] Within the theory by means of which Piaget interpreted his experiments this assumption does indeed play a dominant role. Piaget tried to provide substantial evidence for his belief that mathematical thinking is not predetermined by linguistic structures, but rather by schemata of the coordination of actions.[6] In this interpretive context of the results of empirical research the basic assumption of contemporary theory of mathematics education mentioned above has the status of a rigorous theorem within a developed theory with a variety of empirical interpretations giving it a precise meaning.

If, in the theory of mathematics education, this basic assumption nevertheless continues to remain nothing more than a rule of thumb, then the reasons for this are obvious. A mathematics educator is quite helpless with regard to Piaget's theoretical explications. According to Piaget logico-mathematical thinking was based on the coordination of actions, the latter in turn on biological evolution.[7] His "biological epistemology" did not make allowances for a socio-cultural determination of mathematical thinking. Consequently, Piaget interprets the development of science with regard to its logico-mathematical forms merely as a social reflex of the biological evolution of the individual. By necessity such a theory confers only a marginal function on mathematics education for the development of mathematical thinking.

Although Piaget is frequently quoted with regard to the theory of mathematics education—and in particular the theory of teaching arithmetic—to my knowledge not a single mathematics educator accepts Piaget's biological theory of cognition without reservations. It is not the strict genetic correlation of logico-mathematical structures of thinking as expounded by Piaget that is cited in publications on mathematics education with regard to the interpretation and construction of learning sequences. It is more likely the idea of a cultural transmission of these thought structures in terms of enactive, iconic, and symbolic representations based on the work of Jerome Bruner.[8] Here action, intuition, and language are considered parallel modes of a transmission of mathematical thinking; so far, however, in the framework of this theory no developmental processes have been revealed, represented theoretically, or supported by extensive empirical evidence as explicitly as in Piaget's theory. In view of this situation the thesis of the origin of mathematical thinking in concrete action cannot lead to anything more than the rule of thumb that, especially with young children, lessons should always begin with concrete operations making use of structured materials.

ON THE ROLE OF HISTORICAL CONDITIONS OF THE DEVELOPMENT OF MATHEMATICAL THINKING

Let us assume that mathematical thinking is in fact culturally transmitted by means of representations of mathematical structures. In this case the question raised by psychology as to the conditions of the development of mathematical thinking would be, due to its disciplinary limitation, much too narrow to expect solutions from psychology concerning problems associated with the theory of mathematics education; for, in that case, the development of mathematical thinking takes in fact place in the context of a specific dependence on different, historically determined representations of mathematical structures and their historical development. I am, therefore, of the opinion that it is a necessary consequence of the understanding of mathematical learning processes prevailing in the theory of mathematics education that concept formations of psychology are not sufficient for a theoretical clarification, because they are incapable of capturing the historical dimension of the forms of mathematical thinking.

In my view historical analyses of the conditions of the development of mathematical thinking are needed. Two different developmental processes must be distinguished—that is, the individual process of the development of mathematical thinking under conditions that are subject to change and determined historically and the historical process of the development of ever new forms of mathematical thinking itself. The individual development of mathematical thinking by means of the embodiments of mathematical structures already in existence to which these thought structures are corresponding is not—as suggested by the pedagogical term "discovery learning"—the individual reproduction of the historical process of mathematical discoveries. For the individual developmental process takes place under changing historical conditions engendered by different representations. If the situation of the student with regard to the problem is compared to that of the discoverer, these conditions prove to be additional conditions modifying the historico-genetic process to the individual-genetic process of the development of mathematical thinking.

This has consequences for constructing an overall, explanatory theory of the development of mathematical thinking. In order to gain an adequate understanding of mathematical learning processes it will be necessary to study the development of mathematical thought structures in terms of the simpler conditions of their initial manifestation in the historical process. And only those assumptions can be accepted as basic assumptions of a theory of the development of mathematical thinking that make sense in the context of such an historical analysis.

THE EARLY HISTORY OF ARITHMETIC AS A TOUCHSTONE

These very general thoughts are really in need of comprehensive illustration. In the present context I can merely try to show their soundness and the efficacy of a corresponding research strategy by taking as an example the basic assumption of the theory of mathematics education already mentioned.

If the assumption is to have any validity that mathematical thinking emerges in the course of the active manipulation of concrete objects by internalizing structures of action, then it must be true of the historical development of arithmetical thinking as well. Consequently, it must be possible to prove the existence of objects and contexts of action in the history of arithmetic that correspond in terms of their function to those embodiments of mathematical structures used in mathematics education for the deliberate communication of the forms of arithmetical thinking. However, in this context the fundamental difference between individual and historical development must be observed. While individual development must contend with preexisting embodiments that already assume the historical existence of the thought structures to be developed by making use of such embodiments, the corresponding representations of historical development are to explain the first appearance of these forms of thinking in the course of history. Thus, the invention, creation, and use of such representations must have been possible even without the existence of the thought structures they subsequently represent.

ARITHMETIC AS A SYSTEM OF RULES FOR MATERIAL REPRESENTATIVES OF NUMBERS

While keeping this problem in mind, let us first take a look at ancient Chinese arithmetic. What presumably is the oldest of the extant Chinese books of arithmetic is entitled *Chiu Chang Suan Shu* (Nine Chapters on the Mathematical Art).[9] This book of arithmetic offers insights into the Chinese arithmetic of the first millennium B.C.

It is worthwhile to take a look at the etymology and the semantics of the arithmetical concepts in this book of arithmetic.[10] What is the technique described in the "Suan Shu"? "Shu" is the "method" and "Suan" is a term used for "calculating." But "Suan" refers not only to the activity, but also to the means of this activity—that is, to the counting rods used to indicate the numerals on the reckoning board.[11] Furthermore, it is probable that the hieroglyphic character for "Suan" evolved from the image of the reckoning board.

The indications can easily be multiplied that the technique with which the "Nine Chapters on the Mathematical Art" are concerned is in fact the technique of operating with the counting rods on the reckoning board. The primary subject of this arithmetical technique is not the ideal number, but

Figure 2: Operating with the Chinese reckoning board (according to a late illustration in a Japanese textbook published in 1795). (Source: [15])

PRINCIPLES OF TEACHING ARITHMETIC 155

the actual rod figure representing quantities from the areas of application of this technique on the reckoning board (cf. Fig. 2).

The role of these rod figures is further documented by a special way of rendering numbers—that is, the representation of numbers by rod numerals. These rod numerals were accurate representations of the rod figures arranged on the reckoning board (cf. Fig. 3). By using these representations of the rod figures as numerals the reckoning board defined the structure of numerical

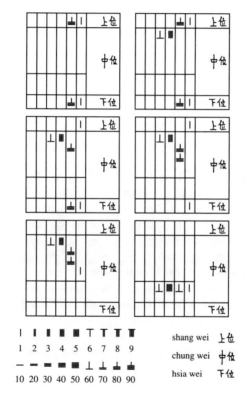

Figure 3: Chinese rod numerals and the use of rod figures for the multiplication of 81 times 81 = 6,561 on the reckoning board. In the case of multiplications the rods are transferred by means of the multiplication table according to an established schema from the 'upper place' *(shang wei)* and from the 'lower place' *(hsia wei)* to an 'intermediate place' *(chung wei)* where they are combined to form a rod figure. The example is following a description in the "Sun Tzu Suan Ching" (Master Sun's Arithmetical Manual) from the third century (cf. [16], p. 63).

representation: It is the representation within a decimal place-value system. The "places" of the numerical representation correspond to the fields of the reckoning board. Thus, in this case a structure of arithmetical thinking is predetermined by the construction of the calculational device.

The strong functional equivalence of the arithmetical knowledge of the Chinese to be inferred from the book of arithmetic and the operations on the reckoning board makes it difficult to distinguish clearly between the arithmetical thought structure and the operation on the reckoning board. Is it appropriate to attribute the invention of the zero to the Chinese arithmeticians when they operate with its material representative—that is, the empty place on the reckoning board—accurately and intelligently or is it premised on the introduction of a separate symbol for the vacant place some 1,000 years later? Did they know the negative numbers when they marked numbers to be eliminated by means of rods of a different color than the normal counting rods and operated correctly on the reckoning board with these rods—that is to say: implicitly following the rules for operating with negative numbers? Did they know the multiplication of matrices[12] when they solved linear systems of equations on the reckoning board according to its rules?

By way of summing up it can be said: The "Nine Chapters on the Mathematical Art" do not constitute a textbook on arithmetic in our sense of the word. Rather, they can be considered as something like a set of instructions for the reckoning board. This does not mean that developed forms of arithmetical thinking could not already be found in this text. However, these forms are always referring back to their origin in the operations with concrete objects—that is, with the counting rods on the reckoning board. The fundamental operations with these counting rods—that is, the additive operations—simulated the actions of composing and decomposing sets of objects and commodities of the context of application represented on the reckoning board by means of the counting rods in the context of a material model. These fundamental operations constituting the starting point of Chinese arithmetic do indeed demonstrate that they could be performed without the prior existence of arithmetical thinking.

THE ARITHMETIC OF CONSTRUCTIVE-ADDITIVE SIGN SYSTEMS

There are, however, early civilizations with a developed arithmetic for which the use of the reckoning board or a comparable calculational aid has not been documented, for example, the Egyptian civilization with its written arithmetical procedures. Did Egyptian arithmetic develop without a material basis analogous to the Chinese reckoning board?

In order to answer this question it is useful to take a closer look at the structure of the numerical representation displayed by Chinese rod numerals. In contrast to our system of numerals the numerical values of these representations of rod figures can be obtained by looking at the written characters themselves. As with the counting rods they represent, the rod numerals require no real understanding of numbers for the performance of elementary additive operations with these numerals. I am calling a system of numerals with that kind of structure constructive-additive, because the iteration of the signs in the rod numerals is identical with the addition of their numerical values, and the result of additions is—as in the case of the counting rods themselves—generated constructively by the iteration of the signs. Thus, there is really no essential difference between the elementary operations with such numerals and the operations with the counting rods on the reckoning board. In this case not even the table of addition up to ten, indispensable for dealing with our present systems of numerals, is necessary.

The question presents itself whether such a constructive-additive sign system could have been introduced and become the basis for the development of arithmetical thinking without previous objective material means of arithmetic like the counting rods; for with regard to the necessary intellectual preconditions there is almost no difference. However, given such an origin of arithmetical thinking, the invention of the place-value system which, in Chinese arithmetic, emerged after all from the use of the reckoning board is not to be expected under such circumstances.

For this merely hypothetical possibility Egyptian arithmetic, as passed on to us in documents from the early second millennium B.C., provides an historical example.[13] The Egyptians made use of two different types of writing. For all painted, carved, embossed, and chiseled inscriptions on buildings or objects of daily life a hieroglyphic pictographic script was used. For writing on papyrus, however, another script was used, hieratic writing, consisting of simplifications of the hieroglyphic characters. These simplified characters preserved their relationship to the original hieroglyphic signs and did not evolve into a separate form of writing. This is especially true of the numerals as well which, in their hieroglyphic form, constitute a strictly constructive-additive system of numerals with a decimal structure. There is no indication whatsoever that the system was going to evolve into a place-value system. The hieroglyphic numerals which were formed by a con-

158   CHAPTER 6

structive-additive process developed into hieratic numerals, written characters for greater ease of writing which, however, preserved their close relationship to the hieroglyphic numerals (cf. Fig. 4). They are nevertheless already documenting a certain—relatively early—stage of the development of arithmetical thinking, since their use presupposes that the results of elementary additions and subtractions were calculated mentally.

In my view the similarity of the numerical representation of Chinese rod numerals and Egyptian hieroglyphs with regard to their constructive-additive basis clearly shows that we must assume a comparable origin of arithmetical thinking. The extent to which, in addition, the Egyptian arithmetical technique continued to be determined by the possibilities of operating with constructive-additive numerals becomes apparent when an Egyptian multiplication is performed with hieroglyphic numerals. An auxiliary cal-

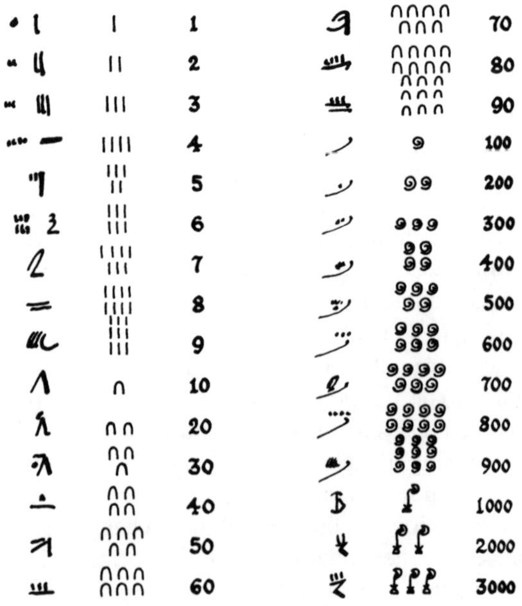

Figure 4: Hieratic numerals from the Rhind Papyrus (17th century B.C., copy of a version from the 19th century B.C.) as simplifications of the hieroglyphic numerals of Egyptian arithmetic. (Source: [8], Appendix 12)

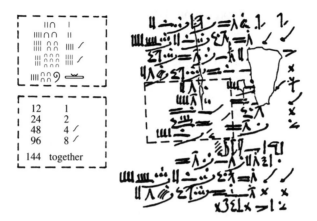

Figure 5: Auxiliary calculation 12 × 12 = 144 (within dotted lines) in an excerpt from Problem 32 of the Rhind Papyrus with hieroglyphic transcription and translation. Direction of writing from right to left. (Facsimile following [2])

culation from Problem 32 of the Rhind Papyrus may serve as an example, because it clearly demonstrates the fundamental problem of the arithmetical technique (cf. Fig. 5).[14] From our vantage point it is the multiplication of twelve times twelve. The Egyptian arithmetician solved the problem additively utilizing the means of the constructive-additive representation of numbers. The multiplicand and an index for the operations performed were graphically doubled until the multiplier from the indices—in the present case from the indices four and eight—could be combined additively.

DEVELOPMENTAL STAGES OF MESOPOTAMIAN ARITHMETIC

Finally, let us take a look at Mesopotamian arithmetic. Because of the greater availability of source material—in Mesopotamia clay tablets were used for writing, and clay is very durable—the development of arithmetic can be traced into the distant past. I shall start with an example from the Old Baby-

Ionian period—that is, approximately from the 19th century B.C. The most important characteristics of the numerical representation of this time are:
1. As in case of the Chinese rod numerals and the Egyptian hieroglyphic numerals the basis of the representation of numbers is a constructive-additive system of numerals.
2. A sexagesimal place-value system is used making it possible to use no more than two basic signs.

The table of reciprocals depicted in Figure 6 may illustrate this structure of the numerical representation.[15] The first two lines of the table are as follows:

$$1; \quad \tfrac{2}{3} \text{ of it} \quad 40$$
$$\text{half of it} \quad 30$$

The following lines with the exception of the last line of the tablet all have the same format:

igi N gal bi M

The column of the numbers N shows how the numerals are formed in a constructive-additive manner using the two basic signs 1 and 10. Thus, the first column on the obverse of the tablet, for example, contains the numbers 3 to 27 in ascending order; some numbers have been left out.

The first lines already reveal the structure of the tablet. If the unit is imagined to consist of 60 smaller parts, the meaning of these lines is as follows:

$$1; \quad \tfrac{2}{3} \text{ of it} \quad 40$$
$$\text{half of it} \quad 30$$
$$\text{the 3rd part} \quad 20$$
$$\text{the 4th part} \quad 15$$
$$\text{the 5th part} \quad 12$$
$$\text{the 6th part} \quad 10$$

The next line reveals the use of a place-value system. As the 8th part of 60 it contains, placed one after the other, the numbers 7 and 30, and one eighth of sixty is indeed $7\tfrac{1}{2}$, thus 7 units plus 30 units of the next place. The entire tablet ends with the reciprocal value of the two-digit number consisting of the 1 and the 21—that is, the number 81. This reciprocal value is stated correctly as the numerical sequence with three digits, consisting of the numbers 44, 26, and 40. The following equation is indeed true:

$$\frac{60}{81} = \frac{44}{60} + \frac{26}{3600} + \frac{40}{216000}$$

Finally, as a special feature of this place-value system it must be emphasized that any designation of the absolute value of the places is lacking; it

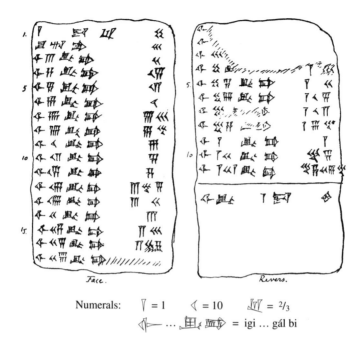

Numerals: $\vert$ = 1 $\triangleleft$ = 10 $\text{\it \vert\vert\vert}$ = 2/3

⟨— ... 𒁉 𒈨𒌍 = igi ... gál bi

Figure 6: Table of reciprocals from the Old Babylonian period (20th–16th century B.C.) according to a drawing by Scheil published in 1915. For an explanation, see text. (Source: [25])

must be obtained from the context. Moreover, missing in the first column are precisely those numbers the reciprocal value of which does not have a finite sexagesimal representation.

If we go back in history by about 200 years to the 21st century, to the period of the IIIrd Dynasty of Ur, we encounter an entirely different situation. At this time, in which the field map depicted in Figure 7 originated, the sexagesimal place-value system was still lacking.[16] The system of numerals had, just like the Egyptian system, a consistent constructive-additive structure with different numerals for every subsequent unit. However, in the crazy quilt of Mesopotamian states a uniform decimal structure was missing which characterized the numerical system of the largely centralized Egypt.

In the context of exploring the origin of arithmetical thinking a further peculiarity deserves attention: At this time different systems of numerals are

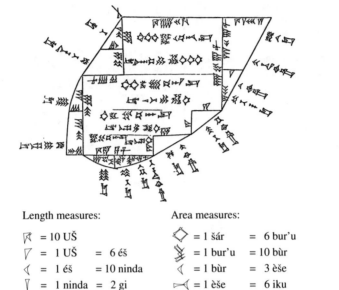

Figure 7: Field map from the time of King Ibbisin (2027 to 2003 B.C.) of the IIIrd Dynasty of Ur, published in 1896 [5], according to a drawing by Thureau-Dangin. (Source: [5a]) The map contains information on lengths and areas calculated in two different ways (written upright or upside down). In addition to the use of special numerals the information about areas is characterized by the semantic indicator gán. Some of the areas at the edge of the field have been marked as 'mountainous terrain.'

used in various contexts, in the example of the field map of Figure 7 different numerals for indicating length and area measures. In their structure the numerals conform to the respective structures of the length or area measures. This permits only one reasonable conclusion: The numerals do not yet represent abstract numbers. Rather, they still represent concrete objects—that is, the standards of measurement that were used.

If we go back another 600 years in history, we find ourselves in the middle of the early dynastic period. A systematically structured table goes back to the 27th century—it represents the oldest comparable find—excavated

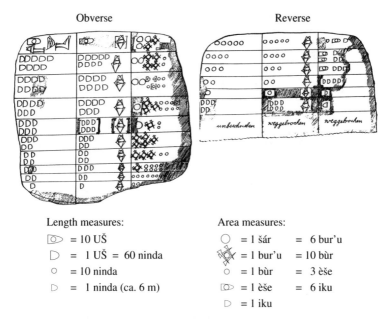

Figure 8: Tablet with information concerning the areas of quadratic fields from the early dynastic period of the city of Shuruppak (27th century B.C.) according to a drawing by Deimel published in 1923. (Source: [4], Vol. 2, p. 75) For a detailed interpretation and a photo of the tablet see Chapter 7.

together with numerous economic texts in the ruins of the city of Shuruppak. It is depicted in Figure 8.[17]

In the third column the table contains the areas of quadratic fields, the lengths and widths of which are listed in the first and second columns. Arithmetically the table is the result of the basic relation of the last line indicating that a field which is one UŠ long—approximately 360 meters—and one UŠ wide has an area of 2 bùr. In contrast to the later cuneiform texts the numerals have been pressed into the clay with a round stylus. With regard to the structure of the numerical representation, however, we once more encounter the same situation: constructive-additive numerals that, with regard to their arithmetical structure, correspond to the structure of the standards of measurement.

The numerals that were pressed into the clay with a round stylus are to be found already on the oldest written tablets unearthed by archaeologists to date, lists of products from the Sumerian city of Uruk. They are believed to

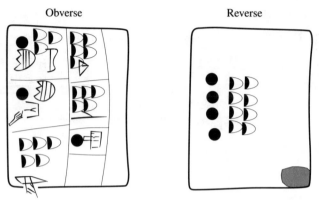

Figure 9: Archaic written tablet (W 6720,a) from the city of Uruk with notations on inventories of products on the obverse and the sum of the records on the reverse (beginning of the 3rd millennium B.C.). (Source: [6a])

come from the transition of the 4th to the 3rd millennium.[18] The simple additions of these oldest written records of the world hardly have anything in common with arithmetic. If the technique of bundling a number of units within a larger unit has been mastered, the additions can be performed quite mechanically.

In Figure 9 such a written tablet has been reproduced. On the obverse of this tablet six notations can be discerned. In each case the recorded amount is indicated by punch marks making use of two basic signs and a designation of the recorded product by an incised pictograph is appended. Obviously, the sum of the notations of the obverse has been recorded on the reverse of the tablet. It can be concluded from the recorded sum that the basic round sign has ten times the value of the basic conical sign.

With the oldest written records we have presumably reached the end of the journey into the early history of arithmetic. However, due to a recent discovery—which, curiously enough, did not occur in the field but in the repositories of the museums—we are able to trace the prehistory of the numerals far into the Stone Age: the discovery of clay symbols designating certain products (cf. Fig. 10). The meaning of these symbols of which there were thousands in the museums was only recognized when the archaeologist Denise Schmandt-Besserat noticed the frequent occurrence of these clay objects and the regularity of their forms while inspecting the holdings

PRINCIPLES OF TEACHING ARITHMETIC

Figure 10: Opened clay sphere (W 20987,8) and clay symbols which were originally kept sealed in this sphere. (Scale 2:3) Place of origin: the city of Uruk (4th millennium B.C.). The surface of some of such clay spheres bear punch marks that, with regard to their number, correspond to the symbols contained inside. (Photo: M. Nissen, courtesy of the Deutsches Archäologisches Institut, Baghdad)

of clay objects from the Near East in a number of museums and—this is the really important discovery—the identity of many forms with pictographs of the oldest writing tablets (cf. Fig. 11).[19] The system of the clay symbols presents us with a mechanical system of accounting that was widespread in all of the Near East since the beginning of the 9th millennium. The use of this system clearly did not presuppose a developed arithmetical thinking. The economic transactions were simply emulated by using the symbols that represented the objects of the transactions. Sealed clay spheres which contained a certain number of such symbols served as tamperproof certification. They could, for instance, accompany products transported by ship as a 'bill of lading,' allowing the recipient to check the accuracy of the shipment. In order to make it possible to ascertain the content of such a clay sphere without the need of breaking it, the symbols were sometimes pressed into the still soft surface of the sphere before the clay had dried, or their shapes were impressed with a stylus. Constructive-additive numerals in their most primitive form were generated in this way. There can hardly be any doubt that the lists of products of the oldest written tablets were the

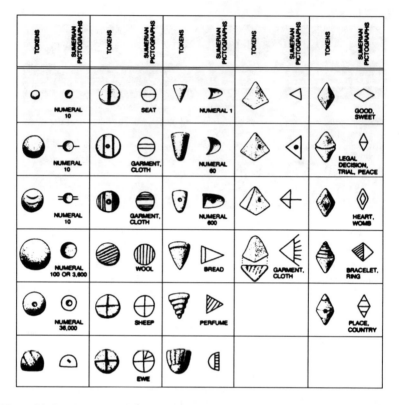

Figure 11: Some examples illustrating the similarity of symbols used to designate certain products in the 4th millennium B.C. and pictographs of the archaic written tablets from the beginning of the 3rd millennium B.C. and interpretations alleged by D. Schmand-Besserat (Source: [27])

result of the subsequent development of this technique of a mechanical form of accounting.

By way of summing up it can be stated: In Mesopotamia the history of the constructive-additive system of numerals can be continuously traced back to a time when arithmetic was only implicitly contained in the operations with clay symbols. There probably is no better evidence for the assumption that the constructive-additive character of the systems of numerals of early arithmetic points to the origin of arithmetical thinking in the operations with concrete objects.

THEORETICAL PERSPECTIVE

The above review of the origins of arithmetic was motivated by a consideration that grew out of a theory of learning arithmetic. Let us first briefly recall this starting point. If mathematical thinking, in accordance with a basic assumption prevalent in teaching strategies for arithmetic, arises from operations with concrete objects by internalizing structures of action, then it must be possible to show such objects and frameworks of action in the early history of arithmetic as well. These objects and frameworks of action must have had a function for the cultural and historical development of arithmetical thinking that is similar to the function that representations in mathematics teaching are currently believed to have for the development of arithmetical thinking in ontogenetic learning processes. However, in contrast to the means of mathematics teaching these objects and frameworks of action to be identified historically cannot have emerged under conditions where the arithmetical thought structures to be developed by manipulating them were already in existence historically.

As a matter of fact such objects with rules of action implicitly exhibiting arithmetical structures indeed played an important role in the early history of arithmetic. Clay symbols, counting rods, and constructive-additive systems of numerical signs represented objects and products of work processes and the distribution of products in the form of a material model. As such they embodied as material representatives more general structures of action such as fixing, ordering, composing, and decomposing sets of objects. Such structures of action evolved in the context of goal-oriented activities without being determined by the specific objectives and without being initially mediated by already preexisting arithmetical structures of thinking. In terms of their functions and their internal nature these structures of action correspond exactly to the seriations, compositions, and decompositions in Piaget's theory of the development of the number concept in children.[20] Historically they represented a general structure within goal-oriented activities. They became the basis of a process of abstraction that established this general structure cognitively and prior to linguistic explication.

Such an understanding of the historical origin of arithmetical thinking challenges the implicit assumption of a naive historiography claiming that number words, numerals, and the means of arithmetic must per se always be considered the result of previously existing arithmetical thought structures.[21] In view of such an understanding of their origin any material representation of arithmetic which can be historically identified poses the problem of determining, on the one hand, the intellectual preconditions necessary for using them and, on the other hand, the intellectual outcome of their usage. Tenets of the history of mathematics presumed to have been established are called into question. What is gained, however, is that beliefs in the history of mathematics become compatible with the results of psychogenetic reconstructions of the ontogenetic acquisition of arithmetical concepts introduced by Piaget and his successors.

The examples presented above can be no more than an initial step in this direction. They are limited to the identification of objects and frameworks of action constituting the starting point of the process of abstracting the structures of arithmetical thinking. This identification does, however, provide a promising perspective supplying new answers to problems of ontogenetic acquisition. If, as in the more recent research most consistently postulated by Jerome Bruner,[22] the cultural context of thought structures is achieved only in terms of their representation by the instruments of social practice, the possibility of uncovering strict genetic associations of these representations presents itself.

Such a genetic association of the thought structures must by necessity elude the theoretical reach of psychology.[23] The domain of material representatives cannot be adequately structured theoretically by means of psychological categories. In Piaget's investigations this domain is, for instance, represented by the arrays of his experiments; however, their relationship to the real environment where the process of socialization takes place, remains unclear and, what is more, cannot be elucidated by psychological explanatory patterns. Maybe this dilemma also explains Piaget's recourse to biological theories. The fact that psychology did not take Piaget quite seriously in this regard can probably be attributed to the circumstance that, due to the predominant psychometric redefinition of Piaget's experimental arrays, psychology no longer reached the scope of Piaget's theoretical endeavors. The dilemma is no longer perceived.

Certainly, this critical evaluation applies to Bruner only in a qualified sense. His theory of cultural transmission by means of enactive, iconic, and symbolic representations takes this dilemma as its very point of departure.[24] Nevertheless, this theory remains within the psychologically constricted perspective. This is the crux of Bruner's theory: If the material representatives on which the development of thinking is based have indeed a strict, genetic relationship that can be theoretically defined, but that cannot be reduced to a logical relationship, then the material representations of mathematical structures are surely not those representations by means of neutral, ahistorical media that action, intuition, and symbolism seem to be according to Bruner's theory.

Detailed reconstructions of genetic connections between the different material representatives of thought structures in the historical development of thinking are required to overcome traditional preoccupations of historians of mathematics which suggest the supremacy of the development of the human intellect in the development of mathematics. The theory of mathematics education would also benefit from such investigations persuing an historical interpretation of Bruner's concept of representation. They would provide a perspective on a constructive theory of developmental connections between didactic representations in the context of mathematical learning.

NOTES

1. Cf. the autobiographical note in [30], p. 255ff.
2. [29], p. 56f.
3. From "Letters from München-Buchsee", quoted from [10], p. 64.
4. Pestalozzi is considered the founder of an education through "intuition" ("Anschauung"). In view of the notorious "exercises by rote" typical of his pedagogy it must be stressed, however, that he accorded language a central role in the learning process. According to Pestalozzi, it is only language that makes abstractions from intuition possible: "Thousands of experiences were bound to convince me that the ability to express precisely what the child experiences through its senses is the first thing it should learn." ([19], p. 64) Cf. his remarks on the abstraction of the number concept by means of language in his preface to the "Anschauungslehre der Zahlenverhältnisse" prepared as a teaching aid by his co-worker Krüsi [18].
5. Cf. [20]; this investigation played a prominent role for the reception of Piaget's work by German teachers.
6. Cf. Piaget's investigation on the development of logical thinking [22] and his lecture on "language and intellectual operations" published in [7], p. 121ff.
7. Piaget's book *Biology and Knowledge* [21] is particularly important for an understanding of his theoretical orientation. Cf. Chapter 1.
8. Cf. [1], pp. 1–67.
9. German translation: [28].
10. For the translations cf. [28], p. 106ff., 142ff., and the references given there. Cf. also [16], p. 3ff.
11. The technique is described in [11], p. 16ff. and in [16], p. 62ff.
12. Cf. [28], p. 80ff. and 130ff.; [16], p. 115ff.
13. For an overview of sources and interpretations see [8]. For the particular approach proposed here see [3], p. 6ff.
14. See [2], Vol. 1, p. 73, and the facsimile and literal translation in Vol. 2.
15. On the Old Babylonian table of reciprocals cf. [17], p. 4ff. and [3], p. 82ff.
16. Cf. [5] and for the applied measures [23].
17. Cf. [4], Vol. 2, p. 26*ff. and p. 75 (drawing). For the interpretation see [3], p. 64ff.
18. Cf. [6].
19. Cf. [26] and [27].

[20] [20].

[21] Note that such an understanding of the historical origin of arithmetical thinking concerns also the debate about constructivist interpretations of the nature of the subject of arithmetic. Constructivism has tried to interpret the origin of arithmetical thinking in constructive-additive symbol systems as a kind of a priori constitution. According to this interpretation arithmetic is based on a "proto-logic" of manipulating such symbol systems; see [13], p. 50ff. and [14], p. 81ff. German constructivists have even drawn educational consequences; see [9]. Others have objected that a single aspect of arithmetic has been overemphasized by this interpretation of the origin of arithmetical thinking. Conceptual development in mathematics had rather to be conceived as a balancing of construction and description; cf., e.g., [24], p. 33ff.; [12], p. 103ff. Careful examination of the historical genesis of arithmetical thought structures and constructive-additive symbol systems indeed provides evidence for such a dialectics of conceptual development. Processes of construction are, on the one hand, constitutive for arithmetical thinking, on the other hand, however, dependent on historical conditions.

[22] Cf. [1].

[23] Cf. Chapters 1, 2, and 4.

[24] Cf. [1], p. 319ff.

BIBLIOGRAPHY

[1] Bruner, J. S. et al.: *Studies in Cognitive Growth: A Collaboration at the Center for Cognitive Studies*. New York: Wiley, 1966.

[2] Chace, A. B. (Ed.): *The Rhind Mathematical Papyrus*. 2 vols., Oberlin, OH: Mathematical Association of America, 1927/1929. (Reprint Reston, 1979.)

[3] Damerow, P. & Lefèvre, W. (Eds.): *Rechenstein, Experiment, Sprache: Historische Fallstudien zur Entstehung der exakten Wissenschaften*. Stuttgart: Klett-Cotta, 1981. [Cf. the selected articles in this volume, Chap. 7 and Chap. 12.]

[4] Deimel, A.: *Die Inschriften von Fara*. 3 vols., Leipzig: Hinrichs, 1922/1923.

[5] Eisenlohr, A.: *Ein Altbabylonischer Felderplan*. Leipzig: Hinrichs, 1896.

[5a] Thureau-Dangin, F.: "Un cadastre Chaldéen." In *Revue d'Assyriologie*, 4 (1), 1897, pp. 13–27. (New publication of the text published in [5].)

[6] Falkenstein, A.: *Archaische Texte aus Uruk*. Berlin: Harrassowitz, 1936.

[6a] Englund, R. K.: *Archaic Administrative Texts from Uruk: The Early Campaigns. Archaische Texte aus Uruk*. Bd. 5 (ATU 5), Berlin: Mann, 1994. (New publication of the texts published in [6].)

[7] Furth, H. G.: *Piaget and Knowledge*. Englewood Cliffs, NJ: Prentice-Hall, 1969.

[8] Gillings, R. J.: *Mathematics in the Time of the Pharaohs*. 2nd ed., Cambridge, MA: MIT Press, 1972.

[9] Inhetveen, R.: "Naive und konstruktive Mengenlehre in der Schule." In *Der Mathematikunterricht*, 17 (1), 1971, p. 7ff.

[10] Jänicke, E.: "Geschichte der Methodik des Rechenunterrichts." In Kehr, C. (Ed.): *Geschichte der Methodik des deutschen Volksschulunterrichts*, Vol. 3. 2nd ed., Gotha: Verlag von E. F. Thienemanns Hofbuchhandlung, 1888.

[11] Juschkewitsch, A. P.: *Geschichte der Mathematik im Mittelalter*. Basel: Pfalz, 1966.

[12] Keitel, C., Otte, M. & Seeger, F.: *Text, Wissen, Tätigkeit. Das Schulbuch im Mathematikunterricht*. Königstein: Scriptor, 1980.

[13] Lorenzen, P.: *Metamathematik*. Mannheim: Bibliographisches Institut, 1962.

[14] Lorenzen, P.: *Methodisches Denken*. Frankfurt a.M.: Suhrkamp, 1968.

[15] Menninger, K.: *Number Words & Number Symbols: A Cultural History of Numbers.* Trans. by P. Broneer. New York: Dover, 1992.
[16] Needham, J.: *Science and Civilisation in China, Vol. 3: Mathematics and the Sciences of the Heavens and the Earth.* Cambridge: Cambridge University Press, 1959.
[17] Neugebauer, O.: *Vorgriechische Mathematik.* Berlin: Springer, 1934.
[18] Pestalozzi, H.: "Anschauungslehre der Zahlenverhältnisse. Zweytes Heft. Vorrede 1803." In Buchenau, A., Spranger, E. & Stettbacher, H. (Eds.): *Sämtliche Werke, Vol. 16: Schriften aus der Zeit von 1803–1804.* Berlin: de Gruyter, 1935, p. 99ff.
[19] Pestalozzi, H.: "Über das Wesen, den Zweck und den Gebrauch der Elementarbücher." In Buchenau, A., Spranger, E. & Stettbacher, H. (Eds.): *Sämtliche Werke, Vol. 14: Schriften aus der Zeit 1801–1803.* Berlin: de Gruyter, 1952, p. 59ff.
[20] Piaget, J.: *The Child's Conception of Number.* Trans. by C. Gattegno and F. M. Hodgson. London: Routledge & Kegan Paul, 1952.
[21] Piaget, J.: *Biology and Knowledge: An Essay on the Relations Between Organic Regulations and Cognitive Processes.* Chicago, IL: University of Chicago Press, 1971.
[22] Piaget, J. & Inhelder, B.: *The Early Growth of Logic in the Child (Classification and Seriation).* Trans. by G. A. Lunzer and D. Papert. London: Routledge & Kegan Paul, 1964.
[23] Powell, M. A.: "Sumerian Area Measures and the Alleged Decimal Substratum." In *Zeitschrift für Assyriologie, 62,* 1972, pp. 165–221.
[24] Ruben, P.: *Philosophie und Mathematik.* Leipzig: Teubner, 1979.
[25] Scheil, V.: "Notules." In *Revue d'Assyriologie, 12* (4), 1915, p. 193ff.
[26] Schmandt-Besserat, D.: "An Archaic Recording System and the Origin of Writing." In *Syro-Mesopotamian Studies, 1,* 1977, pp. 31–70.
[27] Schmandt-Besserat, D.: "The Earliest Precursor of Writing." In *Scientific American, 238* (June), 1978, pp. 38–47.
[28] Vogel, K. (Ed.): *Chiu Chang Suan Shu: Neun Bücher Arithmetischer Technik. Ein chinesisches Rechenbuch für den praktischen Gebrauch aus der frühen Hanzeit (202 v. Chr. bis 9 n. Chr.).* Braunschweig: Vieweg, 1968.
[29] Türk, W. von: *Leitfaden zur zweckmäßigen Behandlung des Unterrichts im Rechnen für Land-Schulen und für die Elementar-Schulen in den Städten. Erster Theil.* 3rd ed., Berlin: Späthen, 1819.
[30] Türk, W. von: *Erfahrungen und Ansichten über Erziehung und Unterricht.* Berlin: Natorff, 1838.

CHAPTER 7

THE DEVELOPMENT OF ARITHMETICAL THINKING:
ON THE ROLE OF CALCULATING AIDS IN ANCIENT
EGYPTIAN AND BABYLONIAN ARITHMETIC
[1981]

PRELIMINARY REMARKS: RECKONING BOARD
AND ROD NUMERALS

The early forms of thinking can only be reconstructed on the basis of mere speculation. Up to now the detailed analyses of the ontogenetic development of thinking presented by developmental psychology are without a phylogenetic counterpart of comparable complexity. This situation is unsatisfactory, all the more so because the explanatory models provided by developmental psychology have far-reaching implications with regard to phylogenetic development. Taking a position in the controversy within developmental psychology concerning the existence of cognitive universals is at one and the same time a decision regarding the historical or the unhistorical nature of certain forms of thinking. A choice between competing explanatory models adapted from biology and socialization theory and applied to the development of language and thinking in children *ipso facto* determines the delimitations between the evolution predating the emergence of human beings and the historical development of the human species. In this situation progress is to be expected only to the extent that the speculative theories are increasingly supplemented by reconstructions of the early forms of thinking guided primarily by the available evidence of the past and corroborated by the latter.

In this context it is of particular interest that, in the search for positive evidence of early forms of thinking, there is at least *one* confirmed cultural achievement generally considered to be a certain indication of the fact that the separation of idea and reality had already taken place: the achievement of a developed arithmetic. The following preliminary considerations may make it clear from the outset from the beginning why this seemingly logical assumption is to be discussed once again, this time in the context of an investigation of the development of arithmetical thinking.

What presumably is the oldest of the extant books on arithmetic of Old China is entitled *Chiu Chang Suan Shu*,[1] translated into German as the *Neun Bücher arithmetischer Technik* (Nine Chapters on the Mathematical Art). It affords insights into the Chinese arithmetic of the first millennium B.C. Perusing this book certainly reinforces the notion that something akin to universal arithmetical ideas must indeed exist—at least as long as we stick to the German translation. Many of the problems might just as well be found in an arithmetical textbook of this century, and the rules applying to the solutions of the problems seem to be derived from the same arithmetical preconditions that determine today's arithmetical rules. Arithmetic seems to be ahistorical.

However, this impression proves to be superficial, for it is due to the conceptual structure of the translation. Concepts like "number," "addition," "multiplication," "fraction," "numerator," and "denominator" represent certain arithmetical ideas and operations of thought. In the text, however, these concepts are translations, and each one of them is really in need of a detailed philological commentary for its justification; for, by necessity, translations demand a transformation into other forms of thinking. The timelessness of arithmetic is merely an illusion—that is, a reflection of the preconditions that prevailed at the time when the arithmetical content of *Chiu Chang Suan Shu* was reconstructed in order to make it accessible to our modern understanding. Thus, the question concerning the arithmetical preconditions of the rules of this arithmetic goes once again unanswered. The evidence to be pursued leads to the semantics and the etymology of the Chinese words from which the translated text, by necessity, was forced to deviate.[2]

What is the technique at issue in the *Chiu Chang Suan Shu*? "Shu" is the "method," the "rule," the "skill." "Suan," the term most frequently used for "calculating," has a variety of meanings. "Suan" refers to predicting the future by means of the number oracle just as much as to the unit of taxation comprising 120 individuals. However, first and foremost, and this is the first solid evidence: not only does "Suan" refer to the activity of calculating, but at the same time to the means of this activity—that is, the *counting rod* that is arranged on the *reckoning board*. The hieroglyphic character for "Suan" is most probably based on the image of the reckoning board.

The indications can easily be multiplied. The "arithmetical rules" of the text usually begin by stating, "Chih ...," "Set up ...," followed by the initial number of the calculation to be arranged on the reckoning board using the counting rods. "Ho," the term indicating addition, signifies the act of "putting together" the counting rods, and "Ch'u," a term covering the operations of subtraction, division, and reduction, signifies their being "taken away." Among the numerous expressions for the object of arithmetic—we would call it "number"—we encounter "Chi," the "piling up" of the counting rods, and "Wei," the place on the reckoning board on which they are arranged.

Thus, the technique with which the *Chiu Chang Suan Shu* is concerned is the technique of operating with the counting rods on the ancient Chinese reckoning board![3] *The primary object of this "arithmetical technique" is not the ideal number, but the counting rods which were piled up in order to represent the quantities* from the areas of application of this technique on the reckoning board.

The role of these rod figures is further documented by a special type of numerical notation used in addition to the hieroglyphic representation— that is, the representation of numbers by *rod numerals;* the latter are accurate representations of the rod figures arranged on the reckoning board. The reckoning board thus determines the structure of numerical representation; it is the representation within the context of a decimal place-value system as presently used. Thus, in this case a structure of arithmetical thinking is predetermined by the construction of the calculating device and by the effect of this construction on the representation of numbers.

With regard to other examples the functional equivalent almost becomes indistinguishable. Does anybody think of the "zero," if he treats its objective representative, the empty space on the reckoning board, correctly and intelligently? Is it acceptable to credit the Chinese with the invention of "negative numbers," when they execute operations on the reckoning board with counting rods of a contrasting color according to specific rules, precisely those rules applicable to operations with negative numbers? Are the Chinese familiar with the multiplication of matrices, when they are solving linear systems of equations according to its rules on the reckoning board? How are we, in light of all this, to make a theoretical distinction between a textbook on arithmetic and a set of instructions for the use of a calculating device?

The example of ancient Chinese arithmetic sheds a new light on the theoretical model so successfully introduced into developmental psychology according to which the logico-mathematical structures of thinking result from the reflection on objective actions.[4] Only if the question is couched in historical terms, will secure explanations for the development of certain forms of thinking be obtained. For the issue here is not just the acquisition of these forms of thinking in the process of socialization, but their actual genesis.

This may serve to indicate the direction of the following investigation. The main question to be explored concerns the problem whether the conclusions to be drawn from the Chinese example can be generalized. Two arithmetical systems will be investigated which are distinguished from today's arithmetic by distinct characteristics and which most certainly did not result from using a reckoning board: ancient Egyptian and ancient Babylonian arithmetic. And this investigation will be guided by the question: *Can the characteristics of these arithmetical systems be explained in terms of operations with the objective means of the particular arithmetical technique just like the characteristics of Chinese arithmetic can be explained through operations with the reckoning board?*

STRUCTURAL CHARACTERISTICS OF ANCIENT EGYPTIAN ARITHMETIC

Sources
Our knowledge of ancient Egyptian arithmetic is essentially based on two sources: a mathematical manual, the so-called "Rhind Papyrus," and a text which appears to be something of a test, the "Moscow Papyrus."[5] Both papyri originated at the time of the Middle Kingdom, around 1800 B.C.

In the middle of the 19th century the "Rhind Papyrus" was acquired by the archaeologist A. H. Rhind in Egypt and, after his death, became the property of the British Museum. The papyrus begins with introductory remarks on the content and purpose of the manual indicating that the text was copied during the reign of one of the pharaohs of the Hyksos dynasty (c. 1650 B.C.) by a scribe named Ahmes from a document dating back to the reign of the pharaoh Amenemhet III (1842–1797 B.C.). The following text begins with table-like compilations of systematic calculations. The main part consists of 87 practical problems and their solutions arranged according to their subjects.

The "Moscow Papyrus" was also purchased. It was the Egyptologist W. Golenishev who bought it towards the end of the 19th century. His collection was later incorporated into the Museum of Fine Arts in Moscow. On the basis of the paleography and the orthography of the text V. V. Struve concluded that the text had been copied by a rather clumsy scribe in a school of scribes of the 13th dynasty (1785–1650 B.C.). The original also came from a school of scribes and was written during the time of the 12th dynasty (1991–1785 B.C.). Presumably it represents an examination with corrections by the teacher. The text contains a mixture of 25 problems from different areas and their solutions. Regrettably the auxiliary calculations which generally provide such ample clues for the reconstruction of arithmetic were omitted.

Both papyri were thoroughly examined and analyzed with regard to their structure in the 1920s.[6] The following reflections are based on the results of these studies.

The Egyptian Arithmetical Schema

If the two papyri are examined with regard to their implicit arithmetical structures, two peculiarities become immediately apparent that are characteristic of Egyptian arithmetic: First, there is the curious way of dealing with the fundamental operations of arithmetic, and second, the consistent representation of fractions as sums of unit fractions.

An auxiliary calculation in connection with *Problem 32* from the Rhind Papyrus offers a particularly simple example concerning the use of the basic types of fundamental operations.[7] It contains the following numerical schema (in contrast to the following written from right to left in the original) as part of an extensive calculation:

1	12
2	24
/ 4	48
/ 8	96
total	144

It is easy enough to decipher the meaning of the schema. Four and 8 equals 12, 48 and 96 equals 144, and that equals 12 times the 12 from the first line. Consequently, we are dealing with the *multiplication* of 12 times 12 which equals 144, and the procedure consists in first duplicating the first line containing a 1 and the second factor until the value of the first column gets close to the first factor, and then choosing those lines from the remaining lines that complement the approximate value with regard to the first factor. If the values of the second column of these lines marked by a slash are then added, the value of the product is obtained—from our own viewpoint due to the distributive law.

The following example comes from *Problem 45* of the Rhind Papyrus. In the description of the solution there is the following passage:

Calculate by adding, starting with: 75 up to times 20; becomes it: 1500.[8]

The following numerical schema associated with it is to be found among the numerical calculations at the end of the text.

1	75
10	750
20	1500

This example indicates that the linguistic representation of the multiplication (20 times 75) closely followed the arithmetical procedure. It also shows that the procedure could be abbreviated by introducing a multiplication by 10.

The third example consists of a calculation from *Problem 69* of the Rhind Papyrus. The numerical schema is as follows:

1	80
/ 10	800
2	160
/ 4	320
total	1120

The picture is familiar by now. What is at issue seems to be the multiplication of 14 times 80 which equals 1120. However, this time the text of the appropriate solution reads:

> Calculate by adding, starting with: 80 for the finding of 1120.

Although we encounter the same numerical schema as in the two previous examples, we must, in accordance with our understanding of the fundamental operations of arithmetic, interpret the numerical schema differently—that is, as a schema for the division of 1120 by 80 which equals 14.

Problem 27 from the Rhind Papyrus, reproduced here in its entirety, may serve as an example for demonstrating how different calculations where combined with each other

A quantity, plus $\frac{1}{5}$ of it. The result is 21.

1	5	/ 1	6	/ 1	$3\frac{1}{2}$
$\frac{1}{5}$	1	/ 2	12	2	7
total	6	/ $\frac{1}{2}$	3	/ 4	14
		total	21		

The quantity is $17\frac{1}{2}$
$\frac{1}{5}$ of it $3\frac{1}{2}$
total   21

From our perspective the solution of the equation $1\frac{1}{5} x = 21$ is determined here. For this purpose we would divide 21 by $1\frac{1}{5}$. The Egyptian procedure avoids such a complicated division. Instead, in order to find the result of $17\frac{1}{2}$, the arithmetical schema is applied three times.

THE DEVELOPMENT OF ARITHMETICAL THINKING 179

The end of the calculation is the easiest to understand. It is already known that $\frac{1}{5}$ of the quantity equals $3\frac{1}{2}$. From our own viewpoint the calculation concludes with the multiplication of 5 times $3\frac{1}{2}$ in order to obtain the whole quantity. The intermediate numerical schema can easily be identified as a division. Twenty-one is divided by 6 and the intermediate result is $3\frac{1}{2}$, which is the value representing $\frac{1}{5}$ of the unknown quantity.

The first numerical schema from which the number 6 is obtained is the most interesting. Formally it seems to represent the multiplication of $1\frac{1}{5}$ times 5. Our modern conception of the problem does not suggest an operation such as this. It also appears as if in this case—in contrast to the previous examples—a division by 5 occurs within the schema. This goes to show the inappropriateness of interpreting the Egyptian arithmetical schema in light of our modern understanding of the fundamental operations of arithmetic. The first of the three calculations is in fact yet another typical application of the Egyptian arithmetical schema which makes use of this schema in order to relate the number n to the unit fraction $\frac{1}{n}$ in a way which can be handled by means of the Egyptian calculating technique. By shifting to a unit that is five times smaller, the specific values of the problem $\frac{1}{5}$ and $1\frac{1}{5}$ are represented by a relation between a unit and its multiple—that is, the relation 1 to 6.

The two previous examples of "divisions" were special cases in that the divisions were without remainder. If this is not the case, the Egyptian arithmetical schema really ought to end without result, because the dividend cannot be composed by adding figures of the previous lines. The procedure to be applied in this case is demonstrated by the following example, *Problem 24* of the Rhind Papyrus. The nature of the problem and the problem-solving procedure are identical to those of Problem 27. The difference is that the "division" to be performed as the intermediary step in the calculation (19 divided by 8) does not yield a integer:

A quantity, plus $\frac{1}{7}$ of it. The result is 19.

1	7	1	8	/ 1	$2\frac{1}{4}\frac{1}{8}$
$\frac{1}{7}$	1	/ 2	16	/ 2	$4\frac{1}{2}\frac{1}{4}$
total	8	$\frac{1}{2}$	4	/ 4	$9\frac{1}{2}$
		/ $\frac{1}{4}$	2		
		/ $\frac{1}{8}$	1		
		total	19		

The quantity is $16\frac{1}{2}\frac{1}{8}$
$\frac{1}{7}$ of it $\quad 2\frac{1}{4}\frac{1}{8}$
total $\quad 19$

Thus, if a "division" has a remainder, the duplications are replaced by successive halvings.

Furthermore, the last example leads us to the second characteristic of Egyptian arithmetic: the representation of the fractions as sums of unit fractions—fractions with the numerator 1—with the unit fractions, which are the additive components of a fraction, simply written one after the other. A number of basic considerations about the procedure are in order.

Mathematical Excursus: Decomposition of Fractions into Unit Fractions
Every proper fraction $\frac{a}{b}$ can be represented as a sum of different unit fractions. Such a decomposition can, for example, be constructed by means of the Euclidean algorithm—that is, the division with a "remainder." The construction proceeds as follows: If b is divided by a, the result is q with the "remainder" r. The relationship of these quantities can be expressed by the following equation:

$$b = qa + r \qquad q > 0 \qquad 0 \leq r < a$$

If q, in case the division leaves a remainder r differing from 0, is substituted by a value b_1 exceeding q by 1 (i.e.: $b_1 = q + 1$) and if the difference between r and the divisor is called a_1 (i.e.: $a_1 = a - r$), then the relationship can also be represented as follows:

$$b = b_1 a - a_1 \qquad b_1 > 1 \qquad 0 \leq a_1 < a$$

If a is determined from this equation and the value is inserted into the original fraction, then this fraction reads as follows:

$$\frac{a}{b} = \frac{1}{b_1} + \frac{1}{b_1} \times \frac{a_1}{b} \qquad b_1 > 1 \qquad 0 \leq a_1 < a$$

The troublesome remaining fraction $\frac{a_1}{b}$ has a smaller numerator than the original fraction $\frac{a}{b}$. Thus, if the whole procedure is applied again to this remaining fraction as well as to any subsequent remaining fraction that is created as long as the division leaves a remainder, then the remaining fraction will at some point equal 0; the original fraction is then represented as follows:

$$\frac{a}{b} = \frac{1}{b_1} + \frac{1}{b_1 b_2} + \frac{1}{b_1 b_2 b_3} + \ldots + \frac{1}{b_1 b_2 \ldots b_n} \qquad b_i > 1$$

This is the desired representation of the fraction $\frac{a}{b}$ as a sum of different unit fractions. If this procedure is applied to the fraction $\frac{6}{7}$, for example, then the decomposition by means of unit fractions is as follows:

$$\frac{6}{7} = \frac{1}{2} + \frac{1}{4} + \frac{1}{12} + \frac{1}{48} + \frac{1}{336}$$

Structural Problems of Calculating with Unit Fractions in the Egyptian Arithmetical Schema

The fact that every proper fraction can be expressed as a sum of unit fractions indicates that it is possible in principle to do fractional arithmetic using only unit fractions. This is exactly what is happening in ancient Egyptian arithmetic. The example of the "division" in Problem 24 of the Rhind Papyrus demonstrates that the decomposition into unit fractions which seems so complicated to us does not normally cause any problems in Egyptian arithmetic. This is true, *because the Egyptian arithmetical procedure does not as a rule allow the emergence of any fractions other than unit fractions.*

There is, however, one exception to this rule. If there are already fractions in the right column of the arithmetical schema—in the case of the "division" by a fraction, for example—then the unit fractions must be doubled in each step of the calculation—that is, two equal unit fractions must be added up. If the denominator is even, it only needs to be divided in half for this purpose; if, however, it is uneven, then—from our modern point of view—a fraction with the numerator 2 emerges. In this case the problem of the decomposition by means of unit fractions is no longer solved automatically by the arithmetical procedure. Let us sum up: *In Egyptian arithmetic the decomposition of fractions with the numerator 2 and an uneven denominator into unit fractions is a problem which played an exceptional role due to the arithmetical technique.*

A second problem is pointed out by the following argument: While it is true that every proper fraction can be decomposed into a sum of different unit fractions, this decomposition is not unique. Apart from the decomposition already cited, the fraction $\frac{6}{7}$, for example, can also be divided in the following way:

$$\frac{6}{7} = \frac{1}{2} + \frac{1}{4} + \frac{1}{14} + \frac{1}{28}$$

If, however, a fraction can appear in a variety of representations, then it may become difficult to recognize it in its different representations as an identical fraction.

This problem can, however, be mitigated by giving certain decompositions by means of unit fractions canonical status according to some criterion, just as, in modern fractional arithmetic, there is one representation for every fraction that is canonically standardized by the demand that fractions must be reduced as far as possible. In Egyptian arithmetic certain decompositions by means of unit fractions are indeed given preference. This does not, however, solve this problem, since the Egyptian arithmetical schema does provide different decompositions of the same fraction as well. The latter would have to be transformed into canonically standardized decompositions using a procedure comparable to our method of "reduction."

Due to the lack of such a procedure it is particularly difficult to evaluate differences or even to determine them correctly. In contrast to a reduced fraction it is, as a rule, not immediately apparent whether a sum of unit fractions is smaller or larger than the integer 1. This poses a technical problem, since the arithmetical procedure invariably demands that certain numbers—especially integers—be combined from the lines of the arithmetical schema by means of addition. Therefore the second point to be made is the following: *A further mathematical problem of Egyptian arithmetic which played an exceptional role due to the arithmetical technique consists of complementing a decomposition into unit fractions to obtain an integer.*

The Duplication of Unit Fractions
What were the solutions to these problems provided by Egyptian arithmetic? The table of calculations at the beginning of the Rhind Papyrus provides an answer to the first problem, the decomposition of fractions with the numerator 2 and an uneven denominator.[9] This table contains the "divisions" of the number 2 by the uneven numbers from 3 to 101. It thus represents a list of the decompositions of the fractions with the numerator 2 and with an uneven denominator between 3 and 101. This list obviously served to master the duplication of unit fractions technically, for the same decompositions are consistently found in the calculations, even in the Moscow Papyrus which is independent of the Rhind Papyrus.

What this means is that this list also served to give certain decompositions canonical priority. In Egyptian arithmetic, for example, the following is always the case:

2 times	$\frac{1}{5}$	equals	$\frac{1}{3}$	$\frac{1}{15}$			
2 times	$\frac{1}{7}$	equals	$\frac{1}{4}$	$\frac{1}{28}$			
2 times	$\frac{1}{9}$	equals	$\frac{1}{6}$	$\frac{1}{18}$			
2 times	$\frac{1}{11}$	equals	$\frac{1}{6}$	$\frac{1}{66}$			
2 times	$\frac{1}{13}$	equals	$\frac{1}{8}$	$\frac{1}{52}$	$\frac{1}{104}$		
2 times	$\frac{1}{15}$	equals	$\frac{1}{10}$	$\frac{1}{30}$			
		.					
		.					
		.					
2 times	$\frac{1}{97}$	equals	$\frac{1}{56}$	$\frac{1}{679}$	$\frac{1}{776}$		
2 times	$\frac{1}{99}$	equals	$\frac{1}{66}$	$\frac{1}{198}$			
2 times	$\frac{1}{101}$	equals	$\frac{1}{101}$	$\frac{1}{202}$	$\frac{1}{303}$	$\frac{1}{606}$	

The Auxiliary Number Algorithm

With regard to the second structural problem of Egyptian arithmetic—complementing a decomposition into unit fractions to obtain an integer—the available sources do not provide such a clear answer. The question as to whether the Egyptians were adept in mastering this problem technically must remain open. While it is true that there are a number of calculations in a special section of the Rhind Papyrus dedicated to this problem—their designation could be translated as "complementary calculation"—and that the problems contain numerous examples concerning the solution of this problem, the calculations unfortunately belong to text passages that are especially difficult to interpret. Consequently, opinions concerning their interpretation diverge.[10]

An "auxiliary number algorithm" was used for solving the problem, a system of auxiliary numbers that were placed under or next to the numbers in question—usually in red ink. I will give an example to illustrate the basic principle of this technique—that is, the solution pertaining to *Problem 33* of the Rhind Papyrus which will be reproduced in its entirety below. (In order to facilitate things, I have numbered each line. In the original the bold text is in red.)

184 CHAPTER 7

(1)	A quantity, $\frac{2}{3}$ of it, $\frac{1}{2}$ of it, $\frac{1}{7}$ of it is added to it. The result is 37.				
(2)	1	1 $\frac{2}{3}$ $\frac{1}{2}$ $\frac{1}{7}$			
(3)	2	4 $\frac{1}{3}$ $\frac{1}{4}$ $\frac{1}{28}$			
(4)	4	9 $\frac{1}{6}$ $\frac{1}{14}$			
(5)	8	18 $\frac{1}{3}$ $\frac{1}{7}$			
(6)	/ 16	36 $\frac{2}{3}$ $\frac{1}{4}$ $\frac{1}{28}$			
(7)		28 10$\frac{1}{2}$ 1$\frac{1}{2}$			
(8)	1	42			
(9)	/ $\frac{2}{3}$	28			
(10)	$\frac{1}{2}$	21			
(11)	/ $\frac{1}{4}$	10 $\frac{1}{2}$			
(12)	/ $\frac{1}{28}$	1 $\frac{1}{2}$	total 40 remainder 2		
(13)	$\frac{1}{97}$	$\frac{1}{42}$	1		
(14)	/ $\frac{1}{56}$ $\frac{1}{679}$ $\frac{1}{776}$	$\frac{1}{21}$	2		

In the original, this text is followed by the "proof by verificaton," in the course of which—taking the result of the calculation: $16\frac{1}{56}\frac{1}{679}\frac{1}{776}$ as the starting point—the number 37 is re-calculated in an even more voluminous calculation. The following auxiliary calculation is also part of the solution. Presumably it wound up at the end of Problem 38 by mistake when the papyrus was copied:

1	42
$\frac{2}{3}$	28
$\frac{1}{2}$	21
$\frac{1}{7}$	6
total	97

Finally, in connection with this problem it must be noted that, in Egyptian arithmetic, there is an exception to the rule of only using unit fractions: The fraction $\frac{2}{3}$ is used and treated like a unit fraction.

How should the solution be interpreted? In this case the "division" by a complicated fraction cannot be circumvented in the same way as in Problems 24 and 27 discussed earlier. The solution, therefore, consists in the "division" of the number 37 by the number $1\frac{2}{3}\frac{1}{2}\frac{1}{7}$. That is to say that by successively duplicating this number, lines of the arithmetical schema are generated which must be added to obtain the number 37.

Up to line (6) the calculation is easily interpreted. The duplication of the unit fractions is accomplished by using the table discussed in the previous section. At line (6) a number is reached that comes very close to the desired value 37. At this point, the structural problem of the Egyptian arithmetical technique mentioned earlier arises: Which part must be complemented so that the number 37 is reached and how is this part to be obtained from the initial line?

The following line (7) contains three seemingly mysterious auxiliary numbers written with red ink and placed underneath the fractions to be complemented in order to add up to the number 1—that is, $\frac{2}{3}$, $\frac{1}{4}$, and $\frac{1}{28}$. The subsequent lines from (8) to (12) obviously represent an auxiliary calculation for determining these red auxiliary numbers. Starting with the number 42 which, in line (8), has been related to the number 1, the auxiliary numbers are determined for the problematic fractions of line (6).

The function of the auxiliary numbers can be inferred from the second auxiliary calculation which accidentally slipped into Problem 38. In this case—starting with the same initial line—auxiliary numbers are calculated as well, this time, however, the auxiliary numbers of line (2) which was the starting point for the entire calculation. The idea behind the auxiliary number algorithm appears to have been the desire to perform the calculation at strategically important points and parallel to the main calculation with a unit 42 times smaller with many of the occurring fractions turning into integers. In light of our modern arithmetical technique the procedure for the first line of the arithmetical schema can be considered a transformation for obtaining the "common denominator 42"—this thought does not, however, contribute a whole lot to our understanding of the Egyptian auxiliary number algorithm.

The auxiliary calculations raise a few more questions as well. The construction of $\frac{1}{28}$ in line (12) or of $\frac{1}{7}$ at a comparable position in the auxiliary calculation deviates from the rules of the arithmetical schema. However, the missing auxiliary calculations of the "division" by 7 or by 28 are so simple that their absence is not to be considered serious. In contrast, the question as to how the number 42 was found is entirely open. The two available sources do not adequately answer the fundamental question concerning our understanding of the auxiliary number algorithm—that is, whether the original number was obtained algorithmically or heuristically.

Finally, let us consider the conclusion of the solution. The final formula of the auxiliary calculation in line (12) provides a key information: "total 40 remainder 2." Which means: The number 1 has the auxiliary number 42;

as indicated by the auxiliary calculation only 40 units have been reached with line (6). The "remainder" is 2. Therefore the line to be added in order to complete line (6) must correspond to an auxiliary number 2!

How can the main calculation ending with line (6) be continued in such a way that the result will be a line corresponding to this auxiliary number? We have already encountered the logic of the solution of this problem with regard to the first arithmetical schema of Problem 27—that is, as a logic of the transition from n to $\frac{1}{n}$ in the Egyptian arithmetical schema. Applied to the complicated case at hand the result is as follows:

If, as shown by the second auxiliary calculation, the first line has a total auxiliary number of 97, then, by making the transition to $\frac{1}{97}$, a line with the auxiliary number 1 is obtained; however, on the basis of the same pattern of argument, such an auxiliary number happens to be associated precisely with the number $\frac{1}{42}$. Thus, line (13) continues the main calculation with exactly these two numbers, and the auxiliary number 1 has been placed at the end of the line as the third number. Duplication—by making use of the table—results in the final line with the auxiliary number 2, the desired line, which completes line (6) so that the integer of 37 is obtained. Therefore the solution reads: $16 \frac{1}{56} \frac{1}{679} \frac{1}{776}$!

The Prevalence of Additive Operations

Even if, as demonstrated by the few examples presented here, operations making use of the ancient Egyptian arithmetical schema could assume very intricate forms, it was nevertheless based essentially on only four basic operations:[11]

– finding the sum of certain numbers,
– completing certain numbers to obtain a predetermined sum,
– duplicating a number—that is, finding its sum with itself, and
– halving—that is, finding a number that, added to itself, results in a predetermined sum.

These basic operations have one thing in common, their additive character. The multiplication table up to ten—the basis of multiplication in our system—is not needed in the Egyptian arithmetical technique. Therefore the hypothesis is obvious that can provide a surprisingly simple answer to the peculiar use of the fundamental operations: *The arithmetical thinking of the Egyptians was dominated by additive numerical operations.* In Egyptian arithmetic numbers only had one single algebraic structure—that is, that of additive composition.

The second peculiarity of Egyptian arithmetic is consistent with this hypothesis as well—that is, the representation of fractions as sums of unit fractions. As impractical as this representation may seem in retrospect, it is extremely functional in the framework of the Egyptian arithmetical technique inasmuch as it permits the application of the additive arithmetical schema to fractions as well. At this point a basic connection between the use of the fundamental operations by the Egyptians and the peculiar representation of fractions becomes apparent: "Multiplications" and "divisions" are performed with fractions as well as integers always in the same way and within the same arithmetical schema. Furthermore, they are performed in such a way that no basic operations other than additive operations are necessary.

However, the hypothesis that the arithmetical thinking of the arithmeticians of ancient Egypt was dominated by the additive numerical structure not only permits the conclusion as to the expediency of the representation of fractions as sums of unit fractions; for if this hypothesis is correct, the Egyptians were not even in a position to develop a representation of fractions comparable to our modern representation. A representation of fractions in terms of "numerators" and "denominators"—however modified— is after all a multiplicative representation. Without the multiplication and the division of integers such a representation of fractions can be determined neither conceptually nor is it technically usable for performing the elementary arithmetical operations.

Reasons for the Structural Peculiarities of Ancient Egyptian Arithmetic
The result of the deliberations so far is as follows: Ancient Egyptian arithmetic exhibits two peculiarities—that is, an unusual use of the fundamental operations and a representation of fractions as sums of unit fractions. The two peculiarities are related to each other if we assume that the arithmetical thinking of the Egyptians was dominated by additive numerical operations. Accordingly, the Egyptian arithmetical schema would have to be interpreted as a universal procedure of solving multiplicative problems by additive means; and the representation of fractions as sums of unit fractions would have to be understood as a form of representation which permits operations with fractions that are exclusively additive in nature.

It is within the scope of structural considerations to uncover such relationships; however, the question why certain possibilities were accessible to the thinking of the Egyptians while others were not cannot be answered in this context. Why is it that the Egyptians were, on the one hand, capable of such sophisticated additive thinking, the developed techniques of which enabled them to deal even with multiplicative problems, but that, on the other hand, they were unable to understand multiplication as a fundamental arithmetical operation? What is it that determined the specific form of their arithmetical thinking? Since it is impossible to answer these questions by means of structural investigations, the means used in ancient Egyptian arithmetic will be analyzed next.

THE MEANS OF CALCULATION IN ANCIENT EGYPT

Arithmetic, a Technique of the Scribes

The introductory description of the available sources already indicated that there was one "professional group" with a special relationship to ancient Egyptian arithmetic: the group of the "scribes." The term "scribe" is somewhat misleading.[12] While writing was the most important tool of the "scribes," their activity was by no means limited to taking dictation or to copying older texts. Rather, the scribes of ancient Egypt were—to a large extent high-ranking—royal officials with broad responsibilities for the planning and administration of the affairs of state. Their training was extensive. The future scribes, mostly children from the aristocratic upper class, were educated at first in exclusive court schools where they not only learned how to spell but the elaborate calligraphy and style as well. Higher education took place in the "houses of life" ("per-anch") as well as in the training academies for specialized administrative tasks which were part of the central administrative offices. In these schools for scribes the knowledge was passed on that constituted the basis for an effective centralized administration of the large ancient Egyptian empires.

Calculating in particular was among the activities of the scribes. There is evidence for this in a papyrus where a scribe is reprimanded for his lack of ability with regard to arithmetic:

Let me tell you what it means when you say: "I am the supervising scribe of the army." You are given the task to dig a lake. You come to see me to get information about the provisions of the soldiers and say: "Figure it out for me." You are neglecting your office, and it falls upon my neck that I must teach you how to discharge its obligations. ... For you must understand that you are the experienced scribe at the head of the army. A ramp is to be built that is 730 cubits long and 55 cubits wide; it contains 120 cases and is filled with reed and logs; it has a height of 60 cubits at the top, 30 cubits at the middle, with a (.?.) of 15 cubits and its (.?.) is 5 cubits. The generals are asked how many bricks will be needed and all the scribes are assembled without any one of them knowing anything. They all count on you and they say: "You are an experienced scribe, my friend. Therefore, make a quick decision for us in this matter. You are famous." ... Do not allow anyone to say of you: "There are things that you do not know."[13]

Problems similar to this one with which the supervising scribe of the army is confronted here, constitute the central part of the Rhind Papyrus as well as the Moscow Papyrus. There can be no doubt that these two "mathematical papyri" do not constitute a particular class of texts. Rather they document—just as other papyri—the aids for and the results of the activities of the scribes. The problems they contain give evidence of the administrative responsibilities of the scribes, and the solutions they contain document the techniques used by the scribes. Thus, the procedures of ancient Egyptian "mathematics"—particularly the arithmetical techniques—were most likely developed by the scribes as well and therefore have their origin in the problems associated with planning and administration confronting the officials of the centralized Egyptian state.

As a rule, this state of affairs has led to extremely one-sided conclusions. For it is the *objects* of concern to the administration of the ancient Egyptian empires that were examined almost exclusively, while the available *means* for coping with the administrative problems were hardly ever analyzed. While the role of the scribes with regard to the development of Egyptian mathematics is always mentioned, it is hardly ever really examined. Instead Herodotus' comment about his impression that geometry must have been invented by the Egyptians, since they had to do a new survey of their fields every time the Nile flooded,[14] is recounted in countless variations allegedly as a key information about the origin of geometry. However, as a matter of fact, this comment explains next to nothing. Where a problem arises, the means for solving it may not be available at all. In order to find out why a problem was solved in a certain way, it is necessary in particular to analyze the means available for solving it.

Ancient Egyptian arithmetic does indeed exhibit a peculiarity with regard to its means: *The only means used to perform arithmetical operations is writing.*[15] It is not difficult to understand this fact as a consequence of the socio-historic context that gave rise to arithmetic. The special status of the scribes within the official hierarchy was based on the importance accorded to writing as the central tool of the administration of the Egyptian state. If Egyptian arithmetic had its origin in the problems of planning and administration, then it is only logical to assume that the arithmetical problems were solved as far as possible by using the most important administrative tool—writing.

This explains why written arithmetical systems were developed and used in ancient Egyptian arithmetic some 2000 years before they generally became the dominant calculating aid. In order to better understand the methods of this early use of written arithmetical systems, it is appropriate to take a closer look at the nature of this means.

Excursus: The Representation of Numbers by Means of Written Signs

Every written arithmetical system is based on the representation of numbers by means of signs or sign combinations. To calculate by writing, therefore, means: Transformation of the sign combinations according to fixed rules relating directly only to these sign combinations and only indirectly to the numbers represented. In particular the various arithmetical systems are distinguished from each other in that they are based on different forms of the representation of numbers by means of signs. The cognitive preconditions necessary for performing the operations depend on the mode of this representation, and thus it determines the entire character of an arithmetical system.

The simplest form of numerical representation is the representation of a number by an *individual sign*. In our modern system of numerals, for example, the numbers from zero to nine are represented by such individual signs. Structurally this representation corresponds to the designation of numbers by means of individual numerical names. As long as numbers are represented by signs only in this fashion, written calculations are basically not yet possible; what is possible is merely a recording in writing of results of arithmetical operations performed elsewhere. Rules referring to the representatives can only be formulated when these representatives are constructed in such a way as to represent, in terms of their objective physical nature, structures of the numbers which can be abstracted from them.

The question as to which numbers in a system of numerals are represented by individual signs is of prime importance for the arithmetical technique. As individually coined written signs representing names of numbers they usually belong to the oldest written representatives of numbers in a system of numerals, for there is no arithmetical operation using written signs that is involved in their development already as a presupposition. In the course of the development of the written arithmetical technique or a technique supported by writing, the representatives of numbers are increasingly constructed by means of rules with reference to arithmetical operations, and the use of individual numerical signs is adjusted to the requirements of the arithmetical technique. Thus, the development of numerals is functionally separated from the general development of writing.

Alternatively, it is possible to use the physico-geometrical characteristics of signs in order to construct objective models of numbers and to denote the numbers by means of these models. In the simplest case configurations are created by iterating a sign, the sign for the unit one. The designated number can be determined by counting these iterated signs. Other individual signs are subsequently defined or redefined in terms of operations of substitution which determine their numerical meaning relative to the unit. In a system of numerals strictly constructed according to these rules, the numerical meaning of every numeral can be obtained directly with the help of the operations of counting and substituting. The dominating principle for constructing numerals in such a system of signs is seriation, and here seriation signifies addition. I would, therefore, propose to call this form of representation *constructive-additive representation*. Our modern system of numerals is completely free of this form of representation, it is, however, well known from the Roman numerals.

A variant of importance to the development of place-value systems emerges when certain sign combinations such as seriations or ligatures of signs no longer signify addition but multiplication. This *multiplicative representation* is no longer constructive in the same sense as additive representation. In this case, the meaning conveyed by the representation is already based on developed arithmetical mental operations. The product is not constructed by means of the signs but is already presupposed as "meaning."

The hieroglyphic numerals of ancient Chinese arithmetic which were used parallel to the rod numerals may serve as an example. Hieroglyphs representing the powers of ten are placed after the basic numerals composed of iterated units in order to indicate the order of magnitude they represent, their "position" on the reckoning board.[16] Another early example of the multiplicative representation of numbers is the sign used for the number 600 in the Early Dynastic period of Mesopotamia which was a ligature of the sign for the number ten and the sign for the number 60; after the transition to the Old Babylonian place-value system, this ligature was substituted by a sign which was identical with the sign for the number ten.

Representations of numbers in a *place-value system* are composed of additive and multiplicative forms of representation combined with each other in a specific way. The seriation of the positions signifies addition. However, the positions themselves are not only free positions for numbers but a multiplicative form of representation in which one of the factors is no longer represented by a sign, but by a position instead. In each case this factor must be inferred from the overall construction of the combination of signs. In our modern system of numerals, for example, it is the powers of the number ten that must be inferred in this fashion.

A form of representation with entirely different, more flexible possibilities of representing numbers is the representation of numbers by making use of *symbols representing functions (or operations) of one or more variables* that, when applied to numerals, signify the particular functional value (or the operational outcome). The fraction line, the decimal point, and the plus and minus signs of our modern system of numerals may serve as examples for this.

Finally, a special form of numerical representation should be mentioned here which is important in Mesopotamian arithmetic. The forms mentioned so far are forms of an absolute numerical representation. In this case every (complete) combination of signs represents a specific number. However, in addition there is also the possibility of a *relative representation of numbers* where the meaning of the signs and sign combinations is dependent on the context. Under these conditions they do not represent numbers, but only certain characteristics of numbers; the numbers themselves must be determined from the context. As "numerals" such signs and sign combinations are, therefore, essentially ambiguous.

If the different forms of the written representation of numbers are compared with each other, it becomes apparent that one of them is special—that is, the additive representation. This is due to the fact that, strictly speaking, it is the only constructive form of representation. It is the only form of representation based on a physical-geometric model of numbers that is independent of any previously established meaning of the symbols—comparable perhaps to the representation which is characteristic of primitive pictographic writing or of the geometrical constructions of Euclidean geometry. For this reason it is possible to perform "mechanical" calculations with constructive-additive numerals—that is, without the manipulation with the numerical representatives already presupposing specific arithmetical mental operations.

All other forms of numerical representation are more indirect forms. Their use as means already presupposes other means. Without the tables of addition and multiplication up to ten, for instance, our own written arithmetical procedures would be worthless; neither can be obtained by using these procedures themselves. Thus, with regard to all forms of representation with the exception of the constructive-additive representation, the cognitive presuppositions of the representations of number refer to a developmental connection of the different numerical representations which were used as means of arithmetic; the development of arithmetical thinking is merely its mental reflection.

The Hieratic Numerals

The oldest form of Egyptian writing is a type of pictography written from right to left or from top to bottom, the writing of the hieroglyphs, the "sacred symbols." As early as the period of the Old Empire, this *hieroglyphic writing* was complemented by *hieratic writing* used as a type of shorthand for the preparation of business documents. In this form of writing the external form of the symbols was streamlined and the pictorial character was largely lost.[17]

The nature of this developmental process is not hard to understand. Hieroglyphic writing continued to be used—as the writing used on monuments and other inscriptions on frescoes and valuable objects integral to the royal life-style. It remained in use with regard to all inscriptions that were chiseled, painted, inlaid, carved, or embossed. The "everyday" writing material, however, consisted of the papyrus made from the pressed pith of the stems of the papyrus plant and the black and red ink made of a suspension of soot or vermillion respectively; the instrument for writing was the brush made of rushes. The simplification of monumental hieroglyphics resulting in hieratic writing which was used in books is nothing other than the adaptation of the forms of writing to the natural characteristics of these writing tools. This development did not entail a structural change of the type of writing. Hieroglyphic pictorial writing remained the basis by mediating between the hieratic abbreviation and its meaning.

The two mathematical papyri also use hieratic writing, the numbers are represented by hieratic numerals. Because of their appearance many of these characters look like individual signs for numerals. It would, however, be quite hasty to interpret them as such, for they are easily recognized as the simplified forms of hieroglyphic numerals. The actual form of the numerical representation in the mathematical papyri can, therefore, only be assessed by analyzing the hieroglyphic numerals of which they are the simplified abbreviations that are easier to write (cf. Fig. 1).

The Hieroglyphic Numerals

In their hieroglyphic form the numerals of ancient Egyptian arithmetic constitute a constructive-additive symbolic system having a particularly con-

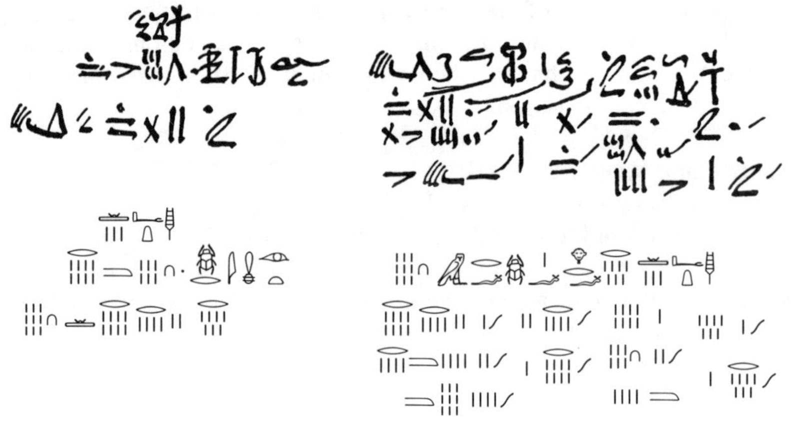

Figure 1: Problem 24 from the Rhind Papyrus (cf. the translation on p. 179).
Above: Original in hieratic writing
Below: Hieroglyphic transcription
Direction of writing from right to left
Hieroglyphic numerals:

⇒ = Individual sign for $\frac{1}{2}$

| = Individual sign for 1 (sign A)

∩ = Individual sign for 10 (sign B)

⇔ = Unit fraction operator

Some of the signs are also used in the text as characters.

The right section contains the problem in the first line (in the original the first part of this line is red). Below it to the far right the first arithmetical schema. To the left of it the second arithmetical schema comprising five lines arranged in two columns is to be found. To the left the third arithmetical schema can be seen. The left section comprises the account of the solution placed below it in the translation. The first line includes the text, "The quantity is." The second line begins with an untranslated formula often to be found before the solution.

sistent structure. There are individual signs for the powers of ten, for the number one it is a vertical line, for the number ten an arch open at the bottom, the picture of a stanchion for fettering animals, for the number 100 the picture of a coiled measuring rope, for the number 1000 the picture of a lotus blossom, for the number 10000 the picture of a vertical finger, and so on. The symbols for the remaining integers are generated without exception by seriation, in descending order from right to left according to the values of the signs.

Similar signs are arranged in characteristic groups so that their number is recognized more easily.

This strictly constructive-additive system of numerals permits the mechanical performance of additions. In order to add two numbers, their signs simply need to be placed next to each other, to be rearranged, and to be substituted by signs with higher values when their number exceeds the decimal limit.

If the individual Egyptian signs for the numbers 1, 10, 100, ... are transcribed using the letters A, B, C, ..., the comparison of an addition in the modern and in the Egyptian system of numerals reveals the fundamental difference (whereby the direction of writing will, in this as well as in subsequent examples, be adapted to our own in order to facilitate the comparison):

	CBBAAA	(= 123)
	CCBAAAAAAAA	(= 218)
total	CCCBBBBA	(= 341)

The addition using hieroglyphic numerals does not require specifically arithmetical mental operations, and, in a modified sense, the same is true of subtraction. Additive thinking is a result of operating with hieroglyphic numerals in this way, not a precondition. To the extent that the additive thinking develops, the hieroglyphic numerals can be simplified and rendered more economically as hieratic characters.

Thus, the written arithmetical system of ancient Egyptian arithmetic has only a superficial resemblance to modern written arithmetical systems. For in Egyptian arithmetic the written numerals have been devised as a mechanical calculating aid to facilitate addition. As far as integers are concerned, the use of individual signs is subordinated to this function as well. In marked contrast to our modern system of numerals, the Egyptian system uses individual signs for precisely the powers of ten—that is, for those numbers represented in the current system solely by their positions. In the Egyptian system, on the other hand, the symbols for the numbers 2 to 9 for which we are using individual signs today, are created out of the sign for the unit in a constructive-additive manner. In our current system of numerals it is the particular use of the individual signs that permits the written arithmetical system we use; it is accepted in turn that the table of addition up to ten which constitutes a fundamental precondition of written calculations must already be known. In contrast, in ancient Egyptian arithmetic the addition table is the very outcome conveyed constructively by the symbolic system.

Apart from this strength of the system of hieroglyphic numerals there is, however, a critical weakness: *the exceptional role played by the additive structure!* It is a lot more difficult to perform multiplications and divisions by means of this system. Only the multiplication or division with a power of ten is simple, because in this case the operation of substitution is identical with the multiplication or division:

> CBBAAA (= 123)
> DCCBBB (= 1230)

This is consistent with the fact that apart from the additive operations the only basic operation known to ancient Egyptian arithmetic is the multiplication and division with the number ten. If the Egyptians wanted to solve the multiplicative problems they encountered in practice they were compelled to develop a more far-reaching means of arithmetical technique.

Relationships Between the Arithmetical Schema and the Numerals
Such a means is the *arithmetical schema* that the Egyptians used for "multiplication," "division," and, in addition, for other purposes as well, such as, for example, the auxiliary number algorithm. The lines of this schema represent multiplicative relationships. If the additive operations are applied to these lines as a whole rather than to the individual numerals, then this corresponds to a constructive creation of new multiplicative correlations of the same relationship. The schema expands, as it were, the heuristic of the anticipation of results of additive operations, a heuristic made possible by the incipient additive thinking. The area to which it can be applied is extended to multiplicative relations without the necessity of introducing a new operation. The constructive-additive representation of numbers remains the only precondition.

The extent to which the ancient Egyptian arithmetical schema remains rooted in this numerical representation becomes especially apparent, if the examples initially provided to explain the arithmetical schema are recalled and the necessary operations with the hieroglyphic numerals are performed "mechanically."

Problem: $12 \times 12 = ?$
Goal of the operations: BAA (= 12) in the left column

A	BAA	(initial line: 1 12)
AA	BBAAAA	($2 \times$ line 1)
/ AAAA	BBBBAAAAAAAA	($2 \times$ line 2)
/ AAAAAAAA	BBBBBBBBAAAAAA	($2 \times$ line 3; B for 10A)
BAA	CBBBBAAAA	(line 3 + line 4)

Problem: 1120 ÷ 80 = ?
Goal of the operations: DCBB (= 1120) in the right column

A	BBBBBBBB	(initial line: 1 80)
/ B	CCCCCCCC	(C for B)
AA	CBBBBBB	(2 × line 1; C for 10B)
/ AAAA	CCCBB	(2 × line 3; C for 10B)
BAAAA	DCBB	(line 2 + line 4)

Thus, "multiplying" and "dividing" require the same operations as adding and subtracting. The difficulty to be overcome in solving the given problems is not due to novel arithmetical operations, but rather to the fact that the result is obtained by means of a heuristic procedure instead of an algorithm: by skillfully combining anticipated results of possible operations to create the desired sequence of signs.

Relationships Between the Representation of Fractions as Sums of Unit Fractions and Calculating with Hieroglyphic Numerals

Let us now explore the means of ancient Egyptian fractions. There were individual signs for the fractions $\frac{1}{2}$ and $\frac{2}{3}$. The original pictorial meaning of the sign for $\frac{1}{2}$—a trapezoid open to the right and also used to signify "side"—is not clear. The sign for $\frac{2}{3}$ is a ligature obtained from the outline of a mouth—also used in the figurative sense of "part" and for denoting a grain measure—and two vertical lines. Originally, there were other individual signs for fractions created as ligatures using the sign for "mouth." However, as the arithmetical technique developed they were considered to be incompatible with it and eliminated from Egyptian arithmetic. Thus, there is a simple explanation for the special status of the fraction $\frac{2}{3}$ mentioned earlier. It constitutes an historical relic from the earliest beginnings of Egyptian fractions.

Apart from these special cases, the Egyptian numerals for fractions—in hieroglyphic as well as hieratic writing—were constructed according to a simple principle: The hieroglyph "mouth"—abbreviated to form a dot in hieratic writing—functioned as an operator converting signs for integers into signs for fractions. More precisely, the operator symbol conferred on the sign for an integer n the meaning of the fraction which is created by dividing the unit into n parts, thus the meaning changes to "$\frac{1}{n}$." For this purpose the unit fraction operator was placed above the numeral or, in the case of longer numerals, above the first sign.

This form of representation automatically results in unit fractions playing an exceptional role. The signs for unit fractions constituted the basis for generating the sign combinations representing the other fractions in a constructive-additive manner by means of seriation—*in exactly the same way in which the integers were constituted using the individual signs of the powers of ten.* Thus, the representation of fractions as sums of unit fractions as a specific characteristic of Egyptian fractions proves not to be an outlandish idea. Rather, it proves to be the trivial extension of the procedure of numerical representation that applies to integers to fractions as well.

The motive for this mode of representing fractions is easily recognized. When the principle of numerical representation is transferred from integers to fractions, the arithmetical systems for integers can be applied automatically to fractions as well. The advantage becomes apparent when the complicated calculation of a sum like that, for example, of the fractions $\frac{5}{6}$ and $\frac{2}{15}$ in our numerical system is compared with the corresponding "calculation" using Egyptian numerals (with the sign X indicating the Egyptian individual sign for $\frac{1}{2}$ and a horizontal line for the unit fraction operator "mouth"):

	X $\overline{\text{AAA}}$	($\frac{5}{6}$ represented as $\frac{1}{2} + \frac{1}{3}$)
	$\overline{\text{B}}$ $\overline{\text{BBB}}$	($\frac{2}{15}$ represented as $\frac{1}{10} + \frac{1}{30}$)
total	X $\overline{\text{AAA}}$ $\overline{\text{B}}$ $\overline{\text{BBB}}$	(result: $\frac{29}{30}$)

The example clearly demonstrates how tempting it must have been in view of the conditions defined by the means of ancient Egyptian arithmetic to take this route of constructing an arithmetical technique for fractions. With hindsight—that is, on the basis of an abstract concept of the fraction that can be represented more or less efficiently—the structural reasons are obvious why this route was bound to lead to an arithmetical dead-end. The structural analysis provided above has shown that the ambiguity of the representation by unit fractions raises problems that are not easily managed. To make the Egyptian form of representing fractions easier to use, a rule corresponding to our reduction into lowest terms would have been needed—that is, a formal rule that would have permitted to coordinate the application of the unit fraction operator with the basic operation of the Egyptian arithmetical technique, namely addition, thus permitting to simplify representations by transforming one representation into the other. However, such a rule of the mechanical transformation of signs was impossible for structural reasons that could not be uncovered by Egyptian arithmetic.

Tables as Calculating Aids
In order to cope with the ambiguity of their representation of fractions the Egyptians introduced a new means into arithmetic. They used a table for ca-

nonically standardizing certain representations of the results of duplicating unit fractions.

In Egyptian arithmetic a second example for the use of a table is known. The method of multiplying or dividing by 10 by replacing the individual signs with signs with a value ten times greater or smaller reaches its limit when numbers between 2 and 9 are divided by 10 because the results are fractions. The Rhind Papyrus also contains the calculation of these fractions in a table following the table for the duplication of unit fractions.[18]

In both cases the table serves as a written record of results for ready use once they have been obtained. In the first case this serves to relieve the constructive-additive numerical representation of the task of having to function as a means for the addition of identical unit fractions as well. In the second case the table serves to close a gap with regard to the application of the operation of substitution. Perhaps these two examples may help to imagine the kind of developmental potential contained in the table as a calculating aid.[19] The domination of additive operations in ancient Egyptian arithmetic did, however, result in the fact that the most important conceivable application of the table remained unused: the use for the systematic organization of the results of multiplicative arithmetical operations in tabular form, that is to say, operations that cannot be translated into simple "mechanical" rules of symbolic transformation by any form of written representation. To elaborate this possibility was reserved for Old Babylonian arithmetic.

THE SOURCES FOR RECONSTRUCTING THE HISTORY OF OLD BABYLONIAN ARITHMETIC

Concerning Chronology
The sources for the arithmetic of the Mesopotamian cultural sphere are far more abundant than those for Egyptian arithmetic. The reconstruction of its development can draw on archaeological sources from several millennia. Thousands of clay tablets with cuneiform inscriptions in particular provide information concerning this development.

Since most of the important sources of a mathematical nature date back to the period of the first Babylonian dynasty, the so-called "Old Babylonian period," it is usually called "Babylonian" mathematics. It would be more appropriate to speak of "Mesopotamian" mathematics, since these documents originate in cultural and socio-historical circumstances that were limited neither to the city state of Babylon nor to the Old Babylonian period. Despite this, the Old Babylonian period is of particular importance for the development of mathematics in Mesopotamia. Therefore, the arithmetic of this period is at the center of the present investigation. However, the reconstruction of its history will have to be based more broadly on the development of the arithmetic of Mesopotamia.[20]

The following chronological synopsis may serve to facilitate the placement of the sources used:[21]

Approx. 9000 B.C.	Establishment of first **settlements** in Mesopotamia
Up to 3000 B.C.	**Prehistoric period**
	Rural communities with slowly developing urban centers (e.g., Habuba Kabira)
	Gradual transition from rain-fed agriculture in the upper valley of the Euphrates and Tigris rivers to irrigation in the delta
3000–2340 B.C.	**Early Dynastic period**
	Development of the first city states, particularly by the Sumerians in the river delta (e.g., Ur, Umma, Lagash, Uruk, Kish, Shuruppak)
2340–2000 B.C.	**Period of the first empires**
	For the first time large areas of Mesopotamia form a political unity (e.g., Dynasty of Akkad, 2340–2159; IIIrd Dynasty of Ur, 2111–2003)
	Sumerian is replaced as the vernacular by the semitic language of the Akkadians
2000–1594 B.C.	**Old Babylonian period**
	Western semitic rulers in Mesopotamia. The first dynasty of Babylon prevails. Largest expansion under Hammurabi, 6th King of Babylon (1792–1750)
16th–12th century B.C.	Kassite rule in Babylon
10th century B.C.	New Assyrian Empire
6th–4th century B.C.	Mesopotamia under Persian rule
4th–2nd century B.C.	Hellenistic period under the rule of the Seleucid dynasty

The earliest written sources date back to the period around 3000 B.C. Two groups of texts in particular provide information on arithmetic: the *economic texts* and the *mathematical texts*.

Despite the relatively favorable situation with regard to source material, the attempt to reconstruct the development of arithmetic is encountering considerable difficulties. To a large extent the *mathematical texts,* which constitute the most important sources, were obtained from unauthorized excavations and were acquired by the museums from dealers. In these cases the determination of the dates is based solely on paleography, on the features of the style of writing characteristic of a particular period. In many cases the place of origin is unknown; invariably it is dubious. Even in cases where the texts come from systematic excavations, dating is often imprecise or questionable.[22] Furthermore, the mathematical texts are not evenly distributed over the millennia; nearly all of them come either from the Old Babylonian and the Kassite periods or from the period of the Seleucid dynasty. The period of at least 800 years without comparable discoveries leads to the conclusion that there are considerable gaps. Only the fact that none of the "mathematical" texts in the narrow sense can, with any certainty, be dated earlier than the Old Babylonian period stands to gain historical plausibility due to the reconstruction of the development of arithmetic. Additional difficulties arise from the status of the way the texts were edited. While the mathematical texts in the narrow sense are quite accessible thanks to compilations and editions, a class of texts of prime importance for the development of arithmetic has hardly been taken into consideration in this context, the metrological texts.

It is true that the dates of the *economic texts* are easier to determine, because, as early as the Early Dynastic period, the year of the reign of the particular ruler was often provided. There are, however, problems with the large number of available texts, because to date they have never been analyzed systematically with regard to the information they might provide on the status of arithmetic at different times. In the present context, this group of texts is used merely by way of example, and many of the statements need to be examined in terms of a systematic evaluation of all available sources. In view of the overall status of the sources an introductory overview of the sources and the archaeological findings seems in order on which the subsequent presentation of Old Babylonian arithmetic and its history are based.

The Belated Discovery of Counters
A reconstruction of the history of the arithmetic of the preliterate period must rely solely on archaeological evidence. A very recent "discovery" is of particular importance in this respect: the "discovery" of clay counters. The history of the discovery of these counters is so unusual that it is worthwhile to recount it here.

Between 1927 and 1931 the city of Nuzi in the area of the upper Tigris river was excavated, a city dating back to the second millennium B.C. Thirty years later, A. L. Oppenheim[23] discovered philological evidence for the use of counters in addition to the usual written arithmetic in the process of analyzing the economic documents found at Nuzi. Upon scrutinizing the excavation reports, he also came across the description of a hollow clay sphere which allegedly contained 48 "pebbles" when it was excavated. The decipherment of the inscription yielded the following text:

> Pebbles for goats and sheep:
> 21 ewes
> 6 female lambs
> 8 rams
> 4 male lambs
> 6 female goats
> 1 billy goat
> 2 female kids
> (seal of Ziqarru)

The "pebbles" were thought to be of such minor importance when they were found that they were lost. Based on his findings, Oppenheim developed a speculative but penetrating theory according to which, in Nuzi, the written economic accounts had been expanded into a system of "double entry bookkeeping" by making use of a system of counters. The obvious assumption that such counters might also have been of considerable importance for arithmetic seems to have been unacceptable to him, since there were no other reports of comparable findings.

Six years later, P. Amiet, working on findings from the much older Elamite city of Susa, came across a similar system of counters. He identified approximately 70 clay spheres containing a variety of geometrically shaped clay pebbles. However, these discoveries were still believed to be merely of local importance.

The archaeologist D. Schmandt-Besserat[24] must be credited with having recognized the real meaning of the clay pebbles. In 1969 she traveled to the museums to find out how the use of clay had been introduced into the Near East. In her research she came across thousands of clay counters from almost all excavations in the Near East which had largely gone unnoticed and had been interpreted incorrectly. The following presentation of the arithmetic of the preliterate period is to a large extent based on this "discovery."

Economic Texts

The oldest written documents known date back to the period around 3000 B.C. and come from the Sumerian cities of Uruk and Jemdet Nasr.[25] These are records of an economic nature which throw light upon the development of writing as well as arithmetic. Somewhat more recent are the clay tablets from the archives of the Sumerian city of Shuruppak (c. 2700–2600 B.C.).[26] In this case we are also dealing mostly with "economic texts," although the percentage of texts of a different nature has increased. These two groups of texts are the most important examples of the so-called "archaic" texts and are used here to represent the oldest forms of written arithmetic in Mesopotamia.

Subsequently, the number of written sources steadily increases. By way of illustration, economic texts from Girsu in the city state of Lagash (c. 2300 B.C.)[27] were examined as an example from the late Early Dynastic period and texts from the IIIrd Dynasty of Ur (c. 2100–2000 B.C.)[28] for the period of the first empires. Up to the beginning of the Old Babylonian period the calculations contained in the economic texts are virtually the only source of information on the development of arithmetic.

Mathematical Texts
Not until the Old Babylonian period do we encounter documents that can properly be called "mathematical" texts.[29] Two groups must be distinguished. They constitute the *table texts*—that is, clay tablets covered with columns of numerals. Their meaning must be inferred from the immanent structure, for aside from these signs they generally contain only abbreviated operational terms. Since the meaning of the latter was determined by these tables, it can only rarely be reconstructed from their etymological roots. Occasionally table texts are placed in the IIIrd Dynasty of Ur, but I do not know of a single example where the determination of this date is beyond doubt. However, one of the table texts is completely atypical; this is the text that belongs to the archaic documents of Shuruppak. This indicates the existence of considerable gaps. The second group, the *mathematical texts* in the narrow sense, are largely collections and solutions to problems. Most of the approximately 200 published texts date back to the Old Babylonian and the subsequent Kassite periods (2000–1100 B.C.) and can only be dated by means of paleography.

For the following period of nearly 1000 years up to the time of Hellenism there are almost no findings of mathematical documents. A relatively small number of documents dates back to the time of the Seleucids.[30] In addition, there is a considerable number of astronomical documents from this period which may provide information on arithmetical techniques. However, in terms of the present account this late period of the development of arithmetic in Mesopotamia was given only marginal attention.

STRUCTURAL CHARACTERISTICS OF OLD BABYLONIAN ARITHMETIC

The Sexagesimal Place-Value System
Old Babylonian arithmetic is characterized by a peculiarity which is as unique in the history of arithmetical systems as the calculation with unit fractions of the Egyptians: *the use of a sexagesimal place-value system.*

The principles of such a numerical system can be explained in analogy to the currently used place-value system of decimals. In the sexagesimal system the number 60 corresponds to the number 10 in the decimal system. Instead of the individual signs 1 to 9 the Babylonians used numerals from 1 to 59 which were formed in a constructive-additive manner using two basic signs, a vertical "wedge" for the number one and a "Winkelhaken" (oblique wedge) for the number ten. In the sexagesimal system, just as in the decimal system, the value of a sign depends on its position in the sequence of the signs. However, in the sexagesimal system the difference between the positions is not the factor 10, but the factor 60.

Thus, in the *decimal system,* for example, the sequence of the signs 140.5 indicates the sum of one hundred, four tens, zero ones and five tenths. In the *sexagesimal system,* on the other hand, each position except the zero can be occupied not only by nine, but by fifty-nine signs. Thus, the number represented in the decimal system by the sequence of the signs 140.5 must be represented in the sexagesimal system by two sixties, twenty ones and thirty sixtieths. If the limits between the positions are identified by commas, the numerical sequence for the decimal number 140.5 written in the sexagesimal system is 2,20;30. If, as in Babylonian arithmetic, 59 different numerals would be used, the identification of the limits between the positions by commas would of course not be necessary.

The Ambiguity of the Babylonian Numerals
The Old Babylonian sexagesimal system differs from the decimal system of modern arithmetic in one essential point: There was no sexagesimal point in Old Babylonian arithmetic. Any conceivable symbol was missing that, in terms of its function, could possibly have been compared to the decimal point. The 59 numerals were simply placed next to each other which means that the numerals were, as a matter of principle, used ambiguously. Babylonian arithmetic was based on a system of *relative numerical representation.* The sequences of signs did not have any absolute numerical meaning.

Thus, the sexagesimal number 2,20;30 (the decimal number 140.5) introduced earlier by way of example was rendered as follows:

2,20,30

Thus, in terms of its representation, it could be distinguished neither from the number 2,20;30 nor from the number 2,20,30 and even less from all those numbers that can be constructed from the sequence of numerals 2,20,30 by placing zeros either in front of it or after it. In terms of the familiar—but in this case entirely inappropriate—decimal representation: The number 140.5 could not be distinguished from the numbers 2.3416666... (equals $\frac{1}{60}$ of 140.5) and 8430 (equals 60 times 140.5) solely on the basis of the form of its representation, nor from any other number that results from multiplying or dividing it by a power of 60.

However, this certainly does not mean that the absolute numerical value was of no consequence in Babylonian arithmetic. In each application of arithmetic the absolute value of a result could no longer be unimportant. However, in terms of arithmetical technique absolute numerical values could also be disregarded only up to a point, because the relativity concerning the factor 60 renders the additive operations ambiguous and therefore meaningless. The numbers 1 and 60, for example, indistinguishable in terms of their Babylonian representation, yield two results if a positive number, for example, the number 2, is added to them: 3 and 62. In the Babylonian system the latter are also different numbers with a distinct numerical representation.

Thus, at the outset we must assume that as a rule the sequences of numerals of Babylonian arithmetic were used unambiguously as well. If the meaning proved to be important for the calculation or for the interpretation of a result, it had to be obtained from the context of the problem at hand and subsequently taken into consideration. In the context of the sexagesimal system such a procedure is facilitated by the fact that an error in the placement entails far more dramatic consequences than in the decimal system. This is so because the erroneous value differs from the correct value not just by the factor of ten, but by at least the factor of 60. However, even if the numerals had always been used with an absolute numerical meaning, *the ambiguity of numerical representation remains a structural idiosyncrasy of Old Babylonian arithmetic,* which is in need of an explanation concerning the conditions of its development as well as its persistence over many centuries.

A Place-Value System Without a Zero?

The case is somewhat different with another idiosyncrasy of Old Babylonian arithmetic—that is, the lack of a sign for the zero. The significance of this characteristic was merely temporary; during the era of the Seleucids, that is to say, about a millennium after the Old Babylonian period which is the focus of the present study, a separate sign for an empty position was used without any further structural changes of the arithmetic.

For a long time, however, operations in Old Babylonian arithmetic were carried out without any identification of an empty position. Instead, 2,0,25, for instance, was simply rendered as 2,25. This may have been due to the fact that the lack of a zero caused far fewer problems than it would have caused with regard to the modern decimal system. As a matter of fact, preceding or succeeding zeros were not at all needed because of the lack of a sexagesimal point, and "internal zeros" are six times more numerous in the decimal system than in the sexagesimal system because of the lower number of numerals. It is, therefore, not surprising that only very few examples of "internal zeros" are to be found in the large number of available documents of the Old Babylonian period. For this reason, however, it is difficult to determine how the problem was overcome.

One of these examples may illustrate the problem. In one of the documents[31] which, according to the style of writing, dates back to the latter part of the Ist Babylonian dynasty and contains examples for solving square and cubic equations, the following calculation is to be found:

$\frac{1}{2}$ of 2,50 break and confront with itself. 2,25 you see. From 2,25 1,10 tear out. 50,25 you see. (Paraphrased as: Square $\frac{1}{2}$ of 2,50. You will get 2,25. Subtract 1,10 from 2,25. You will get 50,25.)

At first this calculation does not seem to make sense, for whatever absolute value of 2,25 and 1,10 is assumed, the result never is 50,25. However, the square of 1,25 is not 2,25, but 2,0,25. In the subsequent subtraction the empty position was taken into account, although it is not expressed in the notation. We must therefore assume that, in order to perform the subtraction correctly, the absolute values represented by the numerals in the various positions were obtained from the context.

Are there Arithmetical Algorithms in Babylonian Arithmetic?

Finally, one more matter must be pointed out: Although a place-value system was used in Old Babylonian arithmetic, there is no evidence in the documents of written algorithms based on this place-value system. It is common knowledge that the development of the modern decimal system is closely linked to the development of such procedures. By contrast, no such correlation is apparent in the case of the sexagesimal system of Old Babylonian arithmetic. In any case, even with regard to more complicated arithmetical operations the result immediately follows the problem.

This situation occasions specific questions and problems concerning the reconstruction of Old Babylonian arithmetic. Do we have to assume that Babylonian reckoners used a calculating aid hitherto unknown whose discovery could finally explain the use of the sexagesimal place-value system?[32] Did the Babylonian reckoners develop quite unusual skills with regard to doing math in their heads for the purpose of dealing with this place-value system thus making the invention of written arithmetical procedures superfluous?[33] These questions also demand an explanation.

Mathematical Excursus: The Structure of Babylonian Arithmetic
If we want to reconstruct the development of Old Babylonian arithmetic in order to better understand its idiosyncrasies, then—for the purpose of defining the questions more exactly—it is appropriate, first, to give a somewhat more detailed account of some of the formal preconditions and consequences of these idiosyncrasies.

Every positive number can be represented in a *sexagesimal place-value system* just as it can be represented in a decimal or any other system. That is to say, every positive integer x can be divided into a sum of multiples of the powers of the number 60 (with their coefficients being integers larger than or equal to zero and smaller than 60):

$$x = a_0 \times 60^0 + a_1 \times 60^1 + a_2 \times 60^2 + \ldots + a_n \times 60^n \quad 0 \leq a_i < 60$$

Such a decomposition can always be constructed by means of the Euclidean Algorithm—the division with a "remainder." For this purpose the number x is divided by 60 thus obtaining x_1 and the remainder a_0 and so on. In the course of this process, the values of x_1 become successively smaller so that the procedure must end up with an x_n that is itself smaller than 60. The following equations result from this process.

$$\begin{aligned}
x &= x_1 \times 60 + a_0 \\
x_1 &= x_2 \times 60 + a_1 \\
x_2 &= x_3 \times 60 + a_2 \\
&\vdots \\
x_{n-1} &= x_n \times 60 + a_{n-1} \\
x_n &= a_n \quad\quad\quad\quad 0 \leq a_i < 60
\end{aligned}$$

208   CHAPTER 7

From these equations the decomposition of x into powers of the number 60 is obtained by means of substitution. This sexagesimal representation is unique, since the assumption that two different decompositions might exist, contradicts the uniqueness of the result of the Euclidean Algorithm. Thus, any finite sequence of integers $a_0, a_1, ..., a_n$ from the interval from 0 to 59 corresponds uniquely to a positive integer x.

This procedure of sexagesimal representation can be generalized by admitting negative numbers as exponents of the base of 60 as well so that sequences of coefficients a_i can also be provided for rational fractions. Let a finite sexagesimal representation of a number, or briefly, a *sexagesimal number* be defined more generally as a finite sum of integral multiples of the powers of the number 60 with exponents which are (positive or negative) integers:

$$x = a_{-m} \times 60^{-m} + a_{-(m-1)} \times 60^{-(m-1)} + ... + a_{n-1} \times 60^{n-1} + a_n \times 60^n \qquad 0 \leq a_i < 60$$

This definition raises the question which of the rational numbers can be represented in this way and which cannot.

If, according to the above definition, x is a sexagesimal number, then this obviously means that there is a power of the number 60—in the case of the decomposition as specified above the number 60^m for instance—that, when multiplied with the number x, results in an integer. And since the number 60 contains only the prime factors 2, 3, and 5, this formulation in turn means that the number x can be turned into an integer by multiplying it by certain powers of these prime factors. This leads to the answer to the above question: Precisely those rational numbers (i.e., quotients of two integers) can be represented sexagesimally, whose denominator after reduction into lowest terms only contains the numbers 2, 3, and 5 as prime factors.

Thus, every number with a finite decimal representation can, for example, also be represented sexagesimally (since it has only the numbers 2 and 5 as prime factors in its denominator), the number $\frac{1}{3}$ can be represented sexagesimally but not decimally, and the number $\frac{1}{7}$ can be represented neither sexagesimally nor decimally. However, just as every rational number can be approximated arbitrarily close by decimal numbers, the same is true of the approximation by sexagesimal numbers. This is true, since every decimal number can also be represented sexagesimally, so that every decimal approximation is at the same time a sexagesimal approximation.

Furthermore it follows from the definition of sexagesimal numbers that sums and products of sexagesimal numbers are also sexagesimal numbers—that is, the set of sexagesimal numbers is closed with respect to the operations of addition and multiplication. The fact is obvious that subtraction cannot be performed without restrictions as long as the concept of number is restricted to positive numbers. However, the question whether *divisions* can be performed with sexagesimal numbers without restrictions is interesting from a technical perspective, because it has immediate consequences for the development of algorithms.

Since the division by a rational number can be defined as the multiplication with its reciprocal, the question is easily answered: The quotient of two sexagesimal numbers (i.e., rational numbers with no prime factors in their denominators other than 2, 3, and 5) will only be itself a sexagesimal number, if either the divisor contains no prime factors other than the numbers 2, 3, and 5 not only in its denominator, but also in the numerator, or if it does contain other prime factors, they must be contained in equal number in the numerator of the dividend as well, so that they can be cancelled out in the quotient. Sexagesimal numbers that only contain the numbers 2, 3, and 5 as prime factors in the numerator as well as in the denominator, which—in other words—can be written as products of powers of the numbers 2, 3, and 5 with integral (positive or negative) exponents, are called *regular sexagesimal numbers*. It follows that the set of the regular sexagesimal numbers is closed not only with respect to addition and multiplication but also with respect to division.

If the sexagesimal numbers are conceived of as numbers of measurements of geometrical and physical magnitudes, then the multiplication with a power of the number 60 can be interpreted as the transition to a measuring unit increased or decreased by this power of 60. The definition of the sexagesimal numbers entails that such a sexagesimal transformation of the unit does not change the sequence of the coefficients $a_0, a_1, \ldots, a_n$ characteristic of the number. Thus, every sexagesimal number can be considered to be characterized by two independent elements—that is, by a sequence of coefficients and by choosing a certain position of this sequence (continued, if necessary, by means of zeros) as the basic unit.

For adding sexagesimal numbers both elements are essential. However, for multiplication the determination of the unit is in a certain sense irrelevant: The two operations—that is, the transformation of the unit and the multiplication are interchangable. The result is the same whether the unit is transformed first and the multiplication is performed second or whether the multiplication is performed first and the result is subjected to a transformation of the unit afterwards. Thus, if a number is multiplied by two numbers which agree in terms of the sequence of the coefficients of their sexagesimal representation and differ only in their basic units, then the two products also agree in terms of the sequence of the coefficients of their sexagesimal representation and differ only in their basic units.

Thus, in the process of performing multiplications, it may make sense for technical reasons to abstract from the respective designated unit. As a result of abstracting from the unit by identifying sexagesimal numbers with the same sequence of the coefficients of their representation, the *relative sexagesimal numbers* of Old Babylonian arithmetic are obtained.

Inquiry into the Reasons for the Structural Idiosyncrasies of Babylonian Arithmetic

Let us recall the results of our examination: Old Babylonian arithmetic—like ancient Egyptian arithmetic—is characterized by distinct structural idiosyncrasies of a highly developed numerical system. However, in this instance the idiosyncrasies are of a totally different nature. Old Babylonian arithmetic is characterized by the following:
– the sexagesimal base of numerical representation,
– the positional form of numerical representation in the place-value system,
– the ambiguity of numerical representation due to the dependence of the meaning of the positions on the context, and
– the lack of distinct arithmetical algorithms.

As in the case of Egyptian arithmetic, structural analysis can detect connections between these idiosyncrasies, but it cannot explain how they came about. The association between relative numerical representation and multiplication appears to be especially significant; inferences from this functionalism to actual developmental correlations must, however, remain purely hypothetical.

In the following, an attempt will be made to gain a more comprehensive understanding by analyzing the means of arithmetic. The situation is somewhat more complicated than in the case of ancient Egyptian arithmetic, since the idiosyncrasies of the latter could invariably be interpreted as consequences of an additive basic structure of arithmetical thinking at the level of structural analysis. A relationship of comparable simplicity cannot be detected between the characteristics of Old Babylonian arithmetic.

However, on the other hand, the external preconditions for an analysis are more favorable due to the availability of better sources. *Thus, with regard to Old Babylonian arithmetic, an attempt can be made to analyze developmental processes in terms of the historical chronology of the sources, processes that could only be tentatively reconstructed with regard to Egyptian arithmetic.*

THE HISTORY OF MESOPOTAMIAN CALCULATING AIDS

Calculating Aids before the Invention of Writing

Number Words

Language presumably is the oldest means of arithmetic. Number words already existed before the first numerical signs. Thus, the etymology of number words and the structure of the counting sequence contain information about the oldest forms of arithmetical thinking.[34]

In many languages steps in the counting sequence indicate former limits in counting, in German (as in English) the counting steps at ten and twelve, for instance. In the Sumerian counting sequence such counting limits of archaic arithmetic seem to exist at five and ten. A limit in counting at five is indicated by the fact that the subsequent number words probably do not have individual names. The etymology of, at least, some of them shows that they were constructed using the number words for the numbers one to five according to the arithmetical operations $5 + 1, 5 + 2, \ldots$. The counting limit at ten is characterized by an individual name which concludes the preceding counting sequence. Another counting limit at three has been suggested based on an etymological association of the first three number words with the concepts "man," "woman," and "many." This identification is, however, not generally accepted any longer.

Another indication of the early development of arithmetical thinking is the fact that in many languages the first three or four number words are treated grammatically like adjectives—that is, they are declined and their gender corresponds to that of the objects counted. This is true of the Indogermanic languages in particular: of Greek up to the number four and of Latin, Gothic, Old and Middle High German up to the number three. There are similar grammatical rules with regard to the first number words of the counting sequence in the Semitic languages to which, for example, the Akkadian language belongs. In the latter, all number words even exhibit a masculine and a feminine form.

Inferences have been made from this grammatical form of the first and therefore oldest number words in particular to the origin of counting. The number word used as an adjective is evidence for the earliest stage of the separation of the number from the object of counting. At this stage the number is nothing but "quantity." It does not yet have an existence independent of the objects of counting; rather, it represents a specific attribute of these objects.

However, this earliest stage of the development of arithmetical thinking is already being transcended, if the language is systematically used as a means for counting by constructing new number words from the preceding ones according to formal rules, for example, the number words greater than ten in German (as in English) and the number words greater than five in Sumerian. The counting principle which is independent of the context allows for the abstraction of "numbers" that are independent of the objects of counting, for *the means of counting makes this independence vis-à-vis the objects of counting into an object of empirical experience.* The counting operation that is independent of the qualities of the objects of counting imparts the experience of the identity of "numbers" with an existence that is independent of the objects of counting.

Counters
The significance of language as a means of arithmetic must, however, not be overestimated. The fleeting nature of the spoken word severely limits operations with the abstracted "numbers." Verbal counting can hardly provide more than a basic ordinal structure of these "numbers." *Counting is not a calculating aid.*

It is thus not surprising that the transition to the formal development of number words already points to extralinguistic means of counting. It is certain that the counting limits at five and ten which, in the examples given, are transcended by using formally constructed number words, were suggested by using the fingers as a means of counting. It must be assumed that, in the course of the subsequent evolution of the counting sequences, the linguistic development merely reflects the development of more efficient, extralinguistic techniques. The construction of number words by making use of arithmetical operations in particular is a sure indication that these number words already refer to enduring objective means of arithmetic in which the numbers are no longer represented merely in temporary-successive, but rather in spatial-simultaneous ways.

With regard to the sphere of Mesopotamian civilization the clay pebbles are such a means.[35] Thousands of them have been found in this area according to the statistics of D. Schmandt-Besserat. In Jarmo in Iraq alone almost 1500 have been recorded. Approximately 1100 of them are shaped like spheres, around 200 like discs, and around 100 are conical in shape. The oldest finds come from two excavations in the Zagros region of present-day Iran. They must be dated back to the 9th millennium B.C. and thus come from the period of transition leading to the emergence of settlements. The counters measuring about one to two centimeters were in use for approximately 5000 years until they were apparently superseded by written numerals. They came in four basic shapes—sphere, cone, cylinder, and round disc—as well as a number of variations of these basic shapes. These basic shapes remained largely unchanged throughout the entire 5000 years in which they were used as counters.

How the counters were actually used can only be inferred from subsequent developments. All statements concerning the arithmetical techniques of this preliterate period of arithmetical thinking are therefore characterized by a high degree of uncertainty.

It is likely that there was a close association between the archaic technique using counters and the economic organization of the rural communities. As a matter of fact the phase of the remarkable persistence of this technique exemplified by the similarity of the shapes of the counters over long periods of time coincides exactly with the period between the transition to farming and animal husbandry and the disintegration of the tribal organization in the early urban civilizations of the 4th millennium. Even at its most primitive level agriculture requires more long-range economic planning than the economic structure of the preceding cultures of hunters and gatherers. The seasonal nature of agriculture required the storage of foodstuffs. It also demanded foregoing the consumption of a certain amount of the harvest set aside as seeds for the next season. Furthermore, in order to avoid losses the number of domestic animals had to be kept under control at all times. Mastering these tasks was no doubt facilitated by making use of arithmetical techniques, although the problems always remained within a narrow, constant framework defined by natural circumstances.

Presumably, the technique of using the counters simply consisted in representing the resources to be managed by a corresponding number of counters and in simulating all changes with regard to the resources by using these counters. The different shapes of the counters then simply corresponded to different goods. Thus, the technique seems to amount more to some form of *mechanical bookkeeping* than a numerical registration. In contrast to counting with number words, the operations with counters in such a technique are already based on additive numerical relationships. However, these numerical relationships remain implicit, for operating with the counters is after all merely a simulation of real processes and modifications of resources.

Tools and Units of Measurement

The number of animals could be controlled directly by using counters. However, numerous agricultural products are not divided naturally into such distinct units so as to permit without difficulty a structural correspondence with regard to the counters. In these cases a measuring tool and the determination of suitable units as standards of weights and measures was needed.

In the development of the tools of measurement the administration of grain, the most important agricultural product of the early rural communities of Mesopotamia, played an eminent part. Inferences as to the original meaning of capacity measures are still possible in part: "sìla" (Akkadian *"qa"*)[36] was the amount of grain constituting a daily ration, "dug" means "jug" and thus indicates an easily managed tool of measurement. Field area measures as measurements for sowing and administration were derived from these grain capacity measures. Thus, "sìla" did not only indicate the daily ration, but sometimes the amount of land which could be seeded with 1 sìla as well. Evidence also exists for the use of the grain capacity measure *"imeru"* (Akkadian) as a field area measure. Because of the etymological association with the word for "donkey" it has been speculated that this measure originally designated an area that could be cultivated with a donkey in one day or which could be sown with a load of seed a donkey could carry.

The meaning of some length measures also indicates the cultivation of fields. Thus, the unit "ninda"—a length unit of about six meters—and in particular the unit "gi" (reed) can be interpreted as designations of measuring instruments. By contrast other length measures were probably used in the context of rural crafts: "šu-si" the width of a finger, "kùš" the cubit. In addition, a measuring rod found in Nippur exhibits measuring units for the width of a hand of four fingers and for a measure of bricks which served for standardizing the sizes of bricks in the manufacture of clay bricks.[37]

The measures "danna" and "gú" also provide information on the evolution of standards of measurement in the context of the objective work processes and contexts of action. "gú" was a weight of 30 kilograms; that is approximately the load an adult human being is capable of carrying. "danna" was a length measure as well as a measure of time, a combination which suggests a measure for travelling on foot—comparable to an "hour's walk."

As a length measure 1 danna corresponded to the approximate distance of ten kilometers and as a measure of time to two hours.

These few examples cannot replace a systematic study of the tools of measurement. However, they already provide information with regard to an essential component of their development. The evolution of tools of measurement presupposed neither the conceptualization of the quantities to be measured nor a developed number concept. *The early standards of measurement did not yet represent context-independent dimensions of reality with an internal arithmetical structure.* An hour's walk and the width of a finger were not yet related arithmetically to each other. Their definition as length measures is the result of later developments. On the other hand, measurements were dependent on and related to each other in ways determined by actual work processes and contexts of action in which they occurred, but which were at odds with their subsequent integration into context-independent dimensions, for example, the close relationship between grain capacity and field area measures.

The technique using counters was adequate for maintaining control of the reiteration and subdivision of context-dependent measuring units within the numerically limited scope of practical work processes as well as for recording the results of the measuring processes based on these operations. The logico-arithmetical structure of the ranges of measurement created in this way has been aptly called the "core" of a measurement system. Such a core consisted of one or at most a few measuring units surrounded by a host of elementary integral multiples and fractions.

The Technique Using Counters as an Administrative Tool in the Early Temple Economy
Throughout the 4th millennium B.C. the archaeological findings indicate a remarkable shift in the technique of using counters. With regard to the oldest finds (9th millennium) approximately 20 different kinds of counters can be identified, and this number of different types remained fairly constant throughout the millennia. In the 4th millennium, however, the number of the types of counters in use rose sharply. For the end of this millennium a statistic[38] of 363 specimens from six different excavations documents 15 main categories and approximately 250 subcategories. This diversity is due in particular to a differentiation of the forms and to a larger diversity of incised signs and patterns. The fact that several categories now occur in two different sizes is also significant. Finally, the fact that numerous counters— about 30 percent in the above statistic—have now been pierced is also new.

The proliferation of forms indicates a considerable functional differentiation. A group of objects, the oldest specimens of which must be dated back to the end of the 4th millennium, furnishes remarkable evidence that, in addition, the counters acquired new functions: closed clay spheres containing a number of counters in their inaccessible interior. Many of these clay spheres, of which about 350 have been found to date, bear seal impressions on the outside leaving no doubt that the clay spheres were used for the tamperproof safekeeping of counters documenting economic transactions.

These archaeological findings document that, starting in the 4th millennium, the possibilities inherent in the technique using counters were used to a larger extent. Chronologically and in terms of location the expansion of this technique coincides with the development of urban centers. The reasons for the expanded use are therefore obvious.

In the rural communities the scope of the problems of economic organization, determined by the periodic uniformity of simple reproduction, was quantitatively limited. The evolution of urban centers,[39] on the other hand, was premised on the production of an annual surplus. The basis for this was created by the transition from rain-fed to irrigated agriculture. The administration of this surplus as well as the consequences of producing it—specialization of skills, the expansion of trade, especially of long-distance trade, transition to mass production, and centralized deployment of labor for the purpose of constructing and maintaining the irrigation system—*raised novel problems of economic organization that exceeded the quantitative framework of the technique using counters of the rural economy.*

At first, the decision-making power concerning collective work was vested in the institutions of the tribal communities. However, towards the end of the 4th millennium forms of the concentration of political and economic power in the hands of individual rulers are increasingly to be found. In the urban centers the first large temples were erected. Their priests administered the surplus of the expanding economy and organized the collective deployment of labor.

It is fairly certain that it was the priests of these temples who developed the archaic technique using counters into a comprehensive instrument for recording and controlling economic processes. The new variety of forms exhibited by the counters of this late phase of their employment corresponds to the fact that goods for consumption and trade were accumulated in the storehouses of the temples in such a variety and at such a rate unknown in rural communities.

The administration of the temple was charged with the collection of tribute and the economical allocation of the resources of the storehouses. Numerous craftsmen were directly associated with the temple and were compensated directly from the supplies of the temple. In the case of the collective deployment of workers the temple assumed the responsibility of compensating the workers. Thus, the temple became the center of a large-scale economy that had to be administered without the existence of a universal currency. All economic transactions had to be accomplished by means of barter.

This helps to clarify some of the functions which the technique using counters had to accomplish in the temple economy. While accounting for the inventory and the economic transactions did not necessitate a new principle, the larger quantities demanded a more rational representation of larger numbers. The way in which the first numerals were subsequently formed suggests that the use of counters of two different sizes served this purpose by specifying that a certain number of smaller counters corresponded to one larger counter. Presumably the holes drilled through many of the counters served the purpose of tying together a large number of counters. Finally, the innovative forms of the tamperproof safekeeping of counters are explained by the fact that the various impersonal contractual relationships of a large-scale economy not only necessitated the recording but also the authentication of economic transactions. Thus, it has been assumed, for example, that in case of a routine shipment by boat a sealed clay sphere with counters accompanying the shipment was intended to reassure the recipient that the type and amount of the goods delivered agreed with the shipment dispatched by the sender.

From Counters to Writing
Around the turn of the 4th to the 3rd century B.C.[40] one of the oldest writing systems known to date emerged in Mesopotamia, an ideographic script, the signs of which were later given in part a new meaning as characters representing syllables. As early as the 3rd millennium this system of writing was simplified to form the so-called cuneiform writing which became the universal writing system of the entire Near East and served to render numerous languages in written form.

The oldest texts[41] are clay tablets with clearly distinguishable "numerals" and pictographs. Larger tablets are divided into "cases" by means of straight lines, each containing one notation.

Originally, the number of the signs in use amounted to approximately 2000. However, in the course of the evolution of cuneiform this number was reduced considerably. Among the pictographs there are naturalistic representations of animals, plants, containers, etc., as well as abstract symbols without any immediately obvious reference to real objects. In the oldest texts the "numerals" are evenly arranged, while the pictographs are distributed irregularly over the available space. It was not until their use as syllabic characters or determinatives—that is, signs that, placed before or after a term, restrict the meaning of that term to a specific domain—that these signs were arranged sequentially according to the spoken language. Originally, this sequence was arranged from top to bottom, later from left to right.

According to D. Schmandt-Besserat, there are close correlations between the counters and the written tablets composed in ideographic writing. She identifies the "numerals" and some abstract signs among the pictographs as representations of counters. In her opinion, the round and conical counters that were used in two different sizes correspond to the circular and conical markings of the "numerals" that were pressed into the clay vertically or obliquely with two round styluses of different diameters. A round disk-shaped counter with the incision of a cross corresponds to the circle with the incision of a cross identified as the sign for "sheep." The sign for "oil" is the acute sector of a circle with an added chord, probably depicting a small container for the oil. A counter in the shape of an ovoid with a circular incision at the thick end corresponds to it so that the profile matches the sign. All in all, counters for 37 signs have been identified by D. Schmandt-Besserat in this way. Another 43 signs are considered as differing only in details from corresponding counters.[42]

Some of the clay spheres bearing markings on the surface also document a close connection between counters and the written numerals. Their exterior not only exhibits seal impressions, but also information on the number and type of the counters inside. For this purpose the counters were sometimes pressed one by one into the surface before the sphere was closed. In this way their shapes were outlined as markings capable of providing information on the contents of the sphere.[43] The connection between the "numerals" and the pictographs of early writing is even more evident in those cases where rough markings were pressed into the surface with a stylus according to the number of the counters inside.

The connection between the counters and the early written characters served as the basis for a theory of the development of this system of writing. According to this theory a differentiation of the counters took place in the late phase of their use into those indicating quantity and those merely giving a qualitative indication of the type of operation or the goods in question. The system of counters was developed into an inclusive information and documentation system during the second half of the 4th millennium, and the ideographic writing superseding it is the result of transferring this system into a medium of representation which was employed in the same way, although it was easier to use.

The fact that, in particular, the oldest of the recovered writing tablets, the texts from the levels Uruk IV and Uruk III and those from Jemdet Nasr, are to a great extent "economic texts" of the kind described, neatly fits into the picture presented by this theory. Only a small group of texts, in particular somewhat more recent texts from the archives of the city of Shuruppak,[44] are of a different type. They document an increasingly divergent development of the two types of signs resulting from the differentiation of the counters into those for the quantitative and those for the qualitative classification, that is to say, the "numerals" and the pictographs. These texts are lists of signs and sign combinations arranged according to their meaning such as lists of species of fishes and birds, domestic animals, plant species, names of vocations, etc. Presumably, these lists were originally used for teaching the meaning of the signs in the temple schools. However, their canonical fixation actually amounted to a classification of existing knowledge.

The decisive step from pictograph to written character was finally taken when these signs were used as syllabic characters. The desire to represent ideal subjects as well presumably served as the motive for this step. In the context of such an endeavor a writing tablet from Shuruppak with 300 names of gods is striking testimony to the early use of pictographs as syllabic characters.

Context-Dependent Numerals

The "Numerals" of the Archaic Economic Texts

The functional equivalence of the constructive-additive representation of numbers and operations with objective calculating aids was demonstrated using the example of ancient Egyptian arithmetic. Just as with counters, adding and subtracting can also be done mechanically with numerals. In addition, the history of Mesopotamian calculating aids provides an example of an actual developmental link between the technique using counters and the constructive-additive representation of numbers. Pressing or drawing the counters into clay *necessarily leads to the development of a system of numerals with a constructive-additive structure.*

The number of the individual signs making up the "numerals" of the archaic economic texts is extraordinarily large. The texts from Uruk contain approximately 35 different numerical signs. However, only five of these have been positively identified in the somewhat more recent texts from Shuruppak,[45] which in turn contain around 30 numerical signs not evident in Uruk. Since many of these "numerals" occur only infrequently, their exact meaning can for the time being hardly be determined, especially in light of the fact that the rest of the content of the texts has only been partially deciphered. In these texts the five numerical signs common to the texts of Uruk and Shuruppak occur far more frequently than the remaining signs. They occur in later texts as well. These are the signs that have already been described as representations of counters which were pressed into the clay vertically or obliquely with two round styluses of different diameters.

Sign a: small conical marking created with a round stylus slanting to the right
Sign b: small round marking created with a vertically-held stylus
Sign A: large conical marking
Sign B: large round marking
Sign C: ligature of a conical and a round marking

The possibilities of gaining information on the meaning of these signs will be explained with the help of an example. Among the writing tablets of Uruk there is one tablet[46] the obverse of which is divided into six cases; in addition to a few pictographs it contains the following "numerals":

Obverse: aaaaaa
aaaaa
aaaaaaa
aaaaa
aa
a
Reverse: bbaaaaaa

The meaning of the pictographs to be found in the six cases of the obverse in addition to the numerals can be inferred from the subsequent development of writing. They are symbols for different types of sheep and goats. There can be no doubt that the writing tablet is a document about a herd of animals.

It is reasonable to assume that the "numerals" on the reverse indicate the sum of the numerals of the obverse, especially since other texts from Uruk have a similar structure. Since the sign a occurs 26 times on the obverse, it follows that the value of sign b must be ten times that of sign a. Since these two signs are indeed the most frequent "numerals" on the writing tablets of

Uruk, it must be assumed that they represent the numbers 1 and 10. The subsequent use of these signs is consistent with this assumption. Thus, concerning its numerical meaning, the written tablet does not leave any questions unanswered.

There are numerous indications in the archaic texts permitting the interpretation of some "numerals." "Numerals" of a higher value are written to the left of those of a lower value. If the value of a sign is n times that of the sign with a lower value next in line, then the latter can occur at most n–1 times within a sign combination. Each identifiable sum allows pretty far-reaching inferences as to the relative quantitative relations between the "numerals" in question. It is fair to assume that the meaning of those signs still being used at a later time essentially had not changed. If all these indications gleaned from the more than 1000 archaic texts and text fragments are systematically evaluated, then the abundant evidence of the quantitative relations between the five most frequent "numerals" suggests that the five signs represent the basic numbers of a sexagesimally arranged numerical system. The following table provides a summary of the quantitative relations of this system:

B = 6 C	B = 60 A	B = 3600
C = 10 A	C = 60 b	C = 600
A = 6 b	A = 60 a	A = 60
b = 10 a		b = 10
		a = 1

This interpretation is nevertheless false, or at least incomplete. It is true that most of the "numerals" contained in the archaic written tablets are in accordance with the assumption of such a numerical system or, more precisely, they do not contradict it. However, first, such a system would not adequately accommodate the many other "numerals" of the tablets and, second, there are incontrovertible counter examples to this interpretation.

A tablet from Jemdet Nasr[47] is particularly revealing. Its "numerals" will be reproduced below (cf. Fig. 2). Unfortunately, a piece of the tablet has broken off so that it is not clear whether it might not have contained other signs as well. (The sign z represents one of the rarely used "numerals," a ligature of the sign b and two horizontal lines.)

```
Obverse:  BBbbbbbb     bbbbbb
          zz           bbbbbbbb
          b            bbbzzz
          Bbbbbbbbbb   BBb
          Bbbbbb       ...?
          bbbbbbbb
          bbb...?
Reverse:  AAABB
```

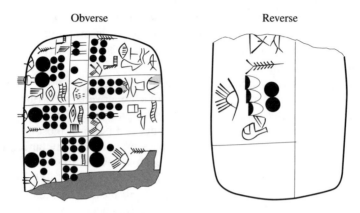

Figure 2: Archaic economic text from Jemdet Nasr. (Drawing by Robert K. Englund.)

"Numerals": ● = Sign b   ◗ = Sign z
            ⬤ = Sign B   ▶ = Sign A

The arrangement of the signs A and B on the reverse and the ninefold occurrence of sign b unquestionably permit the assumption of the relationship A > B > 9 b which is entirely contrary to the assumption presented above concerning the meaning of the "numerals."

Numerous other written tablets corroborate the fact illustrated by this text *that there are uses of the "numerals" that lead to incompatible interpretations of these signs.* These examples have been considered as evidence for the existence of a second system of "numerals" with a modified sequence of the signs and modified quantitative relations between them, a system that is always used with quantities of grain and that represents the following quantitative relations:[48]

$$A = 3\ B\ (?) \qquad A = 300\ (?)$$
$$B = 10\ b \qquad\quad B = 100$$
$$b = 10\ a \qquad\quad b = 10$$
$$\qquad\qquad\qquad\quad a = 1$$

(Addendum to the present edition: These values are incorrect. Recent research has provided overwhelming evidence for the relation b = 6 a instead of b = 10 a. Hence, the correct values are A = 180, B = 60, b = 6, and a = 1. This correction supports my following argument even more. See Chap. 11 and 13.)

For the moment let us simply state that the "numerals" of the archaic texts do not represent numbers in our modern sense, for they do not have a context-independent meaning; their arithmetical function depends on the context in which they are used. In this context the "numerals" apparently are nothing more than substitutions for the flexible use of a technique of representing the qualitative differences of certain objects and quantities of goods using counters. Thus, the same numerical relationship can, on the one hand, be represented by different "numerals" in different contexts, and, on the other hand, the same "numeral" can represent different numerical relations in different contexts.

The "Table of Squares" from Shuruppak

More extensive evidence of the ambiguity of the use of the "numerals" in the archaic texts is provided by the somewhat more recent texts from Shuruppak. They contain more "written" text than the tablets from Uruk and Jemdet Nasr and have largely been deciphered so that the function of their numerals can be better reconstructed. The following table[49] containing three columns is of particular importance in this regard—it probably is the oldest table text found so far (cf. Fig. 3). In addition to the five signs of the older texts discussed so far a further sign c is used in this text, composed of the sign b and four pairs of parallel, crossed, cuneiform markings.

Obverse:	C GAR.DU sag	C	DI	... (?)
	AAAAAAAAA	AAAAAAAAA	DI	BBccccbb
	AAAAAAAA	AAAAAAAA	DI	BBbbbbbbbb
	AAAAAAA	AAAAAAA	DI	Bcccbbbbbbbb
	AAAAAA	AAAAAA	DI	Bcbb
	AAAAA	AAAAA	DI	ccccc
	AAAA	AAAA	DI	cccbb
	AAA	AAA	DI	cbbbbbbbb
	AA	AA	DI	bbbbbbbb
	A	A	DI	bb
Reverse:	bbbbb	bbbbb	DI	bCa
	bbbb	bbbb	DI	CCaaaa
	bbb	bbb	DI	Caaa
	bb	bb	DI	... (?)
	b	... (?)		
	aaaaa			
	... (?)			

(remainder broken off)

There can be no doubt about the formal structure of the table in light of the rapid increase of the numbers in the third column and the formative rule to be inferred from the last two lines of the obverse:

AA	AA	DI bbbbbbbb
A	A	DI bb

A quadrupling in the third column corresponds to a duplication in the first and second columns. If this is a general rule, then it follows that the third column is a sequence of squares generated by multiplying the first two columns.

Given this structural context, the partially damaged lines (with the exception of the last one) and thus all quantitative relations between the "numerals" used (with the exception of the sign a in the last existing line) can be consistently reconstructed:

bb	bb	DI aaaa
b	b	DI a

First and second columns:
$$C = 10\ A$$
$$A = 6\ b$$

Third column:
$$B = 6\ c$$
$$c = 10\ b$$
$$b = 3\ C$$
$$C = 6\ a$$

Figure 3: Obverse of the "table of squares" from Shuruppak.

"Numerals": ○ = Sign B   ▷ = Sign A
             ○ = Sign b   ※ = Sign c

(Photo: Courtesy of the Vorderasiatisches Museum, Berlin.)

If the absolute numerical value for the "numerals" of the first and second columns is assumed to be b = 1 and for the "numerals" of the third column a = 1, then the text proves to be a table of squares:

60 GAR.DU sag	60	DI	3600
54	54	DI	2916
48	48	DI	2304
42	42	DI	1764
36	36	DI	1296
30	30	DI	900
24	24	DI	576
18	18	DI	324
12	12	DI	144
6	6	DI	36
5	5	DI	25
4	4	DI	16
3	3	DI	9
2	2	DI	4
1	1	DI	1
?	?	DI	?

The Metrological Meaning of the Table

However, a number of objections may be raised with regard to this reconstruction. First, the fact is neglected that the signs in the first and third columns are used with different meanings. Second, none of the signs can logically be claimed as a unit. If, for instance, in the first column the value of 1 were attributed not to the sign b, but (in accordance with the examples discussed so far) to the sign a and the value of 10 to the sign b instead, then the same sign a in the third column would have the value of 100. Thus far it can only be said that the quantitative relations between the signs of the first column correspond to the schema most frequently encountered in the Uruk texts. The absolute values of the signs can, however, only be determined arbitrarily and the quantitative relations in the third column define a new schema that does not correspond to either of the schemata reconstructed from the texts from Uruk and Jemdet Nasr.

The meaning of the table is revealed by the first line and by the quantitative relations between the signs. ^{ninda}ninda$_x$ (signs GAR and DU) is an older form of the length measure ninda (sign GAR) corresponding to approximately six meters.[50] According to this, the first two columns represent quantities based on the length unit ninda. Since the end of the first line has been damaged, it does not yield a corresponding unit of measurement for the third column. In this case the fact is helpful that the quantitative relations between the signs of the third column happen to correspond to the quantitative relations between area measures known from later texts:

"Table of squares": B = 6 c   Area measures: 1 šár = 6 bur'u
                     c = 10 b   1 bur'u= 10 bùr
                     b = 3 C   1 bùr = 3 èše
                     C = 6 a   1 èše = 6 iku

Finally, the quantitative relation between length and area measures is also known from later texts: A square its sides being 10 ninda has the area of 1 iku. The next to last line of the table corresponds to this. If, in addition, the well-known structure of the length measures (10 ninda = 1 éš, 6 éš = 1 UŠ) is taken into consideration, then the following interpretation follows from the "numerals" of the table:

First and second columns: Third column:
   C = 10 UŠ   B = 1 šár
   A = 1 UŠ   c = 1 bur'u
   b = 1 éš   b = 1 bùr
   a = 1 ninda (= GAR)   C = 1 èše
                                     a = 1 iku

The interpretation of the first line provided by Deimel is consistent with this interpretation as well.[51] The size of fields was determined either by means of seed or area measures or the lengths of their sides. The sides were, in accordance with our concepts "length" and "width," indicated by the characters "sag" meaning "head" and "UŠ"; "UŠ" in this case did not designate the particular length unit, but the concept "length" (or better "long side") in general. If the sides of a field were of different lengths, all four of them were given. If, however, the opposite sides were of equal lengths, the sign "DI" was added to the "sag" or the "UŠ." Thus, the meaning of "DI" is "to be equal," and Deimel therefore translates the first line of the table as follows:

600 GAR.DU the head, the other sides likewise 600 (GAR.DU);
therefore the area of the field is (200 bùr).

Finally, the last line of the table of which only the sign of the first column is not destroyed is revealing. According to the regular sequence according to which the signs of the first column have been arranged, this place should be occupied by the sign combination for 9 ninda (GAR). However, the line begins with the sign combination for 5 ninda (GAR) instead.

This deviation can be explained if the fact is taken into account that, in the texts from Shuruppak, the smallest unit for measuring areas is the unit

THE DEVELOPMENT OF ARITHMETICAL THINKING   227

iku. However, according to the available reconstruction the next to last line already reads:

> 10 GAR.DU the head ... therefore the area of the field is 1 iku.

Thus, in order to continue the third column, "numerals" are needed that are not the result of a multiplication, but the result of a division of the measuring unit. In the texts from Shuruppak two such signs can indeed be discerned, the signs f and g with the following relationship to the sign a for 1 iku:

$$a = 2\,g$$
$$g = 2\,f$$

Thus, the function of the signs f and g is that of individual signs for the fractions $\frac{1}{4}$ and $\frac{1}{2}$. It must, however, be kept in mind that, contrary to such individual signs for abstract fractions, the meaning of the signs is dependent on the context. This is especially true of the sign g to which the meaning $\frac{1}{2}$ would have to be assigned. In other contexts a much more frequently occurring sign e is used for $\frac{1}{2}$. In conjunction with the grain capacity measure gur both signs have a different meaning with the sign e even occurring with at least two more different meanings (a = 2 e = 4 g = 48 e = 1 gur maḫ).[52] The less frequent sign g which has the meaning $\frac{1}{2}$ only in conjunction with the area measure iku is, in the texts from Shuruppak, also used for designating a single donkey, if donkeys are generally counted in terms of teams of four (sign a).[53] In this case it replaces the sign a otherwise used for the 1 and, depending on the way of looking at it, it would have to be accorded either the value 1 or the value $\frac{1}{4}$. However, in a particular context the meaning of the signs is always clear. Thus, from the overall context of the texts from Shuruppak it may be concluded that two, but no more than two, "numerals" were available for dividing the area unit iku—that is, the signs g and f for the process of dividing iku in half twice. In this case, however, the "table of squares" could only be continued for one more line by dividing the length of the side in half and the area in fourths. This explains the jump from 10 ninda to 5 ninda with regard to the length of the side in the transition to the last existing line, and at the same time it can be concluded that the missing part did not contain any other lines and that the partially damaged last line originally contained only the additional signs aaaaa and f.

Numerals and Measuring Tools

The "table of squares" from Shuruppak is an example revealing conditions which shaped the arithmetic in Mesopotamia following the transition from the technique employing counters to the use of "numerals." The basis was the differentiation of the sign function into a function of quantitative and a function of qualitative designation. This differentiation of sign functions did, however, not lead to a complete separation, for the quantitative designations of signs representing different measures could not be identified with each other due to the incongruent arithmetical structures of the different measuring tools. Even though "numerals" and "characters" were clearly distinguished from the very beginning, the meaning of the "numerals" remained dependent on the objects of quantitative determination. The various measuring tools constituted the points of departure for the development of different "systems of numerals," that is, numerical sign systems, each with a specific arithmetical structure determined by the measuring tools.

Three systems of "numerals"[54] can be identified in the texts from Shuruppak which recur in all economic texts up to and including the Old Babylonian period.

– Sexagesimally arranged *"counting numbers"* and signs for simple fractions which were used for cardinal numbers of finite sets and for some measuring units, for example, for lengths;
– *"area numbers"* arranged in a similar way as the field area measures;
– *"grain numbers"* structured according to the capacity measures for grain which were applied to other agricultural products as well.

In addition to these extensive sign systems there were countless variations for specific applications of measurement, local variations, and special signs for particular purposes, and apart from that a flexible adjustment of the use of the "numerals" to the prevailing arithmetical conditions of specific applications. Despite the otherwise striking structural agreement the system of the "numerals" fundamentally differs from the preceding system of counters. For the written notation made it possible to combine the core measures, which had developed in the process of arithmetical operations using the measuring tools, into comprehensive metrological systems. Now the core measures are no longer isolated arithmetical units. Rather, they have been linked in a variety of ways by means of metrological tables within the types of quantities as well as between the types of quantities. *The metrological table identifies multiples or fractions of standards of measurement of different core measures with each other.* In this fashion numerical relations between the different standards of measurement are established by means of which they are redefined from natural measures as elements of comprehensive systems.

Thus, there are, for example, metrological tables,[55] which relate the "width of a finger" šu-si to an "hour's walk" danna although there is no longer any direct empirical experience that corresponds to the relationship between these measures. Now, they are merely related to each other by a metrological-arithmetical standardization of otherwise only imprecisely determined natural measures. With regard to the area measures the "table of squares" from Shuruppak covers a similar range of measurement, and beyond that it documents the identification of relationships between different types of quantities: According to the conditions given in the initial line of the tablet the length 10 ninda corresponds to the area 1 iku.

The Function of the Sexagesimal Structure of Measurement Systems
The relative inexactness of the natural measures afforded a certain leeway in redefining them in the process of integrating them into a system of measurements. The best-known example is the difference between the ancient Egyptian and the Mesopotamian cubit: The former contains 28, the latter 30 widths of a finger. However, even within the decentralized civilization of Mesopotamian city states this led to a situation where, during a period of transition, quantitative relations between measuring units used at different times or in different places often differed considerably from each other— resulting in a corresponding ambiguity of "numerals."

For example, considering the quantitative relations between "numerals" for grain in the archaic texts from Uruk and Jemdet Nasr one might expect a corresponding subdivision of the grain capacity measure gur in the texts of subsequent periods. However, according to Deimel four different units are defined as gur in the texts from Shuruppak, among them the gur maḫ with 288 sìla, the rest with 36, 72, and 144 sìla.[56] In the Early Dynastic economic texts from Lagash a gur sag+gál with 144 sìla is prevalent.[57] In texts from the IIIrd Dynasty of Ur the gur sag+gál with 300 sìla is used.[58] And in conjunction with these changes of the relationships between the measuring units the meaning of the corresponding "grain numbers" changed as well.

In view of this flexibility of the relationships between the "numerals" it makes sense to consider the efficiency of certain structures, since it is to be assumed that, in the long run, an efficient structure will prevail over an inefficient one. The examples discussed so far have already shown that, in terms of establishing relationships between the "numerals," the sexagesimal structure of the following form was favored:

$$x = 6\,y$$
$$y = 10\,z$$

Indeed, as a relationship between core measures this relationship is especially advantageous, for it relates most simple fractions of a core measure with integral multiples of others: one half, one third, one fourth, one fifth, and one sixth of x, for instance, are integral multiples of z.

There have been numerous speculations regarding the question of how the Old Babylonian numerical system got its sexagesimal structure which differs from the usual decimal structure.[59] In general, the underlying assumption is that the development of a uniform number concept preceded its construction. The present study makes the problematic nature of such an assumption evident. For it is precisely the context dependency of the meaning of the "numerals" over a very long period of time which permitted a structure gradually to take hold that was favorable with regard to the problems of fractions—problems, which the Egyptian arithmetic, after all, failed to solve. Contrary to all theories which seek the reasons for the curious sexagesimal structure of the arithmetic in Mesopotamia in particular conditions defined by local, temporal, or material circumstances it must be stated that, over a period of 1000 years, the sexagesimal structure of the "numerals" merely constituted one structure among many others.

The Relativity of the Unit

From the vantage point of our modern number concept the question arises why the "counting numbers" were not generally applied. Many scholars who interpret the "numerals" merely as representations of a number concept preceding them believe that the "counting numbers" represent the true numerals while all others are merely particular representations of these numbers based on traditional habits.[60] However, such an assumption is already undermined by the fact that, up to the change of the numerical notations at the beginning of the Old Babylonian period, the alleged exceptional role of the counting numbers is not supported by the sources. In addition, according to the interpretation of the sources offered here, the reasons are obvious why the "counting numbers" failed to be generally applied. If it is true that they acquired their particular scaling not as "counting numbers," but, like the "area numbers" and the "grain numbers," from their function as a means for integrating core measures, then the significance of the most important difference between the "counting numbers" and the remaining numerals vanishes: the absolute value of the number one. Hence, the "numeral" for the number one indicates no more than a relative beginning, determined by the exceptional role of a particular tool of measurement, and below the number one the counting sequence can be continued by means of "numerals" for fractions. Thus, in particular the possibility of a strict distinction between fractions and integers is eliminated.

There surely were numerous attempts to standardize the "numerals." This is, for instance, attested to by the introduction of a smallest measure, the še, as $\frac{1}{180}$ of the most important measuring unit under consideration. The term "še" means "grain" or more specifically "barley." As the designation for a single grain of barley it was presumably considered the smallest conceivable "grain measure." When silver increasingly replaced grain as a currency, it endured as a weight measure which was also a unit of value. Later on, as a length measure, it became the one-hundred eightieth part of a cubit (kùš) and, as an area measure, the one-hundred eightieth part of the šár, the square of the length unit ninda.

However, all attempts to standardize the various systems of "numerals" were bound to fail due to the problem of the quadratic relation between "counting numbers" and "area numbers," as long as no better solution had been found for the problem of calculating with fractions than the flexible determination of the meaning of the number "one." Given the conditions of the Mesopotamian "systems of numerals" and their practical use, the precondition for the introduction of a uniform system of numerals was—as odd as this might seem—a uniform solution of the arithmetical problems of working with fractions.

Arithmetical Techniques in the Economic Texts of the Early Dynastic Period
During the entire Early Dynastic period the application of the "systems of numerals" remained limited to problems associated with the organization of the economy. The multiplicity of the "numerals" matches the decentralized structure of the city states of Mesopotamia during this period. In the economic centers of the cities the "numerals" were used according to the particular local conditions. The close association of the arithmetical structures of the "systems of numerals" with the specific use of the tools of measurement prevented these "systems of numerals" from developing into unified symbolic systems, as long as these systems were nothing more than means of economic organization. This function gave rise to them and the arithmetical techniques associated with them were largely sufficient to meet the demands of this function.

Two typical examples may demonstrate the way in which the "numerals" were used. Both examples come from Old Sumerian economic texts from the city of Girsu, the center of the city state of Lagash. Since the later texts, in contrast to the archaic economic texts, in particular also those from Girsu, often bear the date of the reign of the respective king, the age of these texts is quite well known: They date back to the reigns of the city governors or kings Enentarzid, Lugalanda, and Urukagina. They offer insights into the economic organization of the temple of the goddess Baba administered by the wife of the city governor over a period of about 20 years, between 2374 and 2355 B.C. to be precise.

The first example[61] is a record of the tribute of fish paid by marine fishermen (cf. Fig. 4). The text is as follows:

    25 tar fish,
    15 zúbud fish,
2400 Še-suḫur fish,
 960 dried gir fish,
 780 salted gir fish,
  90 turtles (?),
  10 sìla fish oil,
this is the tribute of fish,
 600 Še-suḫur fish,
 360 salted gir fish,
this is the LUL.GU tribute of the previous year,
the second (year),
(from) Nesag.
  55 tar fish
  15 zúbud fish
4680 Še-suḫur fish
 420 dried freshwater fish
 240 salted agargara fish
 130 turtles (?)
  14 sìla fish oil,
this is the tribute of fish,
 840 Še-suḫur fish,
 120 salted ubi fish,
this is the LUL.GU tribute of the previous year,
the second (year),
(from) Lugalšaglaltuk.
These are tributes of fish.
In the month of the feast of "Malt eating
of Ningirsu" by the marine fishermen
delivered.
Eniggal,
(the) inspector,
into the "store-house"
has brought them.
The fish is property
of Baba.
Šagšag,
the wife of Uru-KA-gina,
the king
of Lagaš.
Third (year).

THE DEVELOPMENT OF ARITHMETICAL THINKING

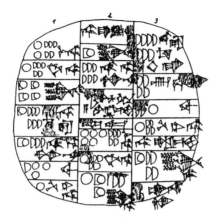

Figure 4: Early Dynastic record of the tributes of fish paid by marine fishermen; drawing by Förtsch ([14]; cf. p. 232).

"Numerals": ▷ = 1 (sign a)   ▷ = 600 (sign C)
             ○ = 10 (sign b)   ○ = 3600 (sign B)
             ▷ = 60 (sign A)
             ⌈ is a sign for a subtractive number representation having the meaning of "minus"

In this text the "numerals" are used solely for documenting the tribute paid. The text does not contain any arithmetical operations. "Counting numbers" are used which exhibit the usual sexagesimal structure—that is, signs for the numbers 1, 10, 60, 600, and 3600. In the case of the fish oil the meaning of the numerical sign is further specified by an added character for the measuring unit sìla, a capacity measure derived from grain capacity measures. Arithmetically, the text belongs to the simpler texts and is only remarkable because of the large numbers it contains.

The second example[62] is a record of amounts of grain distributed for sowing. Among the economic texts from Girsu it is one of the rare texts that contain detailed cost calculations. It reads as follows:

> 8 (bùr) 3 (iku) field area (for) barley,
> plowed by oxen
> (and) seeded.
> Barley (for) feeding (the) oxen during plowing: $24\frac{1}{2}$ gur sag+gál.
> Barley (for) feeding (the) oxen during seed (plowing): $12\frac{1}{4}$ (gur).

> Seed barley: $12\frac{1}{4}$ (gur).
> Bala(?) barley: $1\frac{1}{2}$ (gur).
> 2 (bùr) field area, seeded.
> Seed barley: 3 (gur).
> Barley (for) feeding (the) oxen (during plowing): 3 (gur)
> Total: 1 (bùr'u) 3 (iku) field area (for) barley.
>
> Barley: $56\frac{1}{2}$ gur sag+gál.
> (It is) "royal domain"
> (of the) field "da.UL$_4$.ka".
> 6 (bùr) field area (for) barley,
> plowed by oxen.
>
> Barley (for) feeding (the) oxen: 18 (gur).
> 1 (bùr) field area (was) cross plowed.
> Barley (for) feeding (the) oxen: 3 (gur).
> Seed barley: 9 (gur).
> Barley (for) feeding (the) oxen (during seed plowing): 9 (gur).
> Barley (for) feeding (the) oxen (in order to prepare an) onion (field): $\frac{1}{2}$ (gur).
> Total: $39\frac{1}{2}$ (gur) barley.
> (It is) "royal domain"
> (of the) field "DÙN.UH.ka".
> (The) fields are plowed (and) seed plowed
> (by) "Sagatuka"
> (the) plowing foreman.   Sixth (year).
> Eniggal,
> (the) inspector,
> (has) transferred him the (accounting) tablet.

In this second text the "numerals" serve to document the grain issued and, at the same time, as a means for calculating the total value of separate entries. The translation represents a somewhat inaccurate picture, since the abridged use of signs for syllables and words gives the entire text more the appearance of a table than that of an extensive written text. While all the "numerals" are integrated into the text, they are nevertheless immediately discernible in the rather short listing.

The modern number notation used in the translation for reasons of legibility is also misleading. The original does not contain consistent "numerals" or an absolute definition of units; for this reason the distinction of integers and fractions is lacking as well. "Area numbers" as well as "grain numbers" are used. Since these "numbers" are partly represented by identical signs, but differ with regard to their arithmetical structure, the same sign may denote different numbers within one and the same text.

For instance, in calculating the area of a field of barley of 10 bur 3 iku the following "numerals" are combined:

                    bbbbbbbbaaa
                    bb
            total: caaa

This calculation is based on the "area numbers" with the arithmetical relations which were discussed above in connection with the "table of squares" from Shuruppak:

$$c = 10\, b$$
$$b = 18\, a$$

On the other hand, in calculating the appropriate amount of barley of $56\frac{1}{2}$ gur sag+gál the "numerals" are used as follows:

                    bbaaaagg
                    baag
                    baag
                    agg
                    aaa
                    aaa
            total: bbbbbaaaaaagg

In this calculation the same "numerals" a and b are used as before; this time, however, they represent "grain numbers" with the following arithmetical relations:

$$b = 10\, a$$
$$a = 4\, g$$

It is precisely the specific adaptation of the structure of the constructive-additive system of the "numerals" to the structure of the tools of measurement that simplifies the arithmetical techniques to permit mechanical operations. Each intended meaning of the signs can be derived from the text because semantical indicators specify their meaning, namely, the sign gán indicating "field area" and the sign combination "gur sag+gál" indicating a particular standardization of the grain measures. All other specifications of units of measurement have only been added in the translation to facilitate reading. In the original they, just like the arithmetical relations, must be inferred from the context.

Thus, the constructive-additive "systems of numerals" constituted quite an adequate calculating aid for solving the additive problems associated with running the economy. However, how could problems involving multiplication be solved using these means? This leads us back once again to the "table of squares" from Shuruppak which is about 200 years older. One aspect associated with it has not yet been discussed: How was this table calculated? For the table does not relate length and area measures empirically. Rather, it relates them on the basis of a strict multiplicative relationship between the "counting numbers" of length measurement and the "area numbers" of area measurement.

Unfortunately, given the available sources, it will hardly be possible to answer this question. However, two things seem to be clear. First, the existence of a multiplicative arithmetical procedure does not follow from the existence of the table. It is quite conceivable that the quadratic relation is a result of counting length or area units—by making use of a model for the relationship between lengths and areas and by resorting to small units. Second, it is precisely the tabular presentation of the numerical relation which shows that multiplicative relations were not the subject of the arithmetical techniques of that time, for otherwise the tabular arrangement would not have been necessary. As a matter of fact, the archaic economic texts do not, as far as I know, contain *a single example of a genuine multiplication*.

This situation hardly changed during the entire Early Dynastic period. Evidence for this is provided in particular by the economic texts from Girsu, for, with regard to the use of the "numerals," the two examples presented above are typical for these texts. Of the 195 texts published by Förtsch and systematically edited by Bauer 140 for instance are consistent with the first example. The "numerals" they contain are used exclusively for the purpose of recording or for representing the result of counting objects that were listed individually. Another 46 texts are consistent with the second example. They contain sums computed by making use of the "numerals." All in all, these texts contain approximately 130 additions, most of them exceedingly simple. Of the nine remaining texts five[63] provide weak evidence at best concerning the observance of multiplicative relations; the "multiplications" they contain might just as well be interpreted as repeated additions. Thus, quantities per day are stated, for example; however, in terms of the sum the numbers refer to the entire month (of 30 days). Or the amount of barley per person is stated, but also a total amount that takes the number of persons into account. In one of the examples the equation "the value of 1 ma-na of wool equals the value $\frac{1}{2}$ gur sag+gál of barley" is technically used in the calculation by simply exchanging the relevant "numerals." (Addendum to the present edition: Such numerical relations are sometimes hidden in the text as, for instance, in the second text discussed here. The amounts of barley are implicitly based on simple standardized relations between field areas and barley amounts for seed and fodder.)

Thus, only four texts remain where an arithmetical problem results from multiplicative relations. In two cases[64] the area of fields is deduced from their dimensions—a problem that constitutes the basis for the "table of squares" from Shuruppak as well. In two other cases[65] the amount of barley per area unit iku is deduced from the amount of barley used for planting the fields.

The number of these examples is too small for drawing conclusions as to how the results were arrived at. However, it seems to be fairly certain that a genuine multiplicative arithmetical technique is lacking in these instances as well. The calculation of field areas seems to indicate first and foremost the technical importance of metrological tables. From a later period numerous tables are known in which the areas are given in the case of fields with irregular sides. The reduction of these "multiplications" to special techniques of repeated additions is conceivable as well. The arithmetical problem concerning the two "divisions" is more difficult. However, in both examples the results are not correct. Rather, the figures concerning the amounts of barley per area unit contained in the two texts are merely approximations which are less precise than the "numerals" would permit.[66] This suggests that a non-algorithmic additive method was used for determining the results.

In summing up it must be stated that, in the Early Dynastic economic texts, the arithmetical techniques are dominated by additive operations. Thus, they are entirely consistent with the possibilities that the constructive-additive "numerals" offer to the arithmetical techniques.

The Emergence of Sexagesimal Numbers

Reform of the Numerals and Standard Measures in the First Empires

With the emergence of the first empires[67] the demands on the arithmetical techniques changed. Even if the consequences were not at all dramatic at first, the indications can hardly be overlooked that basic changes became necessary. Thus, it is said that Sargon of Akkad, the ruler of the first state encompassing almost all of Mesopotamia, already tried to introduce a uniform—decimally structured—system of weights and measures.[68] This development became even more obvious in the empire of the IIIrd Dynasty of Ur, presumably the first empire with a tightly organized central administration. The economic texts of this period indicate that the entire economy was now under the direct control of the ruler.

The accumulated surplus made it possible to undertake projects of a scope previously unheard of. Among these are monumental structures like the tower of the temple (ziggurat) of the moon goddess Nanna in Ur and a wall of 26 danna (two hours' walk) as a protection against the raids of the nomadic groups of the Amorites as well as the expansion of agricultural acreage by constructing several large canals fed by the Tigris River.

The expansion and centralization of the economic system as well as the mammoth projects presented the state administration with new tasks that somehow had to be solved with the available means. Scribes of the IIIrd Dynasty of Ur formalized accounting procedures by means of a standardized "form" for recording receipts and expenditures. It has already been mentioned that, during the reign of King Sulgi, the system of grain capacity measures and consequently the arithmetic of "grain numbers" was simplified by replacing the measures designated as "gur," containing 36, 72, 144, or 288 sìla respectively, with the "gur-lugal"—the "royal gur"—which contained 300 sìla.[69] There was even a tendency to express the amounts of grain in terms of sìla and to use "counting numbers" for writing them down.[70]

However, the most important innovation in terms of the development of arithmetic was the *modification of the numerical notation system,* even though this modification did not have any direct consequences. From its early pictographic beginnings writing had evolved into "cuneiform"; in this process it fundamentally changed with regard to both its number of signs, which had shrunk to a few hundred signs, and the increasing abstractness of forms. Toward the end of the Early Dynastic period almost all signs were pressed into the clay as patterns by using a cuneiform stylus. Only the "numerals" continued to be generated with two round styluses—or possibly with the blunt ends of the styluses used for writing.

It is true that, as early as the *Early Dynastic period,* there were in fact attempts to reproduce the "numerals" by means of a cuneiform stylus so that two different kinds of "numerals" were used simultaneously; however, first, the "modern" forms varied considerably and, second, the round numerals continued to dominate. Rather than contributing to the standardization of the written form, the new cuneiform signs were even diversifying the numerical notations and increasing the traditional ambiguity. Thus, they were frequently used in a text in conjunction with the old numerals in order to express a different meaning with regard to the context of application. In the economic texts from Girsu, for instance, entries concerning feed grain are often written with old "grain numbers," while entries concerning the number of animals fed were expressed in terms of "modern" "counting numbers," thus facilitating accurate addition. Or in the case of the statement "$69\frac{1}{24}$ gur sag+gál 3 sìla barley" the signs for the larger unit gur were written using the old form of notation, those for the smaller unit sìla using the "modern" one.[71]

However, in the economic texts of the *IIIrd Dynasty of Ur* the old "numerals" are to be found only in exceptional cases. Even though nothing has changed at this point with regard to the intricate system of specific "numerals" used for particular objects, all signs are now written in a standardized cuneiform style. The cone-shaped oblique impression of the stylus (sign a) became the vertical "wedge," the small round impression (sign b) became a deep oblique impression of the cuneiform stylus, the so-called "Winkelhaken." The ligature of a cone-shaped and a round impression (sign C) was replaced by a ligature of wedge and "Winkelhaken." Other signs, for instance the rarely used large round impression (sign B), were simulated by means of several wedge-shaped impressions.

Later on a simplification resulting from the process of writing was to be of particular importance. It was no problem to simulate rarely used signs like the sign B—representing the number 3600 in the case of the "counting numbers"—with the small stylus. It would, however, not have made much sense to write a sign as frequently used as the sign A—the "counting number" 60—in such a complicated fashion. And since, in any case, the meaning of each "numeral" had to be determined from the context, it became customary to use the same "modern" sign, that is, the vertical wedge, for the signs a and A—the "counting numbers" 1 and 60—which differed only in terms of their size. It is, however, a mistake to confuse, as is often the case, this simplification with the introduction of the place-value system. During the IIIrd Dynasty of Ur when this simplification was introduced and systematically used the entire complexity of the context-dependent "numerals" actually remained intact.[72]

The Operative Denotation of Unit Fractions Associated with Weights for Silver

The weights are among the most recent of the Mesopotamian measures. They are not to be found in the archaic texts, since the only natural weight measure was the carrying load gú, the other weights presupposed the invention of the balance. Therefore, it must be assumed that the system of weights emerged only after the "counting numbers." This would help to explain a peculiarity: The system of weights is the only Mesopotamian system of measures that was strictly sexagesimal in structure from the very beginning. As was the case with other measures, the "grain" še as the "smallest unit" was added later:

$$1 \text{ gú} = 60 \text{ ma-na}$$
$$1 \text{ ma-na} = 60 \text{ gín}$$
$$1 \text{ gín} = 180 \text{ še}$$

When silver was introduced as a standard of value, the system of weights became increasingly important with consequences for the development of arithmetic. First, any goods—whatever the measure used to measure it and whatever the "numerical system" used for calculating—were, by dint of their value, related to the system of weights in linear proportion. The weights thus became a system of greater generality. Second, the natural limitations of the natural measures did not apply to the weights. The value of silver could be divided indefinitely both mentally and actually and thus encouraged the operative representation of fractions.

Such an operative representation of fractions is occasionally found already in the Early Dynastic economic texts in association with the weight unit gín, if it functions as a weight for silver and thus as a measure of value. This is illustrated by the following economic text from Girsu in Lagash:[73]

> 1 gín lal igi 4 gál of purified silver—it is silver of the bar-dúb-ba tribute—from Ursag, son of Amma;
> 1 gín lal igi 4 gál from Il;
> igi 4 gál from Urbaba, the shepherd of the billy goats;
> igi 4 gál from Urbaba, the curator;
> $\frac{1}{2}$ gín from Eta'e, son of Lugalurmu;
> igi 4 gál from Ursag, son of Lugalbad;
> igi 4 gál from Lugalsudduga, son of En;
> 1 gín from Lugalnamgusud.
> Total: 4 gín of purified silver.
> It is silver of the bar-dúb-ba tribute of the shepherds for the third year. AN-bad, the spokesman, made the payment.
> Eniggal, the inspector, weighed it in the month of ... (?) in the "house of the woman."
> They returned it to Baragnamtara, wife of Lugalanda, governor of Lagaš.
> 4th year

"lal" is a symbol for a subtractive number representation introduced as early as the Early Dynastic period for the purpose of abbreviation. "20 lal 1," for example, means "20 minus 1" and frequently represents the number 19 which would have to be written with 10 numerical signs otherwise. The total amount of 4 gín of silver is therefore obtained by $3\frac{1}{2}$ gín plus 2 times "igi 4 gál." Consequently, the formula "igi 4 gál" must indicate $\frac{1}{4}$ gín of silver, and "igi ... gál" is the symbol of an operator which, just like the Old Egyptian unit fraction operator, turns the sign for the number 4 into a sign for $\frac{1}{4}$. For indicating "$\frac{1}{2}$ gín," on the other hand, a particular sign for $\frac{1}{2}$ (sign e) in association with the sign for the measuring unit "gín" was used as usual. The text is also remarkable because of the fact that throughout "igi 4 gál" is used without stating the measuring unit "gín," just as if this established meaning of the expression were self-evident.

Many other instances could be cited as evidence that the operative form "igi ... gál," which diverges from the entire Mesopotamian tradition of emulating natural measures by forming "numerals" in a constructive-additive manner out of individual signs, was used to divide the unit gín of the weight for silver into 3, 4, or 6 small units long before it acquired its specific meaning following the invention of the so-called "table of reciprocals."[74] In addition, a list of weights[75] believed to be a "school text" should also be mentioned here (however, it can be placed into the period under consideration solely because of the style of writing). It runs the gamut from the very small measure of $\frac{1}{2}$ še to the 2.592 million fold measure of 2 gú. The list begins as follows:

$\frac{1}{2}$	(.?.)	še
1		še
$1\frac{1}{2}$		še
2		še
.		
.	etc.	
.		
28		še
29		še
igi 6 gál		
igi 4 gál		
$\frac{1}{3}$		gín
$\frac{1}{2}$		gín
$\frac{2}{3}$		gín
$\frac{5}{6}$		gín
1		gín
1 igi 6 gál		
1 igi 4 gál		
$1\frac{1}{3}$		gín
$1\frac{1}{2}$		gín
.	etc.	

This arrangement of measures is a typical example of a *metrological list*. It displays the characteristics of the use of the expression "igi ... gál." Its composition for the purpose of getting practice is definitely not to be understood purely as an exercise in writing. It highlights the difficulties of structuring an extensive domain of measurement in an arithmetically uniform and simple way with traditional "numerals" that had as yet been modernized only in terms of their written form.[76] It was presumably this difficulty that led to the introduction into arithmetic of an essentially new form for denoting fractions.

The Table of Reciprocals

At the beginning of the Old Babylonian period the invention of an arithmetical tool revolutionized all of arithmetic—that is, the invention of those tables known as tables of reciprocals. The oldest findings of such tables are said to date back to the period of the IIIrd Dynasty of Ur; with regard to the cases I am aware of the determination of this date is, however, uncertain.[77] At this time the new tool was most certainly not yet widely used, since the arithmetic of the economic texts from this period is essentially still based on the additive arithmetic of the Early Dynastic period.[78] The arithmetic of the mathematical texts of the Old Babylonian period, on the other hand, is inconceivable without the development initiated by the table of reciprocals.

Even though the historical development of the table of reciprocals cannot be reconstructed on the basis of findings which can be exactly dated, the history nevertheless provides a number of valuable clues. Originally the operator "igi ... gál" was merely an aid for the construction of new "numerals." It makes sense, however, to use this operator—once it is invented—also to denote those fractions which can actually be expressed already by means of the existing "numerals" as multiples of a smaller unit. In this case it is worthwhile to juxtapose the old and the new denotations in a table. Thus, a table is constructed in which the operator has a novel function. For now it relates a "numeral" that indicates the number of equal parts into which a unit is subdivided and which characterizes the sizes of these parts to another "numeral" that specifies the number of the smaller units which could also be used to represent the sizes of these parts generated by the subdivision. *The technical innovation for denoting fractions has turned into an arithmetical operation* which, due to the strict sexagesimal structure of weights, exhibits simple regularities that could not go unnoticed for long.

A transcription of what is presumably the oldest table of reciprocals found thus far[79] which precisely exhibits this structure will be reproduced below. In this transcription the obverse of the tablet which, although badly damaged, can be entirely reconstructed has already been restored. The modified function of the operator is reflected in a modified terminological form, in the present example in the expression "... igi ... ." Usually an expression is used which is even closer to the initial terminology—that is, the expression "igi ... gál-bi ... ."[80] For reasons of space and readability the

THE DEVELOPMENT OF ARITHMETICAL THINKING   243

"numerals" are transcribed here according to the modern conventions for a sexagesimal place-value representation introduced earlier. After each comma follows the number of the numerals of the next smaller sexagesimal unit. The sign "nu" to be found in the table denotes negation.

Obverse		Reverse	
2 igi 30	16 igi 3,45	29 igi nu	42 igi nu
3 igi 20	17 igi nu	30 igi 2	43 igi nu
4 igi 15	18 igi 3,20	31 igi nu	44 igi nu
5 igi 12	19 igi nu	32 igi 1,52,30	45 igi 1,20
6 igi 10 .	20 igi 3	33 igi nu	46 igi nu
7 igi nu	21 igi nu	34 igi nu	47 igi nu
8 igi 7,30	22 igi nu	35 igi nu	48 igi 1,15
9 igi 6,40	23 igi nu	36 igi 1,40	49 igi nu
10 igi 6	24 igi 2,30	37 igi nu	50 igi 1,12
11 igi nu	25 igi 2,24	38 igi nu	51 igi nu
12 igi 5	26 igi nu	39 igi nu	52 igi nu
13 igi nu	27 igi 2,13,20	40 igi 1,30	53 igi nu
14 igi nu	28 igi nu	41 igi nu	54 igi 1,6,40
15 igi 4			

Upper edge	Left edge	Lower edge
55 igi nu	56 igi nu	59 igi nu
	57 igi nu	1 igi 1
	58 igi nu	

From the history of the development of the operative term there follows an interpretation of the content of the table: The fractions of a given weight unit are expressed as the number of units that are 60 times smaller. From the line "5 igi 12," for example, it can be seen that $\frac{1}{5}$ of a weight unit can also be expressed in terms of 12 units that are 60 times smaller, as for example:

$$\frac{1}{5} \text{ gú} = 12 \text{ ma-na}$$

or: $\frac{1}{5}$ ma-na = 12 gín

If the unit fraction is not an integral multiple of the unit that is 60 times smaller so that it can only be expressed as an approximate multiple plus a remainder, then the procedure is repeated with this remainder using the next smaller sexagesimal unit. Following the usual Babylonian form for writing "numerals," the number of the smaller units is placed to the right of the number for the larger unit (in our modern notation separated by a comma). Thus, the following metrological conclusion can, for example, be drawn from the line "8 igi 7,30":

$$\frac{1}{8} \text{ gú} = 7 \text{ ma-na } 30 \text{ gín}$$

If this method of successively decomposing the unit fraction into smaller units failed because it does not come to an end, the negation "nu" is used to indicate that the particular unit fraction cannot be expressed in terms of a sexagesimally structured system of smaller units.

However, only very few of the approximately 50 published tables of reciprocals, and probably the oldest ones at that, expressly note the unit fractions that cannot be represented sexagesimally.[81] In general these fractions that cannot be represented are simply omitted. Most tablets, on the other hand, do contain an initial line in which the traditional "numeral" $\frac{2}{3}$ is also related to its sexagesimal equivalent 40, and the next line often contains the traditional "numeral" representing $\frac{1}{2}$ instead of the operative form "igi 2 gál."

The table reproduced here is a special case not only because of its early origin, but in other respects as well. It only goes as far as the next sexagesimal unit, whereas all the other tables that have been found are continued up to the value representing one and a half times that of the next unit. As a rule the table of reciprocals ends as follows:

```
      48 igi 1,15
      50 igi 1,12
      54 igi 1,6,40
       1 igi 1
     1,4 igi 56,15
    1,12 igi 50
    1,15 igi 48
    1,20 igi 45
    1,21 igi 44,26,40
```

The value representing one and a half units—that is, the line "1,30 igi 40" was no longer listed. The reason for this is obvious. Once the table had been presented in this formal way, the reciprocity of the relation could hardly be overlooked, the reciprocal relation "40 igi 1,30" is, however, already contained in the table. For the same reason the remaining three lines for which this is true were usually omitted as well so that the table generally ended immediately with the value 1,21 following the value for 1,4. Incidentally, the recognition of the reciprocity of the relation "igi ... gál-bi ..." could also have been the reason for introducing the symmetrical presentation "... igi ... ."

The Revolutionary Development of the Numerals and Their Arithmetic
The introduction of the metrological operator "igi ... gál-bi ..." for the use with weights had far-reaching arithmetical consequences. They did not

merely pertain to weights, but also to the very notation of numbers. None of the tables of reciprocals that have been found specify a particular unit of measurement. This neutrality served to make the relativity of the unit, hitherto already characteristic of the "numerals," an abstract principle by abandoning the determination of the meaning by the context as well. The table explicated the general structure of a strictly sexagesimal metrological structure which, up to this time, had not been achieved entirely for any of the particular tools of measurement. This abstraction from the context is essential for the arithmetical structure and the usability of the table of reciprocals.

The abstraction was not merely a mental process. It was carried out objectively for the "numerals" by no longer writing different units of the sexagesimal structure by using different signs. In any case, this required only a small step, because in the cuneiform notation the same sign was already used for both the numbers 1 and 60. Consequently, only two signs were used for the entire table, the vertical wedge, the cuneiform sign of the old sign a, and the "Winkelhaken," the cuneiform sign of the old sign b. The basic relationship of the two signs has been constant from the outset:

$$b = 10\,a$$

The part of the table reproduced above, for example, using contemporary numerals was in fact written as follows:

```
a         igi a
a aaaa    igi bbbbbaaaaaa baaaaa
a baa     igi bbbbb
a baaaaa  igi bbbbaaaaaaaa
a bb      igi bbbbaaaaa
a bba     igi bbbbaaaa bbaaaaaa bbbb
```

This way of denoting the units decreasing from left to right is based on a strict independence of the purely sexagesimal structure from the actual structure of the tools of measurement from which this structure was abstracted. With its lack of differentiation of the powers of 60 the new style represents this abstracted structure. Therefore, it can be said for the first time that the numerical notation is representing *numbers*. The table of reciprocals thus contains the *first "numerals" in the history of Mesopotamian arithmetic with a context-independent meaning*. It was, therefore, only a matter of time that the arithmetical regularities of the new numbers were deduced from their representations and consciously exploited in dealing with these representations. And the resulting numbers exhibit those structural peculiarities that, at first blush, contribute to the obscure appearance of Old Babylonian arithmetic:

- Their structure is sexagesimal.
- Their unit is relative and consequently the reference of the numerals to the tools of measurement used in the context of their application ambiguous.
- The internal relationship of the numerals to each other follows from their position in the sequence of the signs.

With reference to these numbers the table of reciprocals no longer is a metrological table. Rather, it represents a new arithmetical relationship—that is, the reciprocity with regard to a multiplicative structure of the numbers that can no longer be reduced to addition. It is for this reason that it is called a "table of reciprocals."

The table of reciprocals solved the problems of the operative representation of unit fractions familiar from Egyptian arithmetic at one fell swoop. In contrast to the representation by means of the operator, the sexagesimal representation of unit fractions provided by the table permits adding and subtracting without effort using the familiar means of constructive-additive numerical representation. This fact is the reason for the overwhelming significance of the table of reciprocals for the arithmetic of Mesopotamia and turned it into the wellspring of entirely new arithmetical developments in the Old Babylonian period.

Old Babylonian Arithmetical Techniques

Introduction of Multiplication as an Arithmetical Operation

Since, using the table of reciprocals, unit fractions could be added and subtracted according to the same rules as integers, the possibilities of solving multiplicative problems had changed. As the examples from the older economic texts indicated, multiplicative problems had already been solved earlier as long as they related only to the "counting numbers." For in this case the multiplication could easily be performed as repeated additions, at least with regard to the elementary multiplications found in the economic texts that occurred in practical contexts. In theory these multiplications could have been listed in a table. The result would have been something like a multiplication table, but that would not have made much sense, for the really serious practical multiplicative problems could not have been solved by means of these tables—that is, the problems of calculating the areas of fields which were represented by means of "area numbers" with a different structure. Presumably, such problems were solved using metrological tables like the "table of squares" from Shuruppak. The structural identity of these tables of areas with "multiplication tables" that, while easily generated, were useless, could not be recognized without the common reference of "counting numbers" and "area numbers" to a third shared entity—and that is to say to abstract numbers.

This explains why, almost simultaneously with the invention of the table of reciprocals—or at any rate only a little later—the multiples of the values of the table of reciprocals were listed in separate tables.[82] Because of the neutrality of the new numerals these *tables of multiples* of the unit fractions written in sexagesimal notation represented at the same time *multiplication tables* for the old "counting numbers." The fact that, in addition to the tables of multiples for the values from the table of reciprocals, a table of multiples was, as the only exception, created for the number 7 as well, proves that this structural identity did not go unnoticed. And the number 7 is indeed the only number between 1 and 10 that, due to its not being a regular number, does not appear as a value in the table of reciprocals. If this number is added to the tables of multiples of the reciprocals, all multiplications with the former "counting numbers" can be performed by using these tables.

The following tablet from the temple library of Nippur is a typical example for a table of multiples:[83]

Obverse		*Reverse*	
2,30 a-rá 1	2,30	a-rá 12	30
a-rá 2	5	a-rá 13	32,30
a-rá 3	7,30	a-rá 14	35
a-rá 4	10	a-rá 15	37,30
a-rá 5	12,30	a-rá 16	40
a-rá 6	15	a-rá 17	42,30
a-rá 7	17,30	a-rá 18	45
a-rá 8	20	a-rá 19	47,30
a-rá 9	22,30	a-rá 20	50
a-rá 10	25	a-rá 30	1,15
a-rá 11	27,30	a-rá 40	1,40
		a-rá 50	2,5

The table contains the multiples of the "head number" 2,30—that is, of the reciprocal of 24, resulting from the initial line by means of the successive addition of 2,30. It thus contains the multiples of "igi 24 gál" and must be understood in the original sense of this expression as a table of multiples of $\frac{1}{24}$. However, at the same time it represents the sexagesimal multiplication table of the number $2\frac{1}{2}$, of the number 150, and of any number resulting from multiplying or dividing $\frac{1}{24}$ with a power of 60. Thus, in terms of its function it corresponds entirely to a multiplication table of the type we are using for multiplying decimal numbers and decimal fractions.

Metrological Sexagesimal Tables and the Origin of the Tables of Squares
The old "numerals" had to be replaced by the new numbers, before the new arithmetical tools could be used in the daily computing routine of the temple and palace bureaucracies. This did not require major modifications, since the identification of the signs for the numbers 1 and 60 had already been accomplished. Mainly larger numbers, rarely used in practice, were affected by the change. Because of their practical significance for the area measures iku, èše, and bùr and because the numerical relations between these measures differed considerably from the sexagesimal structure, the "area numbers" remained in use in their cuneiform style. Special signs for the simple fractions $\frac{1}{2}$, $\frac{1}{3}$, and $\frac{2}{3}$ also continued to be used.

This resistance against a change of notations reveals an obstacle to the use of the table of reciprocals: the discrepancy between the traditional structure of the tools of measurement and the sexagesimal structure. The following table from the Old Babylonian temple library of Nippur,[84] probably dating back to the early Old Babylonian period, shows how the problem was solved (cf. Fig. 5).

Obverse			*Reverse*		
m	gán	abbbb	b	gán	bbb
mn	gán	aabbb	bp	gán	bbbb
mm	gán	aaabb	bpp	gán	bbbbb
mmm	gán	aaaaa	bb	gán	a
mmmm	gán	aaaaaabbbb			
mmmmm	gán	aaaaaaaabb			
p	gán	b			
pm	gán	babbbb			
pmm	gán	baaabb			
pmmm	gán	baaaaa			
pmmmm	gán	baaaaaabbbb			
pmmmmm	gán	baaaaaaaabb			
pp	gán	bb			

The left column contains the cuneiform "area numbers" of traditional arithmetic. With regard to their structure they correspond to the structure of the area measures:

$$b = 3\ p \qquad 1\ \text{bùr} = 3\ \text{èše}$$
$$p = 6\ m \qquad 1\ \text{èše} = 6\ \text{iku}$$
$$m = 2\ n \qquad 1\ \text{iku} = 2\ \text{ubu}$$

In the right column the "area numbers" are matched up with the new sexagesimal numbers written with only two signs, the vertical wedge (sign a)

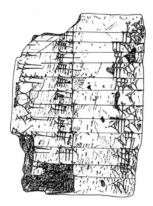

Figure 5: Sexagesimal table for "area numbers;" drawing of the obverse by Hilprecht ([20]; see below).

"Area numbers": "Numerals" of the sexagesimal numbers:

▬ = 1 iku (sign m) | = sign a (vertical "wedge")
▬◁ = 1 èše (sign p) ◁ = sign b ("Winkelhaken")

⌐ = sign for the semantical indicator gán indicating "field area"

for the number 1 and the "Winkelhaken" (sign b) for the number 10. Thus, the table relates the "area numbers" representing specific concrete areas to abstract sexagesimal numbers:

$$\begin{array}{ll} 1 \text{ iku} & = 1{,}40 \\ 1 \text{ iku } 1 \text{ ubu} & = 2{,}30 \\ 2 \text{ iku} & = 3{,}20 \\ 3 \text{ iku} & = 5 \\ & \vdots \\ 1 \text{ bùr } 2 \text{ èše} & = 50 \\ 2 \text{ bùr} & = 1 \end{array}$$

The principle generating the relation follows from choosing a basic unit, in this case the measure šár. The area measure šár is related to the length measure ninda, the preferred unit for measuring fields, by means of a simple ratio: 1 šár is the area of a square its sides being 1 ninda (approx. 6 meters). On the other hand, the unit šár has the following relationship to the unit iku of the table:

$$1 \text{ iku} = 100 \text{ šár}$$
$$\text{or:} \quad 1 \text{ iku} = 1,40 \text{ šár}$$

Thus, if the ratio

$$1 \text{ šár} = 1$$

is used as the principle generating the relation, the actual beginning of the table is the result:

$$1 \text{ iku} = 1,40$$

This generating principle could lead to the conclusion that the table ought to be interpreted as follows:

$$\begin{array}{ll} 1 \text{ iku} & = 100 \text{ šár} \\ 1 \text{ iku } 1 \text{ ubu} & = 150 \text{ šár} \\ 2 \text{ iku} & = 200 \text{ šár} \\ & \vdots \\ 2 \text{ bùr} & = 3600 \text{ šár} \end{array}$$

However, such an interpretation would not do justice to the actual function of the table. There is no sufficient reason in Old Babylonian arithmetic for giving up the flexibility of the relative representation of numbers and for determining the absolute value in any way other than in terms of the substantial context of the calculations. There is furthermore no evidence for the assumption that during this period it was ever attempted to express the absolute value in the numerical representation itself. The obvious assumption that the metrological sexagesimal table is an instrument for establishing an absolute value of the sexagesimal numbers is therefore probably false. We must therefore assume that the table served the sole purpose of *relating concrete area measures to abstract sexagesimal numbers* in order to be able to use sexagesimal numbers instead of "area numbers" for calculating. For this purpose the absolute value of the sexagesimal numbers is however of no consequence.

However, the choice of the basic unit does become significant for the relationship of the differently structured tools of measurement to each other.

Let us assume that we have a similar metrological sexagesimal table for length measures in which the unit ninda is the basic unit. If a table like the "table of squares" from Shuruppak were converted to the new sexagesimal numbers by making use of these tables, it would turn directly into a real table of squares because of the relationship 1 šár = 1 ninda^2. Or, more generally: The relationship within the tables for field areas becomes identical with the *multiplication* of sexagesimal numbers. The metrological sexagesimal table was the tool that turned the possibility of solving the problems associated with calculating areas by means of multiplication, merely theoretically outlined above, into reality.

In the course of the excavations, numerous metrological sexagesimal tables from the Old Babylonian period have been found, in particular tables for grain capacity measures[85] with the choice of the unit gín as the basic unit and for length measures[86] with different basic units. Furthermore, a large number of purely numerical tables of squares from the Old Babylonian period were found which can be regarded as evidence for the validity of the thesis presented here that multiplication as an arithmetical operation resulted from calculating areas and not—as is generally believed—the other way around.[87] The table of squares is not an application of an already existing multiplication technique, but the sexagesimal transcription of a much older type of table—as manifested by the Shuruppak table. Its arithmetical importance was based on the fact that, during the developing phase of the new form of the representation of numbers, constructing the table of squares made it possible to gain insight into the close relationship between multiplications and area calculations.

The Old Babylonian Standard Arithmetical Table

The last of the questions posed at the beginning can finally be answered—that is, the question why there is no evidence of written arithmetical algorithms in the arithmetical and mathematical texts of the Old Babylonian period: The results of arithmetical operations were obtained directly from the arithmetical tables.

The archaeological findings support this assertion. Numerical tablets of the type described in the previous paragraphs have been found in large numbers. The finds mostly come from the Old Babylonian and the following Kassite periods. In general the tablets contain individual tables. However, in addition there are many fragments of texts representing a standard arithmetical table which, in a largely canonical form, combined the tables to form a tool for a comprehensive arithmetical technique.[88] The fragments

that have been found belong to about 40 different tablets. Two of these tablets are almost completely preserved; nearly all the others can also be reconstructed because of the canonical arrangement of the multiplication tables. The structure of the Old Babylonian standard arithmetical table is therefore well known and will be shown below. Nearly all known examples of this standard table roughly follow this schema. However, in one way or another the individual tablets may deviate slightly from the standard form. This standard form consists of the following tables:
- A table of reciprocals of the numbers from 2 to 1,21
- Tables with the multiples (1 to 20, 30, 40, and 50) of the following "head numbers":

50	20	8	3,20
48	18	7,30	3
45	16,40	7,12	2,30
44,26,40	16	7	2,24
40	15	6,40	2,15
36	12,30	6	2
30	12	5	1,40
25	10	4,30	1,30
24	9	4	1,20
22,30	8,20	3,45	1,15

- A table of squares of the numbers 1 to 59
- A sexagesimal table of length measures

The conciseness of cuneiform notation made it possible to fit all these tables onto a single tablet (cf. Fig. 6). The seemingly strange choice of the "head numbers" is due to the fact that, with regard to their origin, the tables are tables of multiples of reciprocals. For the most part the "head numbers" are numbers from the right column of the table of reciprocals. However, the sequence of the "head numbers" does not strictly follow this rule. Rather, it shows several specific irregularities that can probably only be explained by a thorough analysis of the calculations in the economic texts and the mathematical texts where the standard arithmetical table was used.

The tables at the end of the standard arithmetical table indicate its most important application. The table of squares resulted from the problems of calculating areas—in addition the square of the respective "head number" was often placed at the end of each table of multiples as if the calculation of the area of a quadratic field of a given side was of particular importance with regard to the use of the table. Furthermore, the standard arithmetical table often includes a table of length measures.[89] Thus, there is every indication that the table was originally developed as an *aid for the calculations*

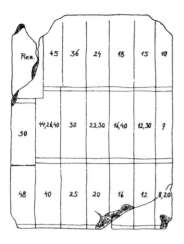 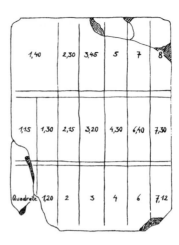

Figure 6: Arrangement of individual tables on an arithmetical tablet (cf. p. 252).
The size of the tablet which is nearly completely intact and was made up of two fragments (museum identifications Ist. A20 and VAT 9734) is 21 cm × 28 cm. It was found in the courtyard of the temple of Assur and, on the basis of a signature on the tablet, it is presumed to have come from the Kassite period. It includes all tables of the standard arithmetical table with the exception of a table of multiples of 2,24 and the sexagesimal table of length measures. (Source: [33], Vol. 3, pl. 65)

of field administration. Moreover, this assumption is quite consistent with the results of the analysis of arithmetical operations in the Early Dynastic economic texts described earlier. It had already become apparent that only the administration of fields raised serious problems of a multiplicative nature that could not be solved satisfactorily by means of the constructive-additive representation of numbers.

Because of the specific way in which the table of reciprocals was combined with the tables of multiples in the standard arithmetical table, the table was suited equally well for multiplications and divisions. The product of one-digit sexagesimal numbers can be obtained directly from the tables of multiples—numbers greater than 20 may require a further addition which was probably performed directly by continuing to count while writing down the constructive-additive numerals. The division of two numbers can be done in the same way if the reciprocal of the divisor is obtained from the table of reciprocals

254   CHAPTER 7

beforehand. Thus, in Old Babylonian arithmetic division was not a separate arithmetical operation but rather the multiplication with the reciprocal.

The situation becomes more difficult when numbers with several digits are involved. Here the *multiplication* requires further additions. For the following reason it is very difficult to determine how these auxiliary calculations were performed: The texts of the Old Babylonian period, however complex they may be with regard to their algebraic argumentation, mostly contain quite simple numerical examples in which there are hardly any multiplications of sexagesimal numbers with more than one digit. For practical purposes the accuracy of the arithmetical table was probably quite sufficient in most cases because, as compared with the decimal system, calculations in the sexagesimal system are six times more accurate given the same number of digits.

The *division* by a number with several digits cannot be directly performed at all with the standard arithmetical table. First, the reciprocal of this number would have to be obtained from the reciprocal table, but here an almost insurmountable limit is set by the number of digits of this table. Finally, the reciprocal table of the standard arithmetical table does not contain any non-regular numbers—that is, numbers the reciprocals of which have no finite sexagesimal representation.

The Unresolved Problem of Division

There have been numerous attempts to overcome this difficulty of the Old Babylonian arithmetical techniques. The fragment of an Old Babylonian tablet of unknown origin[90] contained twelve examples of an arithmetical method developed for this purpose. The most completely preserved example goes as follows (reconstructed):

> 2,13,20 (the *igum*). Its *igibum* what?
> You, by your making:
> The igi (= reciprocal) of 3,20, detaching, 18 you see.
> 18 to 2,10 going (= multiplied), 39 you see.
> 1 appending (= added), 40 you see.
> The igi of 40, detaching, 1,30 you see.
> 1,30 to 18 going,
> 27 you see. 27 is its igibum.
> So is the having-been-made.

(The translation follows the conventions of Jens Høyrup for mathematical texts. Paraphrased as: 2,13,20 [= 2,10 + 3,20] is the *igum*. What is its *igibum?* Find the reciprocal of 3,20. You get 18. Multiply 18 with 2,10. You get 39. Add 1. You get 40. Find the reciprocal of 40. You get 1,30. Multiply 1,30 with 18. You get 27. 27 is the igibum. This is the result.)

THE DEVELOPMENT OF ARITHMETICAL THINKING

The result 27 is the reciprocal of 2,13,20. Thus, this is a procedure for calculating the reciprocal of a sexagesimal number having several digits. The designation of the number as *igum* and of its reciprocal as *igibum* is derived from the word "igi." In the following, the reciprocal will be indicated by a line above the *igum*. The problem is as follows:

$$c = 2,13,20$$
$$\bar{c} = ?$$

The procedure consists in first subtracting the number $b = 3,20$ that can be inverted by using the table of reciprocals from the number $c = 2,13,20$ leaving the number $a = 2,10$ with only two digits.

$$c = a + b \qquad a = 2,10 \qquad b = 3,20 \qquad \bar{b} = 18$$

By using the numbers a, b, and $\bar{b}$, the number c can be decomposed into factors as follows:

$$c = a + b = b\,(a\bar{b} + 1)$$

In this way, the sum

$$2,13,20 = 2,10 + 3,20$$

is converted into a product the factors of which can be inverted by using the table of reciprocals:

$$2,13,20 = 3,20 \times 40$$
$$\overline{2,13,20} = \overline{3,20} \times \overline{40} = 18 \times 1,30 = 27$$

All the arithmetical operations of this example can be performed by using the standard arithmetical table. However, in general this need not be the case, because the second factor need not be contained in the table of reciprocals. In this case an attempt can be made to repeat the procedure with this factor. A succession of decompositions is thus obtained:

$$c_1 = a_1 + b_1 = b_1 \times c_2$$
$$c_2 = a_2 + b_2 = b_2 \times c_3$$
$$c_3 = a_3 + b_3 = b_3 \times c_4$$
etc.
$$\bar{c}_1 = \bar{b}_1 \times \bar{b}_2 \times \bar{b}_3 \times \ldots$$

That this procedure was indeed applied in this way is corroborated by a clay tablet covered with 21 numerical calculations,[91] which presumably also dates back to the Old Babylonian period. One of these calculations will be reproduced below as an example of the high level of development achieved by Old Babylonian arithmetic.

```
        5,3,24,26,40           9
        45,30,40               1,30
        1,8,16                 3,45
        4,16                   3,45
        16                     3,45
                14,3,45
                52,44,3,45
                1,19,6,5,37,30
        11,51,45,50,37,30      2
        23,43,49,41,15         4
        1,34,55,18,45          16
        25,18,45               16
        6,45                   1,20
        9                      6,40
                8,53,20
                2,22,13,20
                37,55,33,20
                2,31,42,13,20
                5,3,24,26,40
```

If the procedure described above is known, then these columns of numbers are not hard to interpret. They represent the calculation of the reciprocal of a sexagesimal number with five digits and subsequently the reverse calculation of the initial number from the result.

$$\overline{5,3,24,26,40} = 11,51,45,50,37,30$$
$$\overline{11,51,45,50,37,30} = 5,3,24,26,40$$

The auxiliary numbers are arranged as shown below:

$$
\begin{array}{ll}
c_1 & \overline{b}_1 \\
c_2 & \overline{b}_2 \\
c_3 & \overline{b}_3 \\
c_4 & \overline{b}_4 \\
c_5 & \overline{c}_5 \\
& \overline{c_5 b_4} \\
& \overline{c_5 b_4 b_3} \\
& \overline{c_5 b_4 b_3 b_2} \\
\overline{c}_1 & (= \overline{c_5 b_4 b_3 b_2 b_1}) \text{ etc.}
\end{array}
$$

However, this elaborate procedure for which a standard arithmetical table is quite sufficient must not obscure the fact that it has by no means solved the problem of division. The procedure is, after all, not an *arithmetical algorithm,* but—similar to the procedures of multiplication and division in Old Egyptian arithmetic—*only a heuristic schema.* One of the weaknesses of the procedure is that it contains an addition, the addition of the number 1. Here the relativity of the representation of numbers causes difficulties, because it is essential in which position the number 1 is added. The absolute value must be inferred from the respective context. While it is true that the procedure could easily be modified in order to avoid this addition, the text of the method cited above shows that the procedure was—at least in part—applied as described.[92]

However, this is a minor difficulty in view of the fact that the whole procedure will be successful only if the separated factors are divisors of the sexagesimal numbers interpreted as integers. If the decomposition is not done skillfully, the number of digits of the numbers to be inverted is increased rather than decreased. How the numbers must be decomposed can only be determined by trial and error. In the end, the procedure fails completely, if the number contains non-regular factors, that is to say, factors the reciprocal of which cannot be represented as a finite sexagesimal number, for example, the factor 7 or the factor 11 or the factor 1,17. In this case, by the way, the entire number is of course also not regular.

In the texts of the Old Babylonian period there are, however, also examples where the reciprocal of a non-regular number would really be necessary for continuing the calculation. Thus, in the context of an extensive calculation of an area there is, for example, the problem of dividing 6,21 by the "area of the GIR," where the value 19,3 had been determined for the "area of the GIR." 19,3 contains the prime number 127 as a factor (in sexagesimal notation: 2,7); therefore it is not regular. In the text[93] this is addressed as follows:

The igi (= reciprocal) of 19,3, the area of the GIR, is not detached. What should be posed that 6,21 has given. 20 should be posed.

(The translation follows the conventions of Jens Høyrup for mathematical texts. Paraphrased as: The reciprocal of 19,3 cannot be found. What must be used so that 6,21 is obtained? 20 must be used.)

The solution seems simple. Instead of multiplying the number 6,21 with the reciprocal of 19,3, a number—maybe by experimenting or by making use of the tables of multiples—is immediately sought that, when multiplied with 19,3, results in 6,21. Here it is the number 20. This simple procedure can of course only be applied here because the numerical example has been constructed in such a way that the dividend 6,21 also contains the troublesome factor 127 (= 2,7). Otherwise the result would also not be regular—that is, it could not be represented sexagesimally at all. Thus, this is a special case without practical relevance. However, virtually all examples of a division by a non-regular number occurring in the available texts of the Old Babylonian period are such special cases.[94] It can therefore be assumed with good reason that the authors of these texts realized that the division by non-regular numbers could technically not be achieved with the means of Old Babylonian arithmetic.

Why did the Babylonians fail to develop arithmetical algorithms that—like the algorithms of our decimal arithmetic—made use of the positional representation of numbers in order to determine at least approximate values? Was the idea of approximation totally alien to them? This was not the case as indicated by the—albeit singular—find of a table of reciprocals which contains the reciprocals of non-regular numbers as approximate values with three or four digits.[95] In addition, a tablet[96] with the following table was found:

igi 7 bi 8,34,16,59	SI-NI-IB
igi 7 bi 8,34,18	dirig
igi 11 bi 5,27,16,22	SI-NI-IB
igi 13 bi 4,36,55	SI-NI-IB
igi 14 bi 42,51,25,40	SI-NI-IB
igi 17 bi 35,17	SI-NI-IB

This is clearly a tablet with approximate values indicating—albeit in one case incorrectly—whether the value is too great or too small. In addition, the last two lines—contrary to the text—do not give approximate values for the reciprocals of the numbers 14 and 17, but for ten times these values. This table nevertheless proves that the idea of approximation was already being systematically developed during the Old Babylonian period.

If the use of approximative procedures was nevertheless systematically circumvented in the mathematical texts of the Old Babylonian period, the reason must be sought elsewhere. The new arithmetical means were no doubt employed to help solve the traditional problems of economic administration. However, in addition, they acquired an entirely different function. They were the basis for the development of an efficient system detached from practical problems, a system of procedures and discoveries concerning the possibilities that were, independent of the motives for their construction, objectively inherent in these means: for the development of Babylonian "algebra."

Arithmetic and Algebra

It would go far beyond the goal of this study to provide even the briefest of summaries of the methods and accomplishments of Old Babylonian "algebra,"[97] especially since a number of the mathematical cuneiform texts constituting the basis of our knowledge of Old Babylonian "algebra" pose problems of interpretation that are difficult to overcome. While the results of the reconstruction of their mathematical content out of the arithmetical procedures used in these texts do provide evidence of the scope and efficiency of Old Babylonian "algebra," they do not yield much information on the way in which the mathematical insights that must be assumed were obtained. Thus, only a final comment on the relationship of arithmetic and "algebra" will be given here. Furthermore, the question, raised by Old Babylonian "algebra" that could not be answered solely by an investigation of the arithmetical means employed, will be stated.

We encounter Old Babylonian "algebra" in a new kind of transmitted texts, that is to say, in the cuneiform texts that are "mathematical" in the narrower sense. In the economic texts that provide information on arithmetical thinking in previous periods, the calculations are always governed by the purpose which was determined by the function of these documents as means of economic administration. With few exceptions the table texts that originate mostly from the Old Babylonian period and only very rarely from earlier periods must be considered to be means for performing arithmetical operations above and beyond addition. They, too, are embedded in the context of the arithmetical techniques generally used in the economic administration. The "mathematical" cuneiform texts, on the other hand, are testimony to a more far-reaching specialization of work. While the problems they discuss still betray their origin in the economic administration of the palaces and temples, they are nevertheless problems of a more theoretical nature concerning the arithmetical techniques used in economic administration. The problems no longer refer to the calculation of this or that field, but rather, for instance, to an arithmetical schema of a series of field calculations in which the given conditions—without reference to practical cases—are systematically varied. The "mathematical" cuneiform texts are definitely not documents of the normal economic administration. Rather, they are evidence of systematic investigations of the techniques employed in economic administration undertaken by specialists.

The "mathematical" cuneiform texts are largely series of problems, often supplemented by the actual calculation that leads to the solution and by a "proof by verification" demonstrating that the solution is correct. Reasons explaining why a calculation must proceed this way and not that are never given. No "theorem" is stated, not even a general rule. No assumption is confirmed by way of "proof." Without exception numbers are calculated from other given numbers using certain implicitly assumed arithmetical relationships.

This has frequently led to the conclusion that the Babylonians—just like the Egyptians—merely provided the material subsequently used by the Greeks to form mathematics as a science. The Babylonians, it is said, merely operated with useful problem-solving schemata that were obtained empirically. The Greeks, however, discovered the mathematical relationships fundamental to these techniques by means of thought.

This assumption is, however, too simple to be true. The formula for solving quadratic and cubic equations is not to be found "empirically." The practically meaningless addition of a length and an area[98] is no longer "useful" in the pretheoretical sense if it serves the sole purpose of constructing a problem with the structure of a quadratic equation. Systematically varied numerical examples in the "mathematical" cuneiform texts demonstrate that it is no longer the particular example that is at issue, but the general context of the algebraic relationships between the individual numbers of the examples. Thus, it is appropriate to distinguish the "mathematical" cuneiform texts from the other economic texts and to consider them as evidence for a Babylonian "algebra" based on but going beyond Babylonian arithmetic.

The "mathematical" cuneiform texts have as their content arithmetical calculations, and it therefore seems to be a truism to say that the Old Babylonian algebra presupposes the arithmetic. However, on closer examination there is a very close connection here, and it is no historical accident that the development of Old Babylonian algebra followed in the wake of revolutionizing the arithmetical means at the beginning of this period.

The new form of numerical representation and the arrangement of multiplicative relationships in the form of tables on this basis were the preconditions for understanding numbers as *context-independent ideal objects* with a rather complicated multiplicative structure. And what is special with regard to the "mathematical" texts is that here it is numbers and their relationships to each other—and in particular the new multiplicative relationships—qua *numbers* and no longer qua representatives of real objects that are made the object of representation. Numerous examples indicate that in these texts the concepts familiar from the economic texts continue only as formal designations of the arithmetical relationships, for example, the "field area" denoting the product, "length" and "width" denoting the factors, the "masses of earth" denoting the volume or, more generally, the product of three factors and so on.[99] For this reason it made sense to the Babylonian mathematician to add an area to a length. In short: The *means of arithmetic* constitute the *object of "algebra"* in the "mathematical" cuneiform texts of the Old Babylonian period.

It was for this reason that the idea of approximation, although quite familiar in terms of the use of the tools of measurement, initially was not incorporated into the "mathematical" texts. It was the very opposite of the newly acquired idea of the number as a concept of arithmetical objects conforming to strict laws.

Historically the examination of arithmetical means as an end in itself had of course very practical consequences. In the Seleucid period the arithmetical techniques developed by algebra provided the conditions permitting the emergence of the first empirical science operating with mathematical means—that is, astronomy. Thus, many of the statements made above about Old Babylonian arithmetic and algebra are no longer valid for this period. It goes without saying that approximations were an integral part of the algebraic tools used by mathematical astronomy at that time. By making use of comprehensive tables of reciprocals with many digits, the problem of division had been solved, if not in theory, then at least in practice.

However, in the Old Babylonian period arithmetic was the precondition for algebra in the strict sense inasmuch as it was the object of the latter and thus determined its fundamental assumptions. In algebra the possibilities afforded by the new arithmetical means were examined. Any deeper study of the genetic connection between arithmetic and algebra in the Old Babylonian period has to find out and to analyze theoretically the means used by the Babylonians for conducting this examination. However, to date this problem has hardly been recognized as the key to our understanding of the development of Babylonian mathematics.

SUMMARY: CALCULATING AIDS AND ARITHMETICAL THINKING

What, if anything, did the investigation of the Old Egyptian and the Old Babylonian arithmetic contribute to answering the question posed at the outset concerning the relationship between the calculating aids and arithmetical thinking? The point of departure of this inquiry was the structural identity of the rules of ancient Chinese arithmetic and the rules of manipulating the rod numerals on the reckoning board. This identity encouraged an attempt to link the explanatory model based on developmental psychology which maintains that the structures of thinking are the result of the reflection on objective actions with the historical approach concerning the development of arithmetical thinking. Consequently, arithmetical thinking in the early civilizations as it can be reconstructed by structural considerations based on the available arithmetical evidence should not be understood as the precondition for the invention of the arithmetical techniques of this arithmetic. Rather, it should be regarded as their result, that is to say, as a result of operating with objective means in the course of mastering arithmetical problems "mechanically." In order to support this assumption it was essential to determine whether the Old Egyptian and the Old Babylonian calculating aids represented the arithmetic in these cultures in the same way as the reckoning board in the case of ancient Chinese arithmetic.

If we try to identify an equivalent to the ancient Chinese reckoning board in the two arithmetical systems of the Near East, it is first and foremost the role of the constructive-additive representation of numbers by means of written signs that must be emphasized. It should be recalled that, while individual designations for numbers are the oldest evidence for an occupation with quantitative problems, in and of themselves they do not constitute evidence for an arithmetic. Arithmetic only begins with the operative manipulation of numerical representations, and the names of numbers alone are not particularly suited to such a purpose. The names of numbers alone do not yet represent calculating aids. However, the numerals of a constructive-additive numerical sign system do, since they permit "mechanical" operations with a numerical representation.

Written signs as objective calculating aids are characteristic of the two arithmetical systems of the Near East. The constructive-additive numerical representation by means of written signs provided a calculating aid comparable to the reckoning board if not in terms of its efficiency, then in terms of its essence. This becomes particularly evident when we remember that this numerical representation in fact emerged in Mesopotamia directly from a technique of mechanical bookkeeping using counters, a technique that did not presuppose any kind of arithmetical thinking and that can be traced archaeologically as far back as the early Neolithic period. Numerals, just like counters, generated by a constructive-additive process represent sets of real objects, and any operation of combining or dividing that can be performed with these objects can in a similar way be performed with their objective representatives as well. These operations with representatives may be called "addition" or "subtraction." However, they do not yet represent mental operations with ideal objects.

Operating in this manner with counters or numerals does not remain without consequences for thinking, since the accumulated experiences make it possible to anticipate mentally the result of operating with the calculating aids. And since operations with the calculating aids themselves already represented the actual combination and division of sets of real objects merely for the purpose of anticipating their result, this incipient arithmetical thinking can even render the objective operations with the calculating aids at least partly superfluous. Thus, even if initially one objective operation is merely substituted for another—namely the operation with real objects for the arithmetical operation with their representation—then the latter operation has nevertheless a special relationship to thinking. For what is essential for the functioning of a calculating aid is not the actual transformation of an objective numerical configuration, but only the possibility of obtaining the meaning of the result from the meaning of the initial configuration by means of such a transformation. And in this nonobjective activity of the transformation of meaning the calculating aids operate as objective means in a way that changes with the development of arithmetical thinking.

One example of this change is the fact that constructive-additive numerals can be relieved of their original additive function by a cognitive structure comprising the fundamental additions up to ten which evolves in the process of calculating. In Old Egyptian arithmetic this evolving cognitive structure encouraged the simplification of the hieroglyphic to the hieratic numerals. Thus, a calculating aid that engenders arithmetical thinking becomes a calculating aid that already presupposes arithmetical thinking, but in turn opens up new, more far-reaching possibilities. Of course, a necessary condition for such a process of the development of a calculating aid is that the experiences acquired in the course of using this means are transmitted from one generation to the next so as to ensure that it is not necessary for each and every person to reinvent tools like the table of fundamental additions in the process of using the calculating aids.

It can be stated as a first result: On the basis of the constructive-additive representation of numbers by means of written signs an arithmetical thinking characterized by additive structures emerged in both of the early civilizations of the Near East. It was transmitted in a specific way by means of the training of the people involved in computing. Subsequently, new calculating aids could be developed that already presupposed additive thinking, but were, in turn, guided by arithmetical goals that could not be achieved by means of the constructive-additive representation of numbers alone.

If we take a look at the goals that were actually tackled under these conditions, another similarity of the Old Egyptian and Old Babylonian arithmetic becomes apparent—that is, the substitution of the individual signs denoting fractions largely unsuitable for performing calculations by a representation of fractions by means of a unit fraction operator. However, initially this innovation merely attests to a growing insight into the structural similarities of actions pertaining to division and multiplication. Dramatic results are not to be expected as long as there are no means for solving the structural problems impeding the application of the additive arithmetical operations to the fractions represented in this way, thus significantly expanding the possibilities of additive thinking. And in fact the result of the investigation of the Old Egyptian and Old Babylonian arithmetic can be stated in simplified terms as follows: The striking characteristics constituting the fundamental difference between these two arithmetics are essentially due to the different means developed for solving this problem.

In *Egypt* the problem arose in the following way: It was necessary to develop the possibilities of performing additions with unit fractions to the extent that the universal arithmetical schema for reducing multiplicative problems to additive problems could be applied to unit fractions as well. The Egyptians were so successful in dealing with this problem that, in practical terms, it was probably solved to a sufficient degree. The experiences relating to the addition of fractions were collected, systematized, and transmitted. The basic operation of duplicating unit fractions which was necessary for applying the arithmetical schema and for which an algorithm was not to be found, was tabulated. This may have been the case with other elementary operations as well. For the remaining structural problems of computing with unit fractions an extensive set of heuristic rules was developed which was used skillfully by Egyptian scribes. However, its use was most certainly dependent on extensive training and broad experience.

Despite this, it must be stated that no means was found in the context of Egyptian arithmetic to solve the problem once and for all. It always remained the central problem of this arithmetic. However, the relative success of the efforts to find a solution proved to be an obstacle with regard to giving serious consideration to problem-solving strategies that would have questioned the existing arithmetical technique in general. Neither its domain of application nor its social organization confronted Old Egyptian arithmetic with problems that could have stimulated the search for possibilities other than those offered by the Egyptian arithmetical technique which, while highly developed, was beset with unsolvable structural problems.

The circumstances under which the problem of using the unit fraction operator presented itself in *Mesopotamia* were entirely different. This problem thus became the starting point of a fundamental discontinuity in the arithmetical development which dates back to the beginning of the Old Babylonian period. Certainly, any comparison with the development of Old Egyptian arithmetic is severely limited by the fact that, due to the paucity of source material, the chronological development of the latter cannot be reconstructed nearly as accurately as that of Mesopotamian arithmetic. Nevertheless, it seems clear that the cause of the different development in Mesopotamia must be seen in the circumstance that Mesopotamian arithmetic was relatively backward at the time when the problem of the unit fraction operator emerged. There is every indication that the Egyptians were already in possession of their efficient arithmetical schema at this time which permitted them to reduce multiplicative problems to additive problems. However, the present investigation shows that in Mesopotamia the unit fraction operator "igi ... gál" was introduced at a time when an entirely different problem still occupied the center of the arithmetical efforts. This was the much more fundamental problem of unifying the context-dependent "numerals" with their diverse arithmetical structures determined by the tools of measurement—a problem of standardizing measuring units and numerical notations which presumably never arose to the same degree in the centralized Egyptian state as it did in the crazy quilt of Mesopotamian city states. By comparison, multiplicative problems were still of minor importance at this time—that is, before the Old Babylonian period; there is in any case no evidence in the sources for an algorithmic arithmetical procedure for multiplications that, as in Egypt, would have had to be rendered compatible with the unit fraction operator. Initially the issue was solely the introduction of a consistent representation of fractions into the metrological tables with their diverse arithmetical arrangements. And it was precisely this circumstance that, almost by necessity, led to the invention of the table of reciprocals, an invention amounting to the discovery of an arithmetical regularity in the metrological tables into which the unit fraction operator had been introduced. And at one fell swoop the table of reciprocals solved the arithmetical problems for which the Egyptians needed their entire elaborate technique of fractions, because it made it possible to add and subtract fractions without any problems by employing the familiar means of the constructive-additive representation of numbers.

If, therefore, the development of Old Egyptian and Old Babylonian arithmetic was so different despite the similar point of departure from the constructive-additive representation of numbers and the additive thinking determined by this calculating aid, the reason for this difference must be sought in the dissimilar conditions in which the particular arithmetical problem of dealing with unit fractions evolved in either case. This result of the investigation raises a new problem—that is, the question as to the significance of external circumstances for the development of arithmetical thinking otherwise determined by the calculating aids. However, with this question the investigation confined to the domain of the history of science reaches its limits, because the external circumstances must appear to be merely accidental unless the socio-historical background of the development is systematically included in the investigation. I will thus have to limit myself to a few concluding comments concerning this question.

The early Neolithic history of arithmetic is almost completely unknown, because etymology is virtually the only source of information. However, this much seems to be certain: The early history of arithmetic is characterized by a high degree of consistency. In all cultures comparable problems in connection with the step-by-step construction and the structuring of an infinite counting sequence are to be found. This consistency is no doubt due to the universal use of language as a means of counting which limited possible alternatives. However, we know that the calculating aids, so successfully developed later, were already known in principle during the Neolithic period. They were employed in the rural economy, in the crafts and trade; their potential for arithmetical discoveries, however, was not explored at that time.

Obviously, "external" circumstances were responsible for this. There was no reason whatsoever to see representations of numbers in the numerical figures obtained by using calculating aids; there was no reason to regard them as anything more than the representations of the material objects of the practical context of action. The use of the calculating aids did not generate an autonomous arithmetic, for such an arithmetic would not have been passed on, since it would not have had a place in the context of work and reproduction of primeval society.

This explanation may become more plausible, if we consider the close temporal and local connection between the subsequent change of the situation and the emergence of states. The time of the early, largely isolated civilizations and states is also the time of the evolution of the first highly developed arithmetic—arithmetic that, because of its distinct characteristics, exhibited a degree of inconsistency never to be encountered again in the subsequent history of arithmetic. And it was invariably the officials of state hierarchies who developed this arithmetic. Thus, the development of arithmetic as a technique removed from the immediate context of the process of labor is a direct result of the social division of labor into physical and mental work.

This background also explains why external circumstances could engender such a high degree of diversity of the early arithmetical systems. Despite the isolation of the highly developed cultures the arithmetical goals were very similar; this is supported by the similarity of the problems in the available arithmetical texts. However, the development of the means for solving them was dependent on the type and extent of the division of labor prevailing under the particular natural and social conditions. For it was the professionalization of doing arithmetic in the context of the administrative activity of the state that was the precondition for exploring the possibilities of the available calculating aids and for developing them into instruments which could no longer be employed without special training.

It was, therefore, the social organization of the early civilizations that created the conditions for the development of an autonomous arithmetic. However, it also limited this development in a specific way. In the context of central administration and labor organization arithmetical thinking already referred to calculating aids isolated from the context of physical labor. The object of arithmetic already appeared as an ideal object, since it could no longer be identified with any one of the concrete objects from the various applications of arithmetic. However, because of the isolation of the early civilizations from each other, the experience was not yet possible that the object of arithmetic is even more general than the particular calculating aids which were developed by the diverse civilizations for mastering arithmetical problems. Thus, the ideal object of arithmetic, the number, continued to appear as the object of reflection on the possibilities afforded by the particular calculating aids in a given context.

The terminology used confirms such a limitation of the arithmetical thinking associated with the early arithmetical systems. In the Old Egyptian "equations" unknown numbers had to be designated which were subjected to certain operations in order to obtain a certain result. The unknown quantities are represented by the Egyptian word "h," in the present context translated as the neutral term "set." This "h" has been interpreted as a precursor of the later concept of a variable. However, it is obvious that a special terminology for the unknown numbers of the "equations" had to be introduced here only because there was as yet no general expression for the ideal object of arithmetic. In a similar way this also applies to the words *igum* and *igibum* in Old Babylonian arithmetic. They were used to indicate the related terms in the context of the investigation of the general characteristics of the relation "... igi ..." determined by the table of reciprocals. No doubt these words also designate the special form of the ideal object of arithmetic in the context of this relation. However, we search the texts of the Old Egyptian and Old Babylonian arithmetic in vain for a word that could, without further ado, be translated as our term "number."

Thus, the final result can be summed up as follows: The development of the calculating aids as means of a centralized administration in the early civilizations did not produce a concept of number, although there can be no doubt that numbers as ideal objects already were the object of arithmetic. Even terms of sufficient generality were still lacking; there was even more of a lack of any kind of definition of the concept of "number."

NOTES

1. [52].
2. With respect to the following translations: [52], p. 106ff., p. 142ff., and the evidence cited; [30], p. 3ff.
3. On this technique see [22], p. 16ff.
4. On this point see my papers in [15].
5. Rhind Papyrus: [11]; [38]; [7]; Moscow Papyrus: [49]. On the other sources cf. [17].
6. Cf. the editions cited as well as [31].
7. All examples according to the numeration of Eisenlohr adopted by subsequent editors. Apart from that the examples follow the edition of Chace et al. ([7]) (unless indicated otherwise).
8. Cf. [49], p. 27f., regarding the translation differing from Chace et al. ([7]).
9. On this point cf.: [31], especially p. 20ff.; [32], p. 147ff.; [7], Vol. 1, p. 13 ff.; [17], p. 45ff.
10. I am essentially following Neugebauer ([31] and [32]). Other interpretations are advanced by Peet ([38], especially p. 15ff.) as well as Chace et al. ([7], Vol. 1, p. 7ff., especially p. 40).
11. Cf. [31], especially p. 6ff.
12. Cf. [6], p. 268ff.
13. The quotation is from Neugebauer ([32], p. 120).
14. [19], II 109.
15. Other views are based on a dubious interpretation of Herodotus ([19], II 36). Cf. also [31], p. 42.
16. Cf. [22], p. 15.
17. Further simplified in the first millennium B.C. to create *demotic* writing.
18. [7], Vol. 1, p. 22f. and 60ff.
19. The use of another table for the "multiplication" by $\frac{2}{3}$ is assumed. Cf. [17], p. 24ff.
20. In the present context the old names are used for locations, in older publications, on the other hand, modern names are often used (Tello instead of Girsu, Fara instead of Shuruppak, etc.).
21. The absolute chronology has been verified only up to about 1500 B.C. In the present context, older dates according to the "intermediate chronology." Cf. [27]; [6], p. 20ff.
22. On this problem cf. [34], p. 59ff.
23. [37], especially p. 123ff.
24. [44] and [45].
25. [13].
26. [8], especially Vol. III.
27. Autographs Förtsch ([14]); translation and edition Bauer ([2]).
28. Primarily Nies ([36]); in addition also [9] and [10]; [12].
29. [5]; [20]; [28]; [33]; [35]; [43]; [53].

30 For example, the text AO 6484, [33], Vol. I, p. 96ff., confirming the continuous tradition.
31 BM 85200, [33], Vol. I, p. 207.
32 Cf. [51], p. 24.
33 Cf. [3], especially p. 229ff.
34 On the etymology of number words: [29]; [50]. With regard to Mesopotamia: [18]; [41].
35 [44] and [45].
36 Following the accepted norm, Sumerian words are rendered in normal type, Akkadian words in italics and text parts with unknown reading in caps. On the etymology of the designations for measures: Neugebauer ([32], p. 100ff.); Lewy ([26]). Both authors, in particular Lewy, with conclusions differing from the view presented here.
37 Cf. [40], p. 7ff.
38 [45].
39 Cf. [23]; [47], especially p. 122ff.; [48].
40 The determination of the date is uncertain. For the finds from Uruk IVa radiocarbon-14 analysis established 2815 ± 85 B.C. ([27]). No writing tablets have been found in the lower levels. The finds of Habuba Kabira (middle of the 4th millennium) contained numerical signs, but also no written characters ([48]). However, for proto-Elamite written tablets found in Tepe Yahya IVC in the hill country of Iran which are archaeologically dated as contemporary to Uruk III—that is, later than the earliest finds in Uruk radiocarbon-14 analysis established 3560 ± 110 for the level in question ([25]).
41 Cf. [13], especially p. 5ff.
42 [45].
43 Cf., e.g., [48], p. 63f.
44 [8], Vol. II.
45 Cf. [13], sign list; [8], Vol. I, sign list.
46 [13], No. 304.
47 [13], No. 621.
48 Cf. Falkenstein ([13], p. 49) as well as the sign list. Subsequent authors have always referred to Falkenstein. However, his interpretation seems questionable to me.
49 VAT 12593, [8], Vol. II, p. 26*ff. and p. 75 (autograph). Cf. also [33], Vol. I, p. 75.
50 Metrological data according to [33], Vol. I, p. 85ff.; Vol. III, p. 52, complemented by [35], p. 4ff.; [26]; [8], especially Vol. III, p. 13*f.; [16]; [2], especially p. 51. Important new findings in Powell ([39]).
51 [8], Vol. II, p. 27*. Cf. also [2], p. 77f.
52 Cf. [8], Vol. III, p. 13*; [8], Vol. I, sign list.
53 [8], Vol. III, p. 9*ff.
54 [8], Vol. I, sign list. For comparison [46], sign list, p. 123ff. for the IIIrd Dynasty of Ur 800 years later.
55 For example, AO 8865, [33], Vol. I, p. 89f. (Old Babylonian). The exact history preceding Old Babylonian metrology is not likely to be reconstructed until a comparative edition of metrological tables is available. Sachs ([42]) reports that there are 200 unpublished metrological tables in the repositories of the museums.
56 [8], Vol. I, p. 3, Vol. III, p. 16*. (Addendum to the present edition: These values are incorrect. The gur maḫ contained 480 sìla whereas the standard gur contained 240 sìla.)
57 [2], p. 114.
58 [46], p. 128.
59 Cf. the following, in part greatly divergent views: [26], p. 6ff.; [29]; [32], p. 108; [50], p. 30; [51], p. 24.
60 This is probably due to the usual philological treatment of the "numerical signs," e.g., [8], Vol. I; [46], List of Numerical Signs, p. 123ff. Cf. also the more recent handbooks like the standard work by Borger.
61 Autograph Förtsch ([14], No. 24); edition Bauer ([2], p. 374ff.).

⁶² Autograph Förtsch ([14], No. 184); edition Bauer, ([2], p. 118ff.).
⁶³ [14], Nos. 9, 90, 92, 103, and 183.
⁶⁴ [14], Nos. 40 and 156.
⁶⁵ [14], Nos. 16 and 99.
⁶⁶ Similar imprecisions regarding all statements about areas in the documents on surveying published and edited by Fuye ([16]). Therefore, the unstated assumption of this work that the statements about areas are the direct result of calculations seems questionable.
⁶⁷ Cf. [1], p. 35ff.; [6], p. 91ff.; [24], p. 29ff.; [47], p. 151ff.
⁶⁸ [1], p. 54.
⁶⁹ [6], p. 144.
⁷⁰ [46], p. 128.
⁷¹ Common in Girsu for example, cf. [14], Nos. 9 and 77.
⁷² Cf., e.g., the "numerals" in the field plan by Eisenlohr ([12]), depicted also in [54], p. 36.
⁷³ [14], No. 65; [2], p. 508ff.
⁷⁴ Cf., e.g., [36], No. 37 (igi 3 gál), dated back to the second year of Bursin (= 1893 B.C.), No. 179 (igi 6 gál); [14], No. 175; [2], p. 510 (igi 6 gál), dated back to the fourth year of Lugalanda (= 2370 B.C.).
⁷⁵ [28], No. 59, p. 32ff. A parallel text (measures for silver) is MLC 1854, [33], Vol. I, p. 92ff.
⁷⁶ Cf. also metrological lists for other measures, e.g., [20], Nos. 29, 37, and 38.
⁷⁷ The date of Ist.T 7375, [10], p. 43 and pl. 14, is based solely on the style of writing (cf. [33], Vol. I, p. 10). The same is true for determining the date of the table of multiples ([28], No. 61) dating back to the IIIrd Dynasty of Ur (cf. [33], Vol. I, p. 50).
⁷⁸ In the economic texts published by Nies ([36]), for instance, I was unable to find a single example containing indications of a multiplication.
⁷⁹ Ist.T 7375, [10], p. 43 and pl. 14; cf. [9] as well.
⁸⁰ For other terminological variants cf. [33], Vol. I, p. 24ff.
⁸¹ Cf. [32], p. 8. The surveys [33], Vol. I, p. 8ff., and [35], p. 12, only contain one other example.
⁸² Cf. the summaries [33], Vol. I, p. 32ff., and [35], p. 19ff.
⁸³ CBM 10190 (= [20], No. 11); at the time misinterpreted by Hilprecht as a table of multiples of the number 150, because the ambiguity of the numerical signs had not yet been recognized.
⁸⁴ [20], No. 40; parallel text [20], No. 39.
⁸⁵ For example, CBM 19814 (= [20], No. 35), CBM 19820 (= [20], No. 34), CBM 45005 (= [20], No. 33), [20], Nos. 31, 32, and 36.
⁸⁶ Sexagesimal tables of the length measures can be found on the following tablets (basic unit in parentheses as far as can be ascertained): AO 8865, [33], Vol. I, p. 88ff. (GAR); VAT 6220, [33], Vol. I, p. 90 (GAR); [20], No. 41 (kùš), No. 42 (GAR), No. 43 (GAR); CBS 7369, [35], p. 30; BM 92698, [53], p. 318f. (2 tables: GAR and kùš); Ist.T 10994, [33], Vol. I, p. 53; Ist.O 4108, [33], Vol. I, p. 73f. (kùš); W 1923-366, [35], p. 34; CBS 8156, [35], p. 30; CBS 29.13.174, [35], p. 33.
⁸⁷ The summaries [33], Vol. I, p. 68ff., and [35], p. 19ff., contain 28 tables of squares, the majority probably from the Old Babylonian period. In addition to the tables of multiples and of squares tables of the same type, but with the structure and notation to length and area measures were found; cf. [35], p. 6ff., as well as Ist.A 20 + VAT 9734, [33], Vol. I, p. 92, from the Kassite period.
⁸⁸ Cf. the surveys [33], Vol. I, p. 35ff., and [35], p. 24ff.
⁸⁹ Usually only a "metrological table" is mentioned in the literature. As far as I could determine all the known examples concern tables of *length measures,* presumably sexagesimal tables and not just lists. The following are the tables in question: Ist.T 10994, [33], Vol. I, p. 53; VAT 6220, [33], Vol. I, p. 90; CBS 7369, [35], p. 30; CBS 8156, [35], p. 30; CBS

29.13.174, [35], p. 33.
[90] VAT 6505, [33], Vol. I, p. 270ff.; cf. also [43], Vol. I, p. 226ff.
[91] CBS 1215, [43], Vol. I, p. 230ff. Cf. also [3], p. 241f.
[92] Overlooked by Bruins ([3], p. 179), whose criticism of the interpretation of Sachs (in [43], Vol. I) thus becomes invalid. Bruins' thesis the tablet and all other arithmetical tablets are merely "school texts" for a brilliantly handled technique of mental arithmetic shows to my mind the importance of supplementing the historiography with a theoretically sound and empirically confirmed reconstruction of the development of arithmetical thinking.
[93] VAT 7535, front lines 22 and 23, [33], Vol. I, p. 306f.
[94] The 13 known cases are listed in Sachs ([41], p. 212, note 24).
[95] YCB 10529, [35], p. 16.
[96] F. Lewis Collection M 10, [43], Vol. II, p. 152f.
[97] Cf., e.g., [32], p. 175ff.; [51], p. 45ff.; [54], p. 44ff.
[98] For example, in AO 8862, [33], Vol. I, p. 108ff.
[99] Cf. [33], Vol. I, p. 208.

BIBLIOGRAPHY

[1] Awdijew, W. I.: *Geschichte des Alten Orients*. Berlin: Volk und Wissen, 1953.
[2] Bauer, J.: *Altsumerische Wirtschaftstexte aus Lagasch*. Thesis, Würzburg: 1967.
[3] Bruins, E. M.: "Computation in the Old Babylonian Period." In *Janus, 58*, 1971, p. 222ff.
[4] Bruins, E. M.: "Tables of Reciprocals with Irregular Entries." In *Centaurus, 17*, 1973, p. 177ff.
[5] Bruins, E. M. & Rutten, M.: *Textes Mathématiques de Suse*. Paris: Geuthner, 1961.
[6] Cassin, E., Bottero, J. & Vercoutter, J. (Eds.): *Fischer Weltgeschichte, Vol. 2: Die Altorientalischen Reiche 1*. Frankfurt a. M.: Fischer, 1976.
[7] Chace, A. B. et al.: *The Rhind Mathematical Papyrus*. 2 vols., Oberlin, OH: Mathematical Association of America, 1927/1929. (Reprint Reston 1979.)
[8] Deimel, A.: *Die Inschriften von Fara*. 3 vols., Leipzig: Hinrichs, 1922/1923.
[9] Delaporte, L.: "Document Mathématique de l'Epoque des Rois d'Our." In *Revue d'Assyriologie et d'Archéologie Orientale, 8,* 1911, p. 131ff.
[10] Delaporte, L.: *Textes de l'Epoque d'Ur*. Paris: Leroux, 1912.
[11] Eisenlohr, A.: *Ein mathematisches Handbuch der alten Ägypter*. 2 vols., Leipzig: Hinrichs, 1877.
[12] Eisenlohr, A.: *Ein Altbabylonischer Felderplan*. Leipzig: Hinrichs, 1896.
[13] Falkenstein, A.: *Archaische Texte aus Uruk*. Leipzig: Harrassowitz, 1936.
[14] Förtsch, W.: *Altbabylonische Wirtschaftstexte aus der Zeit Lugalanda's und Urukagina's*. Leipzig: Hinrichs, 1916.
[15] Furth, P. (Ed.): *Arbeit und Reflexion. Zur materialistischen Theorie der Dialektik—Perspektiven der Hegelschen Logik*. Köln: Pahl-Rugenstein, 1980.
[16] Fuye, A. de la: "Mesures Agraires et Formules d'Arpentage à l'Epoque Présargonique." In *Revue d'Assyriologie et d'Archéologie Orientale, 12*, 1915, p. 117ff.
[17] Gillings, R. J.: *Mathematics in the Time of the Pharaohs*. 2nd ed., Cambridge, MA: MIT Press, 1972.
[18] Goetze, A.: "Number Idioms in Old Babylonian." In *Journal of Near Eastern Studies, 5,* 1946, p. 185ff.
[19] Herodotus: *The History*. Chicago, IL: Chicago University Press, 1988.

[20] Hilprecht, H. V.: *Mathematical, Metrological and Chronological Tablets from the Temple Library of Nippur*. Philadelphia, PA: Dept. of Archaeology, University of Pennsylvania, 1906.
[21] Høyrup, J.: "Algebra and Naive Geometry." In *Altorientalische Forschungen, 17*, 1990, pp. 27–69 and 262–354.
[22] Juschkewitsch, A. P.: *Geschichte der Mathematik im Mittelalter*. Basel: Pfalz, 1966.
[23] Klengel, H.: "Einige Erwägungen zur Staatsentstehung in Mesopotamien." In Herrmann, J. & Sellnow, I. (Eds.): *Beiträge zur Entstehung des Staates*. 2nd ed., Berlin: Akademie-Verlag, 1974.
[24] Klengel, H.: *Hammurapi von Babylon und seine Zeit*. Berlin: 1976.
[25] Lamberg-Karlovsky, C. C. & Lamberg-Karlovsky, M.: "An Early City in Iran." In *Scientific American, June*, 1971, p. 102ff.
[26] Lewy, H.: "Origin and Development of the Sexagesimal System of Numeration." In *Journal of the American Oriental Society, 69*, 1949, p. 1ff.
[27] Margueron, J.-C.: *Mésopotamie ...* Genf: Nagel, 1970.
[28] Meer, P. E. van der: *Textes Scolaires de Suse*. Paris: 1935.
[29] Menninger, K.: *Number Words & Number Symbols: A Cultural History of Numbers*. Trans. by P. Broneer. New York: Dover, 1992.
[30] Needham, J.: *Science and Civilisation in China. Vol. 3: Mathematics and the Sciences of the Heavens and the Earth*. Cambridge: Cambridge University Press, 1959.
[31] Neugebauer, O.: *Die Grundlagen der ägyptischen Bruchrechnung*. Berlin: Springer, 1926.
[32] Neugebauer, O.: *Vorgriechische Mathematik*. Berlin: Springer, 1934.
[33] Neugebauer, O.: *Mathematische Keilschrifttexte*. 3 vols., Berlin: Springer, 1935/1937. (Reprint Berlin 1973.)
[34] Neugebauer, O.: *The Exact Sciences in Antiquity*. 2nd ed., Providence, RI: Brown University Press, 1957.
[35] Neugebauer, O. & Sachs, A.: *Mathematical Cuneiform Texts*. New Haven, CT: American Oriental Society, 1945.
[36] Nies, J. B.: *Ur Dynasty Tablets*. Leipzig: Hinrichs, 1920.
[37] Oppenheim, A. L.: "On an Operational Device in Mesopotamian Bureaucracy." In *Journal of Near Eastern Studies, 18*, 1959, p. 121ff.
[38] Peet, T. E.: *The Rhind Mathematical Papyrus*. London: Hodder & Stoughton, 1923.
[39] Powell, M. A.: "Sumerian Area Measures and the Alleged Decimal Substratum." In *Zeitschrift für Assyriologie, 62*, 1972, pp. 165–221.
[40] Rottländer, R. C. A.: *Antike Längenmaße*. Braunschweig: Vieweg, 1979.
[41] Sachs, A. J.: "Notes on Fractional Expressions in Old Babylonian Mathematical Texts." In *Journal of Near Eastern Studies, 5*, 1946, p. 203ff.
[42] Sachs, A. J.: "Two Neo-Babylonian Metrological Tables from Nippur." In *Journal of Cuneiform Studies, 1*, 1947, p. 67ff.
[43] Sachs, A. J.: "Babylonian Mathematical Texts" Part I–III. In *Journal of Cuneiform Studies, 1*, 1947, 219ff., and *6*, 1952, p. 151ff.
[44] Schmandt-Besserat, D.: "The Earliest Precursor of Writing." In *Scientific American, 238* (June), 1978, pp. 38–47.
[45] Schmandt-Besserat, D.: "An Archaic Recording System in the Uruk-Jemdet Nasr Period." In *American Journal of Archaeology, 83*, 1979, p. 19ff.
[46] Schneider, N.: *Die Keilschriftzeichen der Wirtschaftsurkunden von Ur III nebst ihren charakteristischen Schreibvarianten*. Rom: Papstliches Bibelinstitut, 1935.
[47] Sellnow, I. et al.: *Weltgeschichte bis zur Herausbildung des Feudalismus*. Berlin: Akademie-Verlag, 1977.
[48] Strommenger, E.: *Habuba Kabira. Eine Stadt vor 5000 Jahren*. Mainz: Zabern, 1980.

[49] Struve, V. V.: *Mathematischer Papyrus des staatlichen Museums der Schönen Künste in Moskau.* Berlin: Springer, 1930.
[50] Tropfke, J.: *Geschichte der Elementarmathematik. Vol. 1: Arithmetik und Algebra.* New edition of H. Gericke et al., 4th ed., Berlin: de Gruyter, 1980.
[51] Vogel, K.: *Vorgriechische Mathematik. Part 2: Die Mathematik der Babylonier.* Hannover: Schroedel, 1959.
[52] Vogel, K. (Ed.): *Chiu Chang Suan Shu: Neun Bücher Arithmetischer Technik. Ein chinesisches Rechenbuch für den praktischen Gebrauch aus der frühen Hanzeit (202 v. Chr. bis 9 n. Chr.).* Braunschweig: Vieweg, 1968.
[53] Weissbach, F. H.: "Die Senkereh-Tafel." In *Zeitschrift der Deutschen Morgenländischen Gesellschaft, 69,* 1915, p. 305ff.
[54] Wussing, H.: *Mathematik in der Antike. Mathematik in der Periode der Sklavenhaltergesellschaft.* Leipzig: Teubner, 1962.

CHAPTER 8

THE FIRST REPRESENTATIONS OF NUMBERS AND
THE DEVELOPMENT OF THE NUMBER CONCEPT
coauthored by Robert K. Englund and Hans J. Nissen
[1988]

Due to early practice and habit we are so familiar with manipulating numbers that we hardly ever stop to think about the intellectual achievement represented by the detachment of the designation for a quantity from the counted object itself. Ethnological investigations of primitive peoples without an abstract number concept in our present sense first indicated that this separation must not be taken for granted. Despite this, archaeologists always assumed the existence of such a number concept when interpreting the oldest known written documents which come from the period between 3200 and 3000 B.C.

Most of these texts were found in the ancient city of Uruk in southern Mesopotamia. We reported in an earlier article ([3]) on our progress regarding the decipherment of these written documents known as archaic texts. It was our goal to show what kind of information the newly acquired knowledge provided with regard to the development of writing. In the following we chose another aspect of our work as the subject of our report: the analysis and interpretation of the numerical signs in these texts and the consequences for the development of mathematical thinking.

CHARACTERISTICS OF THE NUMERICAL SIGNS
IN THE ARCHAIC TEXTS

In fact it turned out that the idea of numbers as general, object-independent magnitudes was by no means established at the time of the emergence of writing as had been rather uncritically assumed. While the evolution from the symbolic representation of a quantity of certain objects to the purely mental construct of the number had been initiated, it had not yet been completed. In that respect the numerical signs in the oldest written documents also shed light on a critical phase of the cognitive development of humankind.

Our analysis was based on a computerized data base containing transcriptions of most of the extant archaic texts. By employing predominantly statistical methods we could, for the first time, conduct a comprehensive investigation of all aspects of how the oldest numerical signs were used. However, before we discuss the more general aspects of this topic, we would like the reader to turn detective so that he or she can, with regard to the crucial points, participate in solving this puzzle.

The archaic texts contain approximately 1,200 signs and variants of signs that differ in terms of their meaning (because of the difficulty of distinguishing such variants of signs from differences in writing the signs, their exact number is hard to ascertain, especially since the so-called Lexical Lists, the main source for the identification of signs, only contain 706 signs). Of these the approximately 60 numerical signs immediately stand out because of the peculiar way in which they are written: Normally they were not—like the other signs—incised into the clay surface with a pointed stylus; rather, they were pressed deeply into it with a round stylus held vertically or obliquely. In addition, some of them bear markings incised with a pointed stylus.

Since all administrative texts contain numerical signs, some of them are by far the most frequently occurring signs of the archaic texts. Due to the character of these texts as economic records the numerical signs are at the same time the main vehicle of information. This prominent role demands that they be given special attention concerning their decipherment.

EARLIER ATTEMPTS AT INTERPRETATION

In contrast to the other written signs some of the numerical signs—and especially those most frequently used—remained in use for centuries almost without any change whatsoever with regard to their appearance. Thus, they are to be found even in relatively late groups of texts that can be almost completely read. They still appear as late as the Ur III period approximately 1,000 years later when they were, however, finally replaced by their cuneiform counterparts.

For this reason the familiar sexagesimal signs for the numbers 1, 10, 60, and 600 in the archaic texts could apparently be deciphered without effort. The few available texts that were in good condition and that, in addition to

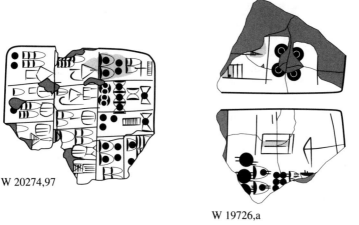

Figure 1: The use of diverse systems of numerical signs for different objects or classes of objects is characteristic of the numerical notations on the oldest known written documents, the archaic tablets of Mesopotamia. It clearly indicates that the number concept had become only partially detached from the objects counted. The text on the left is an account of dairy and textile products. In the first case of the third column 1223 (2×600, 2×10, and 3×1) units of an as yet unidentified product from goats are recorded in a sexagesimal system. In the case beneath 18120 (2×7200, 3×1200, and 1×120) units of another dairy product (presumably cheese) have been recorded in a bisexagesimal system. The fragment to the right comes from a tablet containing information on field areas and quantities of grain. On the obverse (top) an area is recorded in the so-called GAN_2 system comprising roughly 643 acres, an estimate based on the meaning of the signs in subsequent periods. This side presumably contained several entries concerning areas of fields together with the yields harvested. On the reverse (bottom) the total of these yields has largely been preserved. It has been recorded in the ŠE system used specifically for recording quantities of grain comprising 180161 (3×54000, 2×9000, 5×30, 2×5, and 1×1) units of a capacity measure of approximately 1 gallon (that is to say, a total of more than 190210 gallons). Presumably the administration of such quantities which far exceeded the scale of a rural economy played a decisive part in the early development of arithmetical thinking. (Drawing by Robert K. Englund.)

individual numbers, apparently contained their sums as well—in most cases on the reverse—seemed to confirm these identifications; for after inserting the numerical values known from later texts, arithmetically accurate calculations were the result.

This seemed to prove that the numerical signs had their later arithmetical meaning from the very beginning. Their decipherment was achieved with a particular sense of confidence. However, the fact that the apparent certainty only applied to a few of the approximately 60 numerical signs was largely suppressed. With regard to calculations inconsistent with the arithmetical meanings that were postulated, the tendency was to assume fundamental mistakes on the part of the scribes rather than questioning the modern arithmetical meanings ascribed to them.

In addition, the numerous peculiarities that distinguished the archaic numerical signs from modern numerals were thought to be of no fundamental importance. The fact that the sign for 60, for example, had the meaning 300 when the text dealt with grain and the sign for 10 the meaning 18 in the context of field areas, was considered an unenlightening exception at best. Mostly it was disregarded. Nobody expected what we know with certainty today—that is, that particularly the numerical signs occurring most frequently in the archaic texts changed their arithmetical meaning depending on the area of their actual application.

In 1936, the German sumerologist Adam Falkenstein for the first time reported on the archaic texts found during the early excavation campaigns. In his publication he attempted to cope with the large number of peculiar numerical signs by assigning them to three different groups; he interpreted the first as a sexagesimal system, the second as a competing decimal system, and the third as a system of fractions.

However, even then there were many more signs than numbers, and Falkenstein was compelled to interpret many graphically distinct signs as different ways of writing the same number. Despite these obvious weaknesses Falkenstein's list of the archaic numerical signs remained unchallenged as the authentic source for the history of mathematics for almost 40 years.

Not until the middle to late 70s the Soviet sumerologist Aisik Vaiman and the Swedish mathematician Jöran Friberg turned their attention in several publications once again to the problem of the meaningful categorization and interpretation of the numerical signs contained in the archaic texts published by that time. However, to date their investigations have not received the attention they deserve. Although they came to some extent to different results, they made a few important discoveries.

Thus, Vaiman showed that the numerical signs of the archaic texts can be fitted into a much larger number of systems with much greater plausibility than Falkenstein had assumed. Furthermore, Friberg demonstrated that the assumption of a competing decimal system was based on a logical error that can be traced all the way back to the publication of the first proto-

Elamite texts by Vincent Scheil in 1905 (cf. [6]). Because his identification of a decimal system had been uncritically cited over and over again, the error had become firmly established so that, what could have been easily refuted by careful analysis of the summations contained in the texts, was considered sound knowledge.

This was essentially the situation when we decided in 1983 to undertake a completely new analysis of the numerical signs of the archaic texts based on all the tablets that had been excavated. The project was an initial test whether it would pay off to commit modern techniques of data processing for deciphering and interpreting the archaic texts for the first time and on a large scale. As a joint project of the Seminar of Near Eastern Archaeology at the Free University of Berlin and the Max Planck Institute for Human Development and Education a computerized data base had been created at a main frame of the computer center of the university, now available in a preliminary form for the electronic processing of the data.

RULES FOR USING THE SIGNS

We chose a fundamentally different method for the analysis than that commonly used up to that point. People had always attempted to obtain the meanings of numbers as directly as possible from the few texts containing arithmetical operations and to use the values determined in this fashion for interpreting those texts from which the arithmetical meaning could not be inferred directly.

Each arithmetical activity, however, takes place at two entirely different levels. Initially it simply consists in the mechanical application of rules to certain notations. Only then does it include a certain numerical understanding as well by means of which a certain meaning is attributed to the rules and the results of their application. Instead of inquiring into this numerical understanding right away, we first tried to find out as much as possible about the rules for using the signs without worrying about the meaning too much.

In this process the many hundreds of fragments with numerical notations that were lacking any context and therefore seemed to be worthless suddenly proved to be an exceedingly rich source of information. Our procedure for reconstructing the rules for handling numerical notations from the fragments can be compared to a computerized puzzle. Every substantiated or assumed rule was immediately checked against all the available texts by means of the computer. How many examples could be found that supported it? How many exceptions or counterexamples were there?

The ambiguity of the signs, a cause of frequent irritations in the past, hardly caused any trouble in this regard, since only the rules for using the signs were explored, not their meaning.

Three very general rules are characteristic of the representation of quantities by means of archaic numerical signs: We shall call them the correspondence rule, the substitution rule, and the seriation rule. They will be reviewed briefly before we turn our attention to more specific rules.

According to the correspondence rule quantities are represented by sign repetitions. The sign for the counting or measuring unit is repeated as often as this unit is contained in the quantity. However, this is done only up to a certain limit determined by the context.

If the quantity to be represented exceeds this limit, then—according to the substitution rule—a sign with a higher rank replaces a certain number of repetitions of the sign with the lower rank. In this case, the sign with the higher rank is itself treated as a unit and, if necessary, repeated according to the correspondence rule.

Finally, the seriation rule states that the signs of a numerical notation are always recorded in the order of their ranks beginning with the sign of the highest rank.

These rules have been known for so long and apparently have been considered so trivial that a closer inspection did not seem worthwhile. If the signs were interpreted as symbols for numbers, then the rules seemed more than plausible as rules of a system of number notation comprising units of a different order. If, for instance, a certain sign represented the number 1 and another the number 6, then the first sign could by necessity be repeated at most five times.

Other explanations are conceivable however. The first sign could, for example, represent a small measuring cup and the second a larger one with six times the volume. In this case it would be obvious as well why the smaller sign could only be repeated five times.

For the time being we did not try to answer such questions; instead, we focussed on how often particular signs were repeated at most in different contexts. This allowed us to ascertain the value of the sign with the higher rank and to determine how this value depended on the context.

THE FIRST REPRESENTATIONS OF NUMBERS

SUMMATIONS

It is relatively easy to determine the "value" of a sign, if there is a sum in which this sign replaces a certain number of repetitions of the next sign with a lower rank. For instance, on the reverse the text W 20676,2 (Fig. 2, left) there is the sum of the two notations on the obverse. The calculation is as follows:

$$1\bullet 7\flat\ +5\flat\ =2\bullet 2\flat$$

It can be concluded from this that ten ⟩ are replaced by one • or:

$$1\bullet = 10\flat$$

It is, however, not easy to determine the extent to which these results can be generalized; for the arithmetical use of the archaic numerical signs is always ambiguous. Thus, the text W 15897,c21 (Fig. 2, right) contains the sum

$$3\flat\ +3\flat\ +3\flat\ +3\flat\ =2\bullet$$

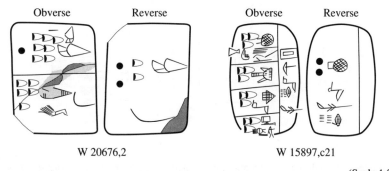

Obverse   Reverse   Obverse   Reverse

W 20676,2   W 15897,c21

(Scale 1:2)

Figure 2: Arithmetical ambiguity is one of the most peculiar characteristics of the numerical signs on the earliest known written documents, the approximately 5,000-year-old archaic texts from Mesopotamia. As indicated by the texts depicted above, the arithmetical meaning of these signs changes with the area of application. In both texts the sum of the entries on the obverse has been noted on the reverse. In the upper tablet where beer jars have been recorded, 17 units and 5 units are added together. The result is 22 units. In this case the sign • indicates ten times the value of the sign ⟩ . In the text below four times 3 units of an unknown grain product are added up. The result is 12 units. Here the sign • has the value of 6 ⟩ . (Drawing by Robert K. Englund.)

Following the previous example the result 1 • 2 › could be expected. The sum 2 • implies that, in the present context, the sign • does not represent the tenfold, but only the sixfold value of the sign › .

However, it is not just due to the fact that the arithmetical meanings of the numerical signs change with the context that the sums cannot always be interpreted as easily as in the above examples. If several complicated numerical notations are summed up and a corresponding number of necessary substitutions is performed, then the problem becomes a tricky numerical puzzle that often has several possible solutions.

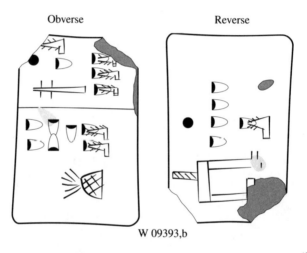

W 09393,b

(Scale 1:2)

Figure 3: Sometimes the relationships between the values of the different signs can be determined even in the case of algebraically ambiguous sums by means of plausible additional assumptions. An example for this is the decipherment of an archaic numerical sign system, the so-called EN system. The text W 09393,b contains the summation

1 • 1 ›   3 ▱ + 2 ›   1 ⁻ 1 ⁻ 2 ▱ = 1 • 4 ›   1 ▱

that, as an equation with four unknown quantities would have many solutions. However, the order of the signs can be inferred from the sequence of the signs in the notations. Furthermore, it can be assumed that the signs with higher values are integral multiples of those with lower values. The maximum number of repetitions of any sign finally indicates the minimum value of the sign next in rank. By means of these three supplementary conditions the relationships between the four signs as specified in Figure 6 can be unambiguously determined. (Drawing by Robert K. Englund.)

For instance, the sum in the text W 9393,b (Fig. 3) only permits us to infer the following equation:

$$1\,\overline{\underline{}} + 1\,\overline{} + 4\,\text{⟨sign⟩} = 1\,\text{⟨sign⟩}$$

This equation contains four unknown quantities and thus, in general, many solutions. However, in this case fortuitous circumstances nevertheless make it possible to determine the quantitative relations between the signs in this special case. The order of the signs can be inferred from the sequence of the signs in the notation. Furthermore we know that, as a rule, the numerical signs with higher values represent integral multiples of the numerical signs with lower values. Finally, the sign ⁻ must have at least four times the value of the sign ⟨sign⟩, since the latter is repeated three times on the tablet. The following quantitative relations can be positively concluded from these conditions:

$$1\,\text{⟨sign⟩} = 2\,\overline{\underline{}}$$
$$1\,\overline{\underline{}} = 2\,\overline{}$$
$$1\,\overline{} = 4\,\text{⟨sign⟩}$$

In general, the possibilities of determining the values of numerical signs by means of sums are very limited, if only for the simple reason that only a few of the archaic texts contain sums; of these only a fraction is preserved well enough to permit the identification of all the notations that are summed up. However, there is another, much more direct way of inferring the values of the signs. A probabilistic experiment may serve to demonstrate this.

It is common knowledge that there are dice that do not have six faces; there are also those in the shape of a tetrahedron with only four faces. Let us assume that someone informs us of the successive results of rolling a die without telling us how many faces there are in that die.

At the beginning we would know little about the number of faces. It would only be certain that it had to be at least as large as the largest number coming up. However, with every roll of the die the probability decreases for the highest number not to come up, and we would ultimately almost certainly know how many faces there actually are in the die. We would even notice it if, under certain conditions, the die were to be exchanged for a different one.

284   CHAPTER 8

STATISTICS AS A METHOD OF DECIPHERMENT

Similarly the numerical notations of the archaic texts can be interpreted as carrier of information about the values of the numerical signs involved. For instance, every fragment containing the repetition of a sign provides information on how many times this sign can be repeated at a minimum in a certain context. With every new fragment the probability increases as well that the maximum number of possible repetitions actually occurs at least once.

This argument, for instance, provides a simple statistical proof that the common assumption is false—that is, the assumption that, in addition to a sexagesimal system, a competing decimal system initially existed in Mesopotamia. In this fashion we obtain reliable and independent confirmation of Jöran Friberg's findings mentioned earlier that were based on sums—that is, obtained by a different method.

The proof is obtained in the following way. The sexagesimal system is based on the quantitative relations:

$$10 \, \text{᠈} = 1 \, \bullet$$
$$6 \, \bullet = 1 \, \text{)}$$

For the supposed decimal system, however, the following quantitative relations would have to be true:

$$10 \, \text{᠈} = 1 \, \bullet$$
$$10 \, \bullet = 1 \, \bullet$$

If the sign • is repeated more than five times within a numerical notation without having been replaced by the sign next in rank, then this numerical notation cannot belong to the sexagesimal system.

However, this does not at all mean that it must therefore represent a decimal system. This hasty assumption is clearly refuted by the statistics. Our data base mentioned above of the archaic texts from Uruk and Jemdet Nasr as well as contemporary texts from other locations lists 46 notations containing the sign • being repeated more than five times together with the sign ᠈ ; in none of them the sign ᠈ is repeated more than five times (Fig. 4). The statistical probability that this is a coincidence is only about 0.02 percent.

Thus, the statistic contradicts the quantitative relations of the alleged decimal system. Rather, it confirms the quantitative relations of a system which we, because of its use in conjunction with grain (represented by the non-numerical sign ŠE), called the ŠE system:

$$6 \, \text{᠈} = 1 \, \bullet$$
$$10 \, \bullet = 1 \, \bullet$$

THE FIRST REPRESENTATIONS OF NUMBERS

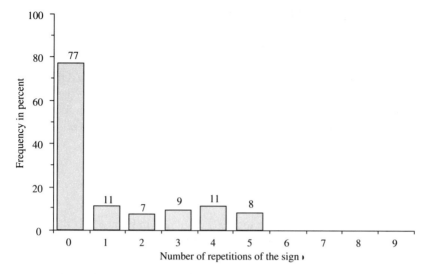

Figure 4: Statistics provide by far the most instructive method for deciphering the numerical signs. While there are only a few tablets with summations yielding results that can be unambiguously determined, all the available texts can be included in a statistical analysis. The statistical proof that the sign ● changed its value according to the context of its application may serve as an example for the efficiency of such analyses. In the sexagesimal system the following relationships are true: 10 ▶ = 1 ● and 6 ● = 1 ▶   Thus, in this system the sign ● can be repeated at most five times. However, on a number of tablets where quantities of grain are recorded this sign is repeated more than five times as well. According to the statistical analysis the result of which is shown in the above chart the lower valued sign next in line ▶ is never repeated more than five times in all these cases, although, in the sexagesimal system, it may actually even be repeated up to nine times. There is only a 0.02 percent probability that this is merely a coincidence. Thus in this so-called ŠE system primarily used for grain the following is valid: 6 ▶ = 1 ● (and 10 ● = 1 ●) The fact that, in texts with more than five repetitions of the sign ●, the frequency of notations completely lacking the sign ▶ is greater than average may be explained by a tendency toward quantities represented by round numbers.

This and similar statistical methods permitted the systematic examination of the extent to which the results secured by analyzing individual key texts could be generalized. Determining the semiotic rules according to which the signs were manipulated from the more than 6,500 known numerical notations turned out to be such a complicated puzzle primarily because a single result does not often permit far-reaching conclusions. Only as the work progressed did the different individual results slowly come together to reveal a consistent overall pattern.

286   CHAPTER 8

THE DECIPHERMENT OF CALENDAR ENTRIES

The decipherment of the calendar entries (U_4 system, Fig. 6) is one example demonstrating how various types of information must be combined with each other to solve such a puzzle. It had long been believed that combinations of the sign U_4 (◊) with numerical signs represented calendar entries. However, hypotheses regarding the precise meaning diverged widely. As we now know, Vaiman was the only one who was correct in his assumption:

$$◊\text{-} = 1 \text{ day}$$
$$◊ = 1 \text{ month}$$
$$\text{-}◊ = 1 \text{ year}$$

We found the confirmation in the texts compiled in Figure 5. If the indications they contain are combined, then it is not only possible to decipher the meaning of the calendar entries, but also to determine that, as early as in the archaic texts, the administrative month was calculated as having 30 days and the administrative year as having 360 days.

Actually, the text MSVO 1, 21 (Fig. 5) could already have yielded the solution, if it had not been the case that one of the signs was written in a very misleading way. This text contains the calendar entry ◊, to be read as "1 month 5 days," although it looks more like "1 month 14 days" because of the sign at the top right that resembles a • more than a ▫. How many days does a month actually have?

The quantity of grain 3 ▸ 2 — 1 ⁼ (ŠE system, cf. Fig. 6) recorded after the statement of the time period provides an explanation. By means of the quantitative relations determined by us it proves to be 35 times the amount corresponding to the sign ⁼. If this sign represented one daily ration, then "1 month 5 days" would have to mean 35 days and the month would have to be 30 days long.

However, as already mentioned, the period recorded in this text was initially misread. The actual decipherment of the calendar entries was therefore a little more complicated. It was closely associated with the interpretation of the numerical functions of the signs |₁ and |▷ that are not related to the calendar entries themselves.

In the text MSVO 1, 27 (Fig. 5) the sign |₁ represents quantities of grain amounting always to 10 percent of the previously recorded amount of grain. This might of course be a coincidence; however, other texts confirm this observation. Among other things this is true of the text MSVO 1, 21

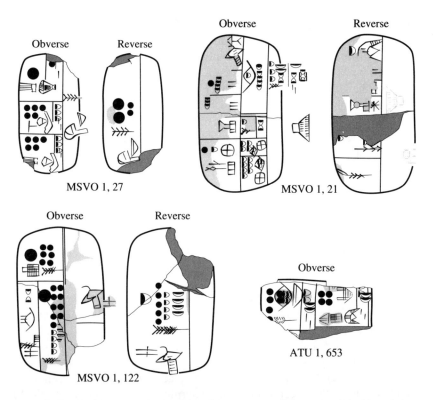

Figure 5: The decipherment of calendar entries was successfully accomplished by solving the puzzle raised by the four texts reproduced above. On the tablet MSVO 1, 27 quantities of grain have been recorded which are added up on the reverse. This text reveals that the sign ╷₁, the exact meaning of which is unknown, always occurs next to a quantity of grain that amounts to exactly 10 percent of the amount preceding it. The badly damaged tablet MSVO 1, 21 possibly is a record of feed grain for sheep. In the first case of the second column there is the statement "1 month 5 days" followed by an amount of grain amounting to 35 times that which is represented by the sign ⁼, followed in turn by a supplement of 10 percent (rounded off). If it is assumed that the sign ⁼ represents a daily ration, then the result is a month of 30 days. However, at first this conclusion was not very firm, since the calendar entry on this tablet is inconclusive and can be interpreted as "1 month 14 days" as well. The tablet MSVO 1, 122, an account of grain for an unknown purpose, contains the statement "3 years" (⇒) in the second case of the first column below the sign ⁼ and in the following case a quantity of grain amounting to exactly 3 times 360 ⁼ plus 10 percent. This indicates that the year was calculated as having 360 days. The text fragment ATU 1, 653 begins with a quantity of grain followed by the statement "24 months" and finally the sign ▷ with the meaning of "(grain) ration." It is known from other texts that this sign represents $\frac{1}{3}$ ⁼ of grain. The amount of grain recorded on the tablet amounts to 24 times 30 times $\frac{1}{3}$ ⁼ which confirms the above assumption of a month having 30 days. Here again the supplement of 10 percent is recorded separately. (Drawings by Robert K. Englund.)

mentioned earlier. The notation 1 — 1 ⸚ 1 ▨ to which the sign |₁ has been attached representing $3\frac{1}{3}$ daily rations is so close to 10 percent of 35 rations that the difference may easily be explained as a mistake in rounding off.

The text MSVO 1, 122 which contains the sign ⇒ denoting "3 years" became the key to the decipherment (Fig. 5). Initially only a very bad copy of it and a photograph were available. On the basis of these sources it seemed possible that the sign ⸚ next to the calendar entry for "3 years" indicated exactly one daily ration and that the amount of grain listed following the notation could be calculated as 3 times 360 daily rations according to which one year would have had 360 days.

One of us (Englund) examined this question in the Iraq Museum in Baghdad on the basis of the original of the text. It appeared that the quantity mentioned in the text was greater thus apparently disproving the assumption concerning the length of the year. However, after carefully cleaning the text he found that the amount of grain is exactly 10 percent higher than the value to be expected—the very 10 percent listed separately in text IM 55582.

Finally, the fragment ATU 1, 653 (Fig. 5) that did not make any sense before provided a splendid independent confirmation of this interpretation. From other texts we already knew that the sign ▷ the image of a ration bowl, represents a smaller ration—typically an amount of grain of $\frac{1}{3}$ ⸚. This probably represented the daily ration of the workers at the lowest level who were dependent on the temple and palace administration, a ration of 0.6 to 0.8 liters that, according to our determination of quantities of the ŠE system, was close to the subsistence level.

According to the calendar entry of the text the latter ought to record such small rations for "24 months" or 24 times 30 days. In the context of the ŠE system 24 times 30 times $\frac{1}{3}$ ⸚ does in fact result in the amount of 4 • , recorded before the calendar entry. Furthermore, the quantity of grain of 2 › 2 — recorded in a separate case together with the sign |₁ does indeed correspond to 10 percent of this amount.

NUMERICAL SIGN SYSTEMS WITH SPECIFIC AREAS OF APPLICATION

To a large extent the rules that had been identified concerning the use of the signs were consistent with the usual attribution of numerical values. However, in addition we found rules that were inconsistent with the usual interpretation of the signs as numerical signs of an integrated system.

There are, for instance, three signs to which the numerical values of 60, 120, and 600 have been attributed. We found the curious rule that the signs for 120 and 600 never occur together in the same notation. If the sign for 120 occurs, it is sometimes repeated more than five times without ever being replaced by the sign for 600. If, on the other hand, a numerical notation actually contains the sign for 600, then the sign for 60 is repeated instead of being replaced by the sign for 120.

The overall result of our puzzle provided a plausible explanation for these and other peculiarities. We found out that the more than 6,500 numerical notations could be assigned almost completely to a certain number of different systems of numerical signs in which the same signs have in part different numerical meanings. Thus, the signs for 120 and 600 never appear together simply because they belong to different sign systems. A thorough analysis has shown that, with few exceptions, the different systems have areas of application that are strictly separate.

The 15 systems that have been identified can be assigned to five basic systems (Fig. 6): the sexagesimal system, the bisexagesimal system, the ŠE system, the GAN_2 system, and the EN system (ŠE, GAN_2, and EN are the conventional readings of signs associated with the systems). For some of these basic systems there are derivative systems with the same arithmetical structures, but with other uses and a modified graphical form of the signs. In addition, there are some systems such as the U_4 system for calendar entries that combine numerical with non-numerical signs thus linking them with specific metrological meanings.

Finally, there are non-numerical written signs and combinations of non-numerical and numerical signs that mainly express certain subjects or situations (and are thus ideographic in character), but in part also possess a numerical meaning. There are, for instance, sign combinations for certain herd animals with numerical signs fusing the name of the animal with the statement about its age to form a single sign. In the archaic texts the transitions between quantity, metrology, and ideography are fluid. For this reason distinctions between numerical signs, denotations of measures, and non-numerical signs always seem to be somewhat arbitrary.

The attribution of the various systems to specific areas of application did not follow obvious general rules. The sexagesimal system and the bisexagesimal system as well as the systems derived from them were systems for discrete objects with the sexagesimal system having by far the most varied application.

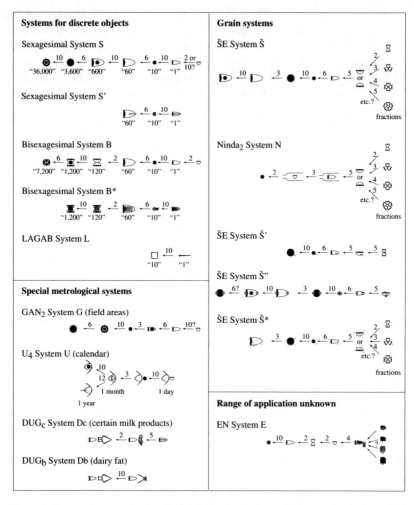

Figure 6: The computer-assisted analysis of all numerical notations occurring in the roughly 4,000 electronically stored texts and text fragments of the oldest known tablets identified 15 numerical sign systems with distinct areas of application belonging to five basic systems. In addition there were indications of at least two more systems for which there is not enough evidence in the available texts in order to reconstruct them. Within the systems some of the same signs are used with a completely different arithmetical meaning.

THE FIRST REPRESENTATIONS OF NUMBERS   291

The bisexagesimal system was used chiefly for grain products as long as the latter were not measured by means of capacity measures and therefore recorded in the ŠE system. In addition, it was also commonly used for a certain dairy product as well as a certain type of fish.

The derivative systems were used for very specific purposes: the system S' derived from the sexagesimal system, for instance, in documents dealing with animal husbandry presumably for recording the animals slaughtered in the course of the current accounting periods of one year as well as in texts on the production and distribution of beer for recording a particular type of beer. The ŠE system and the systems derived from it denoted capacity measures for grain and grain products with the different systems presumably symbolizing different types of grain. The GAN_2 system was used to register field areas. The use of the EN system is unknown.

There are clear indications as to the existence of other archaic numerical sign systems that we do not yet know. The very fact that 25 of the 26 records of the EN system were found at a single location of the same archaeological site should give us pause with regard to the completeness of our knowledge. If by chance there had been no excavations at this place, we would only have a small fragment with the sign ▱, which would not have provided any clues as to the system it was based on.

Ten of the signs of our list of 60 numerical signs do not appear in any of the reconstructed systems. The reasons for this vary. Three of the signs, for example, only occur on a small tablet which we interpret as the clumsy attempts of a pupil largely unversed in writing. In the case of four other signs we are not sure whether they might perhaps not be variants of other numerical signs or whether they are numerical signs at all. However, at least in the case of two of the ten signs, that is, the signs ▱ and ▱ we are, due to the way they are used, convinced that they belong to two other systems. The reason why we do not know very much about these systems is due to the fact that no authoritative texts have been transmitted.

THE LACK OF AN ABSTRACT NUMBER CONCEPT

In light of all this the question arises to what extent these arithmetical notations can be considered representations of "numbers" at all given their complicated rules of application partially limited to specific contents, their numerous substantially ambiguous meanings and the fuzzy boundaries between their operative and their ideographic functions (their functions as numerals and as signs for concrete objects). What is the stage of development they represent in a civilization that, a thousand years later, produced a highly developed mathematics as the first civilization on this earth?

Up to now it was generally believed that the archaic numerical signs were early representatives of the particular notion of numbers which later distinguished the arithmetic in Mesopotamia. This view seems to be quite reasonable. The sexagesimal system which soon became the predominant system survived the archaic period by about 1,000 years without any graphical modifications and another 2,000 years in a modified, but structurally equivalent representation by means of cuneiform signs. In addition, sources containing information about Sumerian number words exhibit an extraordinary structural similarity of the Sumerian sequence of number words to the sexagesimal system (however, these sources date back to a time when Sumerian had not been spoken for hundreds of years).

This was sufficient reason that, to the scientists dealing with later texts, the sexagesimal system necessarily appeared to be the true system of numerals while everything else seemed to be metrology. Further support was added by the subsequent development of the other systems. For, as a matter of fact, the bisexagesimal system was very soon, that is, during the early dynastic period, replaced by the sexagesimal system. The ŠE system even disappeared already at the end of the archaic period. For the EN system there are only texts from the oldest stage of writing at Uruk. Only the GAN_2 system persisted in addition to the sexagesimal system and developed into a comprehensive system of area measures.

The results of our analysis of the archaic numerical signs and the rules for their use, however, contradict this view. It is true that the sexagesimal system was already fully developed during the archaic period; and it occupied a privileged position in as much as it was applied to a particularly large number of object categories. Despite this, less than 50 percent of all numerical notations of the archaic texts belong to this system. Although the same signs were generally used in the remaining notations, they were used as signs belonging to other systems and thus with a different arithmetical meaning.

Thus, particularly the signs of the sexagesimal system apparently representing abstract numbers in reality were especially variable with regard to their arithmetical meaning. The system in which they were used, and thus their numerical value, changed in accordance with their particular context. Only by disregarding the evidence provided by our statistical analyses a fixed numerical value can be attributed to them as their "proper meaning" permitting their interpretation as numerals of an overarching abstract number concept. Paradoxically, it is especially the signs of the derivative systems, that is, signs associated with specific contents, rather than the "abstract" signs that were used with an unequivocally determined arithmetical meaning.

NUMBER ANALOGUES

What is the theoretical alternative to an interpretation of the signs as signs for numbers? In a famous paper on the thinking of primitive peoples published in 1912 the Gestalt psychologist Max Wertheimer gave a very precise description of the psychological nature of arithmetical notions prior to the development of an abstract number concept. On the basis of research reports on primitive peoples he postulated psychological structures which he called number analogues; while they are associated with certain contents and are, therefore, less abstract than our numbers, they nevertheless serve similar purposes.

There are by now numerous ethnological studies confirming Wertheimer's assumptions. They indicate that many preliterate cultures developed a host of context-dependent, proto-arithmetical techniques which neither presupposed nor entailed an abstract number concept.

Numerous characteristics of the archaic numerical sign systems analyzed by us place them in the vicinity of such proto-arithmetical techniques. For instance, the numerical signs actually did not designate numbers, but units. Thus, the number four was not represented by a symbol like our numeral "4"; rather, it was represented downright physically by repeating a sign for the unit four times. This means that there was no formal distinction between the representation of objects and that of numbers. Thus, the rules for applying the numerical sign systems could also be used according to the proto-arithmetical notion that the four repeated signs represented the four depicted objects rather than the number four.

Furthermore, there are no explicit, context-independent arithmetical operations in the archaic texts. In summations the envisioned compositions of sets of objects are simply performed with the signs representing them. Accordingly, there is no operation which inverts an addition—that is, there is no subtraction. The rare multiplicative relationships are conceived as real interrelations—for instance by expressing that certain amounts of grain correspond to certain numbers of breads—and were presumably largely transformed by means of additive operations. Numerous "atypical" examples of notations, that is, notations modified by the context, give the impression that designating the content as a memory aid always had priority over sticking to formal rules for the use of the signs.

However, these proto-arithmetical characteristics do not mean that there is no difference between the proto-arithmetic of preliterate cultures and the arithmetic revealed by the archaic texts. It is true that the numerical notations contained in them show characteristics that were, up to that point, known only from preliterate cultures; but at the same time they display a formal complexity and a capacity for mastering complex arithmetical problems that would be unimaginable in such cultures. It is this that makes these sources so unique for studying the development of the number concept.

The analysis of the operations carried out by means of the numerical signs permits inferences as to the underlying cognitive processes. Thus, we already encounter first abstract operations expressed in terms of corresponding symbolic actions such as already standardized operations of substituting iterated units by higher value units. On the other hand, however, context-dependent mental operations persist which refer to the concrete actions of their areas of application. They determine the variation of arithmetical meanings according to the context in which they are actually applied.

FROM NUMBER ANALOGUES TO THE ABSTRACT NUMBER CONCEPT

Up to now the investigation of the early development of the number concept was always faced with a dilemma: It is true that, on the one hand, the extant preliterate cultures provide insights into very early forms of this development; however, on the other hand, there is an unbridgeable gap with regard to the documents of the developed mathematical thinking of the early civilizations and of classical antiquity.

In the preliterate cultures it is not only certain forms of abstract thinking that are missing, but also the problems and the arithmetical techniques which would allow them to become manifest. By contrast, abstract thinking and abstract symbolic operations have so far always been found together in the literate cultures. The question whether a development took place between the two stages or whether we are merely dealing with the epigenetic, that is, as if biologically determined, unfolding of a universal structure of human thinking, remained the subject of controversial speculations.

The numerical sign systems of the archaic texts now provide the missing link: It proves that the development of the basic structures of arithmetical thinking is culturally determined.

The identification of this missing link permits the exact dating of a crucial stage of development so that we can attempt to reinterpret the available documents of early arithmetical thinking. This leads us more or less to the following overall picture of the development toward the abstract number concept (Fig. 7).

First archaeological evidence indicating the construction of correspondences between objects and counting symbols in the broadest sense can already be found in the late Paleolithic Age (approximately 40000 to 10000 B.C.). However, the counting symbols used hardly exhibit systematic structures that would allow an inference as to their operational use.

This preoperative stage of the development of the number concept is followed by a proto-arithmetical stage where conventions concerning the symbolic representation of quantities were introduced for the first time. In the cognitive domain there was a parallel construction of number analogues. Presumably the basis for this development is the so-called agrarian revolution—the transition to systematic animal husbandry and food production—with its requirements concerning storage and the organization of the division of labor. In the Near East the period for which evidence of the use of counting symbols exists does in fact correspond roughly to the period of the emergence of settlements and the expansion of agriculture and animal husbandry.

An analysis of the numerical sign systems of the archaic texts of Mesopotamia gives us a snapshot of the transition from this proto-arithmetic to arithmetic in the narrow sense. Very soon the peculiar characteristics of the archaic systems of numerical signs disappear, and a context-independent "primary" arithmetic already develops in the course of the third millennium B.C. As was the case in other early civilizations this development is advanced by the administrative officials of the state bureaucracies as well as by the schools of scribes where they were trained.

It is, however, still a long way from this primary arithmetic which, while context-independent, nevertheless remains tied to certain concrete forms of numerical representation by means of symbolic systems, to the intellectual concept of the abstract number. Babylonian mathematics which developed from this tradition and first flourished during the first third of the second millennium B.C. (that is to say, about 1,500 years after the invention of writing) knew neither definitions nor proofs; it did not even have a term comparable in its meaning to our word "number."

It therefore appears that structural statements about ideal objects and thus the development of the number concept had to await Greek antiquity.

Examples	Stages of the Development of the Number Concept	Chronology
	Preoperative Stage: First assignments of symbols to objects, however no systematic structures.	approx. 40000 to 10000 B.C. Early Paleolithic Age
	Proto-arithmetical Stage: Control of quantities by means of one-to-one correspondences between objects and symbols. Transmission of standardized symbolic systems. Introduction of higher value units by means of substitution.	starting around 10000 B.C. Preliterate sedentary civilizations
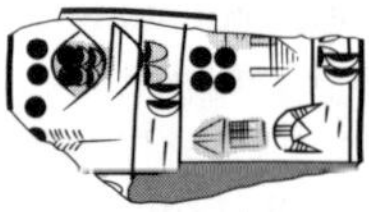	Archaic Arithmetic: Developed arithmetical techniques, however no context-independent meaning of the "numerals" as yet.	around 3200 B.C. Literate sedentary civilizations
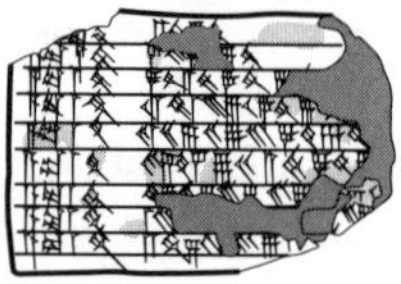	Primary Arithmetic: Universal representation of numbers, explicit arithmetic, however no number concept apart from its representation. No concept formations related to numbers as ideal objects.	starting around 3000 B.C. Highly developed early civilizations
EVCLIDIS ELEMENTORVM LIBRI XV	Abstract number concept: Conceptualization of numbers as ideal objects. Definition of numbers and proof of their arithmetical characteristics.	starting around 500 B.C. Greek antiquity

Figure 7: Historical stages of the number concept. The abstract number concept is the result of a long developmental process with the invention of writing constituting a clearly defined demarcation. This development began with one-to-one assignments of signs to objects. Standardized "means of counting" transformed such assignments into a socially transmitted technique. With the invention of writing "numerical signs" became an efficient second order means of representation, although initially the representation of numbers remained closely associated with the objects counted. It acquired its abstract independence of contexts only slowly. Finally the elaboration and conceptual clarification of arithmetical laws by means of written representation led to the distinction between specific representations of numbers and the number concept itself.

BIBLIOGRAPHY

[1] Damerow, P.: "Individual Development and Cultural Evolution of Arithmetical Thinking." In Strauss, S. (Ed.): *Ontogeny, Phylogeny and Historical Development.* Norwood, NJ: Ablex, 1988, pp. 125–152.
[2] Damerow, P. & Englund, R. K.: "Die Zahlzeichensysteme der Archaischen Texte aus Uruk." In Green, M. W. & Nissen, H. J. (Eds.): *Zeichenliste der Archaischen Texte aus Uruk.* Berlin: Mann, 1987, pp. 117–166.
[3] Damerow, P., Englund, R. K. & Nissen, H. J.: "Die Entstehung der Schrift." In *Spektrum der Wissenschaft, February,* 1988, pp. 74–85.
[4] Englund, R. K.: "Administrative Timekeeping in Ancient Mesopotamia." In *Journal of the Economic and Social History of the Orient, 31,* 1988, pp. 121–185.
[5] Friberg, J.: *The Early Roots of Babylonian Mathematics, Vols. I and II.* Göteborg: Chalmers Technische Hochschule, 1978/1979.
[6] Friberg, J.: "Numbers and Measurements in the First Written Records." In *Scientific American, 250,* 1984, pp. 110–118.

CHAPTER 9

ON THE RELATIONSHIP BETWEEN ONTOGENESIS
AND HISTORIOGENESIS OF THE NUMBER CONCEPT
[1993]

The present study has as its subject Piaget's psychogenetic theory of the historical development of cognitive structures. Using the number concept as an example, the connections between the ontogenetic and the historical development of cognitive constructs, between cognition and culture, postulated by Piaget are examined.

Piaget's constructivist conception of number has two historical roots: Its conceptual background originates in the philosophical tradition of neo-Kantianism, its empirical foundation is based on psychological investigations on the development of the number concept in children. Piaget later generalized the results of his investigations by advancing a psychogenetic theory of the development of cognitive constructs which encompasses the historical development and therefore cannot be adequately validated and verified by psychological means alone. In the present study, this theory will be contrasted with ethnological and historical findings in order to determine its efficacy and its limitations.

Following a short introduction to the neo-Kantian background of Piaget's conception of number (part one), his theory of the development of the number concept in children (part two) and his theory of the historical development of arithmetical thinking (part three) are briefly summarized. Piaget's theoretical interpretation of the relationship of the two developmental processes presented in his genetic epistemology is at the center of this discussion. Psychological investigations on the ontogenetic development of the number concept by Piaget's successors will, on the other hand, not be discussed despite the important revisions and additions that Piaget's theory owes to them (see [34]; [62]; [73]; [33]), since they generally treat the dimension of Piaget's theory dealing with aspects of cultural history only implicitly. Likewise, historical and philosophical investigations are largely disregarded, because they, with few exceptions (e.g., [98]), prove to be indifferent regarding the problems and the empirical findings of developmental psychology.

Following the exposition of Piaget's theory, the relationship of representation and transmission will be discussed and the topic of the present study will be explicated theoretically (part four). Two case studies on the early history of arithmetical thinking constitute the main part of the study. The first study (part five) is dedicated to the techniques for controlling quantities in an extant Stone Age culture without contact to Western conceptions of numbers—that is, the culture of the Eipo in the highlands of New Guinea. The second study (part six) deals with the forms of quantification documented by the archaic systems of numerical symbols of the transition from the preliterate period to the literate culture in Mesopotamia.

Finally, the consequences for Piaget's theory of the development of the number concept are explored by summing up the relationship between ontogenesis and historiogenesis (part seven). It will be demonstrated that, on the one hand, the two studies essentially confirm Piaget's constructivist theory; on the other hand, however, one of its central theoretical tenets must be revised: Motivated by theoretical considerations, Piaget subdivided the historiogenesis of cognitive structures such as the number concept into an epigenetically determined "prehistory" and a proper "history" of mathematical thinking. In the "prehistory" of a concept the development takes place parallel to the child's development of thought. The proper "history" of a concept goes beyond the domain of ontogenetic developmental processes. Its results are open. Agreement of proper historical developments with those of the child's development of thinking concern merely the developmental mechanisms. This subdivision of the historical development of a concept into a "prehistory" and a proper "history" is not supported by the historical evidence.

The scientific findings that constitute the foundation for the conclusions presented here are not derived solely from my own work even if this has not been acknowledged in each particular case, but also from the work of my collaborators: In the first case study I collaborated with Wulf Schiefenhövel (Max Planck Institute for Human Ethology, Andechs near Munich), in the second case study with Robert K. Englund and Hans Nissen (Seminar for the Ancient History of the Near East at the Free University of Berlin). However, the responsibility for the conclusions drawn and especially for any misinterpretations of our results in light of questions pertaining to developmental psychology rests solely with me.

ON THE PRECONDITIONS OF PIAGET'S CONSTRUCTIVIST CONCEPTION OF NUMBERS

Since time immemorial numbers have been considered an undisputed example for the existence of objects of thought the knowledge of which is not based on experience. As early as antiquity they became a paradigm for ideas verifiable by means of pure thought, and the Platonic conception of numbers according to which it is meaningless to speak of the *genesis of the number* concept still has its adherents.

Kant who, with his *Critique of Pure Reason,* attempted to define once and for all the domain of "a priori" knowledge, the cognitive universals of human thought, and to determine the totality of those objects that we can know through reason alone, has had a particularly lasting effect in this respect. It goes without saying that Kant, as did philosophers before him, also held the belief that the knowledge of numbers and of the properties of numbers were part of the fundamental endowment of human beings equipped with such cognitive universals. The proposition $7 + 5 = 12$, which Kant, in his introduction to his *Critique of Pure Reason,* cites as an irrefutable example of nontautological knowledge not derived from experience—in his terminology this is a "synthetic judgment a priori"—became one of the examples cited most frequently for such propositions. It is, however, often overlooked that Kant's broad-based attempt to determine given and unchangeable objects of human cognition became a point of departure for the constructive understanding of such objects and of the concept of number in particular. Kant based the possibility of nontautological knowledge acquired through reasoning on the ability of human beings to make thought itself an object of thinking. According to Kant, numbers are a construction within "pure intuition," based on the reiteration of one and the same mental operation: "Arithmetic attains its concepts of numbers by the successive addition of units in time ..." ([43], p. 27; cf. [42], p. 116ff.). Thus, Kant of all people, the protagonist of a conception of numbers as given and unchangeable objects of thought, contributed to the dissemination of the assumption that numbers are the result of an activity of thinking, that they are constructed in thought by means of reflection.

In the long run, Kant's identification of the reasons for the possibility of cognition by pure reason proved to be more effective than his attempt of cataloging all judgments to be obtained by reason. Hegel, who radicalized the principle of construction by means of reflection (see Chap. 1) and who sought to derive all basic categories of the sciences from this principle genetically, already pointed to the arbitrariness necessarily inherent in the absolute definition of such categories. Neo-Kantianism, which inspired Piaget's genetic epistemology, can be characterized precisely by the fact that it made use of the functions of thought determined by Kant, but dismissed Kant's systematic determination of the types of knowledge to be obtained through reason and not through empirical means, because they had become historically obsolete. In this sense Frege, for example, correctly perceived himself as being in the Kantian tradition although he—contrary to Kant—denied the synthetic nature of arithmetical propositions ([29]) and attempted to reduce arithmetic to logic by making use of strict formalistic means and to construct the numbers with the aid of purely logical concepts. Only a few people were impressed by the view of radical empiricists (e.g., [57]) according to which arithmetical propositions are abstracted from experience in the same way as, for instance, the propositions of physics.

Frege's example clearly demonstrates once more that a constructivist conception of numbers need not necessarily be associated with a genetic interpretation of the construction process in Hegel's sense. The idea of a developmental process generating numbers is basically alien to modern constructivism as exemplified by Lorenzen's writings ([54]; [55]) as well. Constructivist theories of the number concept primarily establish a procedure for the construction of numbers that is considered possible or even necessary. This does not necessarily include the view that the construction must be understood as a temporal process that can be identified in individual thinking or in the historical development. Thus, it was indeed neither philosophy nor mathematics, but an empirical discipline that prompted the development of genetic theories of the number concept—that is, the psychology of child development. "Piaget's genetic epistemology is a radical reformation, in dynamic, evolutionary terms, of Kant's transcendental epistemology." ([96], p. 1)

ONTOGENESIS OF THE NUMBER CONCEPT IN PIAGET'S GENETIC EPISTEMOLOGY

In Piaget's influential study on *The Child's Conception of Number* ([67]), which will be the starting point for the subsequent discussion, the number concept is understood as the result of the construction of a cognitive structure based on experience, a construction which takes place in the thinking of every child over the course of several developmental stages. Piaget seeks to prove this by means of empirical research. The results of his research not only prompted him to abandon the assumption that the number concept is a hereditary schema of thought, but likewise the assumption that the properties of the numbers are abstracted from the real objects of experience.

In his study Piaget counters the assumption of a hereditary schema of thought by showing that the cognitive structure underlying the number concept is not present in the thinking of the child from the beginning; rather, it evolves as the result of a developmental process only at a certain developmental stage. He demonstrates the difference of the thinking of children before the formation of the number concept by showing that, during the early phase of development, the "preoperative" phase, essential prerequisites are lacking in the thinking of the child without which a meaningful conception of numbers is impossible. An example for such a prerequisite is the ability to understand that there is something that remains identical in a set of objects if their placement is changed—however much the appearance may change—as long as no objects are added or subtracted. This prerequisite is not present at the preoperative stage of development. At this developmental stage in the child's thinking there is, for example, no awareness of the necessity that the result of counting a set of objects must always be the same, no matter how the objects are arranged and in what sequence they are counted. Counting does not yet have an arithmetical meaning, as long as the result is merely linked with the act of counting and cannot be ascribed to some entity as a logically necessary attribute. Given the absence of this "invariance of quantity," the child's thinking more than anything else is lacking the structured substratum which, by way of differentiating it from directly perceivable objects, we designate as their "quantity," that substratum which is conceived in developed arithmetical thinking as varying in terms of a more or less when numerical values representing amounts are attributed.

The proof that a child at the preoperative stage does as a matter of principle not presuppose an invariance of quantity does not just demonstrate that the number concept is not based on a hereditary schema of thought. At the same time it becomes clear that such a schema of thought cannot be abstracted from the objects ([66], Vol. 1, p. 132). After all, the mere substratum of the number concept, as yet amorphous and not structured arithmetically, which is only missing at the preoperative stage, already constitutes an interpretive schema that reveals the invariance of quantity only in contradiction to the sensory evidence of the apparent change of quantities still perceived globally and not numerically.

How should we, in accordance with Piaget, conceive of the developmental process in which the child—proceeding from a type of thinking without invariance of quantity and without a space of possible quantities distinguished from the real objects—ultimately constructs that space, structured by logically necessary arithmetical operations, the elements of which are identified in the counting process?

Piaget, like Kant, considers cognitive structures and the number concept in particular to be the outcome of a construction achieved by reflection, more precisely of a reflective process, which he called "reflective abstraction," in which in the process of thinking a structure of the object is constructed based on the coordination of activities of the subject. However, according to Piaget, this process of reflective abstraction originates primarily in concrete actions and not only in mental operations as was assumed in the Kantian tradition (see Chap. 1). According to Piaget's investigations, numbers are the result of a constructive process based on the active relationship with objects. Numbers are abstractions of the coordination of correspondence and comparison activities forming certain systems of actions.

The execution of the actions, from which the number concept is being abstracted, precedes the development of the number concept. While these actions are already mentally present in "prelogical thinking" at the preoperative stage ([66], Vol. 2, pp. 73–75), they are, however, not operative before the number concept emerges; they can be executed neither independently of the object and universally nor reversibly and thus independently of particular situations. Consequently, the actions do not yet constitute those closed systems of actions to which the developed number concept can be applied "a priori."

Piaget explicitly specified these systems of actions and the association with the arithmetical operations and explored their development in the acting and thinking of the child. He based the cardinal properties on the construction of one-to-one correspondences by means of actual correspondence activities (cardinal correspondence), the ordinal properties on the ordering of objects according to qualitative characteristics (seriation), the additive numerical operations on the composition and decomposition of sets or logical classes of objects and groups of objects (additive composition), and the multiplicative numerical operations on the combinatorial coordination of sets and relations (logical multiplication). In the developed cognitive structure of the number concept these systems of real actions are mentally represented by means of reversible cognitive operations constituting a closed system of possible deductive conclusions. In this system additive and multiplicative relations on one hand and relationships based on the logic of classes and relations on the other can be mutually inferred from each other.

The reversibility of the operations constitutes the arithmetically structured substratum of the number concept, a substratum based on the reflection of identical transformations representing the quantity which remains unchanged under these transformations. In this context the concept of the reflective substratum (Reflexionssubstrat; [100], p. 46ff.) is aimed at expressing conceptually the constructive nature of the process of reflection. It denotes the constitution of an ideal object as the invariant of a system of transformations—that is, the determination of an object as a relative of the relations. Abstracted from its genesis, such an object appears capable of axiomatic representation, as implicitly defined by its internal structure.

In this way numbers become ideal objects of thought whose existence is not dependent on the existence of real objects to which they can be applied. Accordingly, they are independent of the numerous forms of representing numbers that vary from one culture to another. Thus, according to Piaget, the cognitive structure underlying the number concept which makes it possible to uncover the properties of the numbers in pure thought without apparent recourse to experience is, in a certain sense, a result of experience and abstraction after all; not, however, of an abstraction from objects, but of a reflective abstraction from actions performed with these objects.

Thus, according to Piaget, the cognitive structure of the number concept is not part of the inherited conditions of mental activity; rather, it is acquired in the course of ontogenesis through action-based experiences. However, if this is true, how do we explain the fact that the properties of the number concept apparently are logically necessary and can seemingly be obtained in pure thought independent of experience?

Piaget assumes that, while the cognitive structure of the number concept is acquired by means of action-based experiences, it is nevertheless not influenced substantially by the content of these experiences; for he views the coordination of actions accomplished by this structure as the result of an endogenous unfolding of inherent possibilities of intellectual development through the interaction with the environment. In the basic cognitive structures of intelligence which, according to his view, include the structure of the number concept as an integrating structure of certain logical operations, he perceives a highly developed form of the biological coordination of behavior exclusive to human beings. While, according to his theory, their endogenous development is not predetermined by heredity, it is nevertheless an epigenetic process—that is, a process determined by internal necessity and determined with regard to its developmental stages and its outcome (see, e.g., [68], pp. 318–321).

Thus, Piaget's epigenetic theory of the relationship of biology and cognition must be seen as an attempt to bring the development of the logico-mathematical structures of thinking into line with their universality. For this reason Piaget shares with Kantianism the rigorous distinction between two different forms of knowledge—that is, empirical knowledge whose contents are determined by the object and logico-mathematical knowledge whose contents represent cognitive universals of the human species. Two distinct modes of cognition correspond to the two forms of knowledge. Empirical knowledge can never lead to anything but empirical certainty that admits conceivable alternatives as well; in contrast, logico-mathematical knowledge yields knowledge with the status of logical necessity that, once acquired by means of the development of appropriate cognitive structures, cannot be shaken by any empirical experience.

In order for the universalism of the logico-mathematical structures to become plausible, this concept must be limited to very elementary structures. Until very recently there has, however, been agreement that, at the very least, the laws of logic were included in the core of universal structures of thinking. Faced with the disparity of actual thought processes and strict logical deductions in an attempt to simulate such processes on computers, modern cognitive science called even this precondition into question (see, e.g., [59]).

In the case of numbers the answer to the question concerning the status of arithmetical propositions reveals fundamental historical changes even earlier than that. In the Platonic tradition only the natural numbers were considered true numbers, and even Kant accepted this assumption without hesitation. In the course of the 18th century a development took place that increasingly led to an expansion and differentiation of the number concept. This development began with the acceptance of "fractions" as numbers. Their interpretation in terms of the Euclidean theory of proportions which distinguishes strictly between a number and a proportional relationship thus became obsolete. In his influential *Eléments de Géométrie* (cf. book 3), published in 1794, Legendre was the first to acknowledge this and consequently, as was commonly done later on, based the theory of proportions entirely on arithmetic.

However, this was only a first step in challenging the understanding of number dating back to the mathematics of classical antiquity. More and more new types of numbers emerged. The very designations of these numbers indicate the growing difficulties of viewing numbers as a priori given and of assuming that they constitute an integral part of the basic universal cognitive endowment of human beings. Numbers that did not fit the traditional scheme were called "irrational," "negative," and "imaginary." With the emergence of hybrid entities such as finite groups, algebraic numbers, quaternions, etc., the differentiation between numbers and other mathematical entities really became problematical.

Piaget's effort to remove the difficulties resulting from this development for his understanding of the number concept as a universal cognitive structure is completely in line with the reflection on the foundations of mathematics in the early 20th century. Since the attempt to reduce all mathematical concepts to the concept of natural numbers seemed largely successful, the idea suggested itself to reduce the natural numbers themselves completely to logic. Frege and—independently, but in a similar manner—Russell and Whitehead constructed the natural numbers as equivalence classes of classes with the criterion of equivalence being defined as the existence of a one-to-one correspondence of the elements. Although Piaget does not share the assumption that the number concept can be completely reduced to logic, he rightfully emphasizes his proximity to this view. In his opinion the numbers can be reduced to logic just as well as, conversely, logic to the numbers. He maintains that this unity of the number concept and the logic of classes and relations is precisely the cognitive structure which, as a universal characteristic of human intelligence, is abstracted from the coordination of actions in the course of the ontogenetic development and subsequently predetermines all experiences ([66], Vol. 1, especially p. 8ff.). At the same time this closed system of thought structures, which encompasses the natural numbers and logic, constitutes the basis for all subsequent forms of mathematical thought.

According to Piaget, the coherence of the developed structure is the reason for the status of logical necessity of mathematical knowledge, since the latter is determined by the system and not by the real objects that are assimilated—that is, interpreted in terms of this system. In this sense mathematical knowledge is a priori knowledge, except that this a priori is not yet present at the beginning of the ontogenetic development, but only at the end of this process as a necessary result of development ([68], p. 269; [66], Vol. 1, p. 137).

Thus, the number concept in Piaget is, just as in the earlier Kantian tradition, limited to a few basic properties, essentially to the properties of the natural numbers. He defines negative numbers, fractions, irrational numbers, etc., as reflective abstractions of the operations of this basic structure and explains in this way that they appear only relatively late in the history of mathematical thought. Their apparent "artificial" character stems from the fact that they are no longer directly related to concrete actions, but only indirectly as constructions of the primary operations already detached from the confines of reality. They are connected to the original actions by a common unique functioning of cognition at all levels of abstraction:

> From the most elementary actions which provide the child and the primitive with the ability to count small collections up to the negative, complex and transfinite generalizations of number seemingly showing no longer any relation to the concrete actions the same operative mechanism is to be found which, in spite of its often irregular appearance due to difficulties of becoming conscious, develops in most continuing and equilibrated way according to its internal logic. ([66], Vol. 1, p. 131)

HISTORIOGENESIS OF THE NUMBER CONCEPT IN PIAGET'S GENETIC EPISTEMOLOGY

Let us now turn to the question concerning the consequences that follow from this "psychogenetic" theory of the development of the number concept for the reconstruction of the historical development of arithmetical thinking. First, it must be stated that the existence of such a development is not at issue, although the arithmetical laws seem to be given a priori. At the very least the forms of representation of numbers and numerical operations and the arithmetical techniques are subject to historical change, and this change clearly exhibits developmental connections—that is, sequences of mutually dependent stages of ever more efficient arithmetical techniques ([56]; [41]). Such connections do not just appear sporadically. Overarching historical developmental processes can be identified, possibly from the first recorded numerical notations up to and including modern arithmetic. Consequently, it can at least not be ruled out that there is a single, unique process of the historiogenesis of all forms of thinking associated with arithmetical techniques.

In standard mathematical historiography there is a tacit understanding about the nature of this development. As a matter of principle, arithmetical techniques are presented in terms of the means of the modern numerical notation and interpreted in light of today's number concept, as if no historical alternatives to our arithmetical conceptuality existed (see, e.g., [35]). From this point of view historical developments do not concern the number concept as a cognitive construct, but only its concious explication and external representation. In light of his psychogenetic theory of development Piaget basically questions such an understanding. While he did not dedicate a separate comprehensive study to the historical development of the number concept itself, numerous remarks in the studies where he discusses the nature of the number concept deal with the implications of his ontogenetic investigations for the reconstruction of the history of arithmetical thinking.

Before we examine these implications in detail, some general comments regarding the psychogenetic reconstruction of developments in the history of science as processes of conceptual development are in order. Piaget analyzed the development of the forms of cognition in the maturing child as a process of the genesis of concepts. A genetic approach seemed to him to be absolutely necessary with respect to the historical development as well. In his view this approach was realized in the historico-critical method, and he identified a relationship between historico-critical method and genetic epistemology ([69], p. 8). From this vantage point the usual preconception in the history of mathematics deserves to be criticized, because it views historical forms of cognition like that of arithmetical thinking as deficient early forms of the outcome of the historical development and fails to understand them on the background of the specific conditions of their development. However, Piaget's criticism of a historiography of science failing to engage in genetic reconstruction has a deeper motivation as well. In his view there are close connections between the process of psychogenesis and the historical development of concepts, parallels between ontogenesis and historical development that went unnoticed by historians of science. The blindness of the historians of science concerning the epistemological dimension of their object of study necessitates a reevaluation of all the facts they assembled; the history of science must be reread from a perspective that differs from that of most histories of science ([69], p. 31f.).

First, we will have to examine the question to what extent such a parallelism between ontogenesis and historical development asserted by Piaget follows from his theory. We are aided in this endeavor by the fact that it was the development of the number concept which Piaget used as his preferred example for illustrating the epistemological meaning of developments in the history of science.

As we have seen, Piaget assumes that, although the logico-mathematical structures of thinking—including the number concept—are acquired through action-based experiences, they are, nevertheless, not influenced substantially by the content of these experiences. Thus, according to his theory no essentially different cognitive structures can have been in existence in the course of historical development than those structures identified by him in the course of individual development. However, the question arises whether the development of these structures in ontogenesis might not be tied to historical conditions so that certain historical periods can be matched up with certain stages of individual development characteristic of this period.

The answer to this question resulting from Piaget's theory depends on the assumptions about the cognitive conditions already present in the so-called "primitive" cultures. In cultures with no or only rudimentary arithmetical techniques the ontogenetic development either leads to the same developmental stages as in our own culture—that is, it reaches the level of formal thinking and, consequently, specifically generates the same number concept, or it does not lead to the stage of formal thinking in such cultures and thus, in particular, not to a developed number concept.

In the first case we must assume an essentially similar logico-mathematical competence in all cultures, and the historical changes of logico-mathematical thinking have no epistemological relevance. Such changes therefore require for their explanation a theory of the transformation of the universal logico-mathematical competence into graduated forms of performance that explains why and how the historically changing situational conditions result in a consistent developmental connection between the performances of individuals of different generations and different historical and cultural periods. We need such a theory, for instance, in order to explain the transformation of the universal cognitive structure of the number concept into the structure of the process of changing representations attested to by the historical modification of arithmetical techniques.

In the second case, on the other hand, the historical development as a whole or the developments of particular concepts divide into two different periods respectively—that is, a period in which the ontogenetic development does not yet or not yet fully produce logico-mathematical thinking, and a second period, in which the cognitive achievements are already based on the developed logico-mathematical thinking. In this case, we need, first, a theory of the dependence of the ontogenetic development on situational preconditions capable of explaining why, under certain historical conditions, the ontogenetic development does not advance beyond a particular developmental level. Secondly, we need a theory of the modifications of thought for the second period that can be observed in this case as well despite the completely developed logico-mathematical competence.

In the second case, for instance, the historical development of arithmetic divides into the periods before and after the evolution of the number concept. The historiogenesis of the number concept would have to have taken place in the first period; furthermore, it would have to have taken place in the same sequence and with the same structural characteristics as the development of the number concept in ontogenesis. It should be possible to identify the transition to a second period historically in which the number concept is generated by ontogenesis in the same way as under current conditions. Again, a theory of a performance of arithmetical achievements developing historically with the assumption of a basically constant number concept would be necessary for this period.

However, whatever the answer to the question concerning the relationship of ontogenesis and historical development within the framework of Piaget's epigenetic theory of logico-mathematical thinking, the development of the cognitive structures which include the structure of the number concept essentially is, according to this theory, an ontogenetically determined process resulting from biological development. Historical developmental processes of logico-mathematical thinking, on the other hand, are merely a phenomenon derived from this ontogenetic universal. There is just the alternative that these structures have either determined the human intellect all along or that they appeared only at a later time in history.

Which of these alternatives did Piaget himself adopt? It seems that Piaget originally leaned towards the first alternative. He apparently assumed that the ontogenetic development proceeding from the sensorimotor stage via the stage of concrete thinking and proceeding to formal thought was—in view of all these stages of cognitive structures—indeed universal throughout history. However, in the course of scrutinizing the studies in the field of the history of science he ultimately made a decision clearly favoring the second alternative.

The posthumously published study *Psychogenesis and the History of Science* (1989), which Piaget coauthored with the physicist Rolando Garcia, contains a theory of the connection of the psychogenesis of logico-mathematical thinking and historical development, which distinguishes between a "prescientific" and a "scientific" period in each scientific field. According to this theory, there are "analogies" ([69], p. 63) between the ontogenesis and the historical development in both periods. They result from the fact that the "mechanisms" of interaction between the subject of cognition and the object of cognition as well as the transition from one developmental stage to the next are in agreement with regard to both types of developmental processes ([69], p. 8). Each cognitive act consists in an *assimilation* of experiences to existing cognitive structures and simultaneously in an *accommodation* of these structures to the new contents of experience. The construction of new structures in the transition to the next developmental stage is brought about by *reflective abstraction* from the actions or—at the higher stages of development—from the cognitive operations to which the objects of cognition are subjected.

However, according to Piaget and Garcia there is an even more direct relationship between the historical development and ontogenesis in the prescientific period. In this period there is agreement not only of the mechanisms of development, but beyond that also of the contents of the concepts so that the historical development of the concepts parallels the development of the concepts in ontogenesis. However, in their study Piaget and Garcia provided only one example—that is, the development of basic concepts of mechanics from Aristotle to Newtonian physics.[1]

They argue that four developmental stages of the concept of impetus can be distinguished in ancient and medieval natural philosophy which correspond exactly to the developmental stages of the thinking of the child about mechanical movement and its causes. Accordingly, the scientific period of mechanics begins only with Newtonian physics. During this period the development does continue according to the same laws of the transition from one stage to the next, without, however, a significant relationship to the development of the concepts in the development of the child which is completed with the formation of the formal operations of logico-mathematical thinking ([69], p. 63). The authors maintain that the reasons for the absence of such a parallelism in the scientific period of a discipline, to which all their analyses with the one exception of pre-Newtonian mechanics are dedicated, are due to the fact that, in this period, the level of abstraction is above the level that is the subject of developmental psychology ([69], p. 30).

One of the problems related to this theory of the association of the development of concepts in the course of ontogenesis with the historical development of concepts consists in the fact that Piaget and Garcia find themselves compelled to locate the time of the transition from the prescientific to the scientific period for different scientific disciplines in completely different historical periods. They argue that the development of the concepts of mechanics proceeded extremely slowly in comparison with the development of mathematical concepts for instance. For this reason, the phase of the parallelism of the historical development and ontogenesis does not occur until classical antiquity and the Middle Ages—that is, a period with a sufficient number of written records permitting the reconstruction of this development. The situation is different with regard to the development of the basic mathematical concepts. Here a comparable parallel between ontogenesis and historical development can only be observed in a very early period. The number of the written documents from this time is too small to make an analysis of the development of the concepts possible ([69], p. 31).

In addition to geometry which had virtually developed into a paradigmatic deductive theory already in antiquity ([69], p. 111), this problem is of particular concern to the number concept. However, in his study on the development of the number concept in children published in 1941, Piaget did not initially deal with this problem and treated the number concept as if it were a universal, culturally independent structure of human cognition ([67]). In any case, in this study the assumptions concerning the sequence of the developmental stages were stated without any qualifications with respect to disparate cultural conditions.

In light of the contemporary discussion on the interpretation of ethnological findings as evidence for culturally specific modes of thinking particularly also in the area of arithmetical thought this emphasis on a universal number concept is, to say the least, astonishing, especially since this interpretation was discussed specifically in the field of Gestalt psychology with which Piaget was familiar. Thus, Max Wertheimer, following the argumentation of Lévy-Bruhl, tried to show that the number concept represents a relatively late developmental stage of area-specific thought structures which he called "number analogues" and which, although less abstract than our numbers and not generally applicable to any particular subject, served analogous purposes ([99], p. 108).

Seemingly such argumentations did in fact prompt Piaget to revise or at least clarify his theory. In his *Introduction à l'Epistémologie Génétique*, published nearly ten years later, there is sporadic evidence that it was primarily the provocatively presented ideas of Lévy-Bruhl on *the primitive mentality* that led him to realize the problem of the cultural dependence of arithmetical achievements and subsequently influenced his theory ([51]; on the arithmetic of primitive cultures: [52]). Piaget adopts Lévy-Bruhl's thesis that the thinking of "primitive peoples" is "prelogical," and identifies it with the preoperative thinking of children. He justifies this identification by attributing insufficient experiences to primitive people in the same way as to the preoperative child in dealing with those actions, which, according to his theory, form the basis for the concrete operations and the number concept in particular. Not until the demise of the "primitive" mentality did humankind achieve the transition to a mode of thought corresponding to the developmental stage of concrete operations ([66], Vol. 2, pp. 72–77). However, compared to the development of the concepts of mechanics, this transition in the case of the number concept admittedly occurred very early. The phase of the parallel development of ontogenesis and cultural evolution with regard to the basic mathematical concepts is essentially complete as early as Greek antiquity. At that time the arithmetical and geometrical concepts already had become scientific concepts.

The question as to how the developmental processes in the second scientific period, following the attainment of the number concept, are to be interpreted, is rarely explicitly stated in Piaget's earlier writings. In his *Introduction à l'Epistémologie Génétique* Piaget is reconstructing the period of the subsequent history of arithmetic, in which the number concept is expanded successively, as a progressive process of "becoming conscious" of the available operations of the cognitive structure of the number concept ([66], Vol. 1, p. 261). "There is only a reflection upon an effectively earlier element, a reflection, however, which enriches it by its becoming reflected." ([66], Vol. 3, p. 301) By means of reflective abstraction, more and more general arithmetical structures are created that are only indirectly related to the systems of actions constituting the basis for the number concept.

This interpretation of the development of expanded number concepts is not entirely consistent with the argumentation of Piaget's study on historical processes coauthored with Garcia, because in the latter the process of the development of scientific concepts is understood as a much more open process than in his earlier writings. However, here the development of the number concept no longer is a subject for discussion. Only from the analysis of the development of algebra presented in this study, we can infer that, at this time, Piaget no longer considered the subsequent development of the number concept to be merely a form of the reflection of elementary numerical operations to the extent he did in his earlier writings.

SYMBOLIC REPRESENTATION
AND HISTORICAL TRANSMISSION

Piaget's psychogenetic theory is designed to bridge the gap between a psychologically naive historiography of science and a psychology that generalizes its conclusions in an uncontrolled manner to historical processes. It offers two theoretical models for parallels between ontogenesis and historiogenesis. On the one hand, cognitive constructs can be epigenetically determined, so that ontogenesis and historiogenesis merely reflect the same epigenetic process. According to Piaget, this is true in the case of the elementary number concept. On the other hand, parallels between ontogenesis and historiogenesis can be based on psychological mechanisms of construction equally effective with regard to both. Both models are unsatisfactory in one respect: They do not take into account that ontogenetic and historiogenetic developmental processes of cognitive constructs are of a totally different nature. The former are psychological processes in the particular individual, the latter are processes between individuals based on their interaction.

Ontogenetic developmental stages consist in the ability of the individual to actually activate the acquired cognitive constructs in different situations, an ability which is ultimately based on the cognitive identity of the consciously acting individual. In contrast, historiogenetic developmental stages presuppose the reproduction of cognitive constructs constituting a certain developmental stage between individuals. Such developmental processes, therefore, raise a specific theoretical problem—that is, the problem of how to explain the occurrence of culturally specific forms of ontogenetic processes. A psychogenetic theory of the historiogenesis of cognitive constructs must be capable of explaining how, in the historical process, a developmental sequence of individual cognition can at the same time appear as a developmental sequence of the cognition of different individuals succeeding each other historically.

A consistent historical developmental sequence of psychogenetic constructs presupposes that historically acquired cognitive constructs characteristic of specific stages can be reproduced socially. If we accept Piaget's assumption that these constructs rest on action-based experiences, then a theory of the historiogenesis of logico-mathematical structures must be able to explain how culturally specific conditions can influence the actions of the maturing individual in such a way that specific developmental goals are established for the ontogenetic development at the different stages of historical development. It must, furthermore, explain why they would be the very goals corresponding to the individual developmental stages and mechanisms. Evidently, this problem arises even if, like Piaget, we consider the historical developmental stages of cognitive structures as ultimately determined by the epigenesis of the individual or by universal mechanisms of cognitive construction.

Basically, three different levels must be taken into consideration at which culturally specific conditions of interaction between individuals can influence the action-based experiences of the individual with regard to the development of particular cognitive constructs.

– First, culturally specific systems of actions can be reproduced between individuals by direct simulation or indirectly by pedagogically motivated processes of interaction.
– Second, such systems of actions can be based on culturally specific means of action (tools, instruments, machines, etc.), the adequate, interactively transmitted use of which results in the development of particular cognitive constructs.
– Third and last, systems of actions may refer to culturally specific representations of cognitive constructs in actions, images, and symbols, and thus merely indirectly to actions involving the material objects of a culture. Such systems of actions performed within culturally specific symbolic scenarios represent products of reflection the meaning of which is reconstructed by the individual in the socialization process.

The specific problem of a historical analysis of the development of such cognitive constructs follows from this mediation of the social reproduction of cognitive constructs by actions, material means of actions, and representations of cognitive prerequisites for actions. While psychological analysis, represented by the Piagetian research tradition in particular, addresses conditions of action determined by the cultural environment and embodied in systems of actions, means of actions, and media of representation merely as marginal conditions of the ontogenesis with the mere role of providing various possibilities for performing the actions on which the cognitive constructs are based, the historico-psychological analysis is specifically concerned with processes concerning the modification of these conditions and with their effect on the cognitive constructions to be accomplished in ontogenesis.

In the context of historical analyses it makes sense to look also for substantial influences of cultural, historically changing conditions on basic cognitive constructs like the number concept, although, according to Piaget's theory, such influences ought not to occur ([3]; [4]; [15], [16]; [86]). In fact, even in the circle of those committed to constructivist approaches in the Piagetian research tradition there have been fundamental revisions of his theory of the epigenetic determination of the structures of logico-mathematical thinking in order to facilitate such an analytical perspective; for this determination excludes the possibility that the factors that otherwise seem to be associated most closely with revolutions in thinking can have a significant influence on the development of its logico-mathematical structures. These factors include objective representations of cognitive structures in actions, images, and symbols as well as social interactions conveying the meaning of such representations.

The writings of L. S. Vygotsky, on the one hand, and Jerome S. Bruner, on the other, became particularly important in this respect. Despite numerous differences with regard to specifics, their theories offer alternatives to Piaget's epigenetic conception inasmuch as the representations of such structures are believed to have a substantial influence on their development ([95]; [5]; [6]; [1]). This, in particular, addressed the implication of the epigenetic conception that there is a fundamental difference between the basic logico-mathematical structures of thinking and cognitive structures in general—that is, the difference between universal structures independent of experience and area-specific structures dependent on experience. However, Piaget's basic constructivist idea according to which such structures cannot be directly obtained from reality remains unaffected by this approach.

If such a revision of Piaget's theory is adopted, then cognitive structures are to be interpreted in general as socially transmitted and historically changeable value and interpretation systems; and (according to Bruner) their representations are to be seen as models of reality in the media of actions (enactive representation), images (ikonic representation), and conventionally attributed signs (symbolic representation) which, as "amplifiers" of the basic human capacities of action, perception, and cognition, coordinate the psychogenetic mechanisms of the construction of cognitive structures in any cultural setting in particular way. Culture is the embodiment of cognitive structures in the means of thought and action which can be described theoretically as representations in these three media. Culture begins with anthropogenesis. When human beings emerge the biological evolution is complemented by cultural evolution as a new type of evolution based on the transmitted means that make the ontogenesis of certain cognitive constructs possible. In the form of models of reality culture supplies ontogenesis with instruments for the interpretation of reality which are adopted spontaneously or initiated by interaction in the process of dealing with these models and which are subsequently adapted to individual objectives. Bruner was incidentally well aware of the basic criticism of psychology implied by such a theory of the relationship between culture and cognition. He accused psychology of giving "lip service" to ideas on the relationship between culture and personality designed to cover up its inability to explain the social reality of human beings by means of its resources ([5], p. 321).

There is a close intrinsic relationship between the representations of cognitive structures and the psychogenetic functions used to explain the development of cognitive structures in Piaget's theory. Assimilation and accommodation, for example, do not merely change the cognitive constructs, but (as collective processes of a social unit) the meanings of culturally determined actions, images, and symbols as well. Thus, representations mediate between the cognitive activities of the individual and the collective interpretations established in social interaction.

Even more important for the correlation of ontogenetic and historiogenetic developmental processes is the relationship of representation and reflection. With regard to representations, Hans Aebli introduced the concept of a "second order" behavioral system (sekundäres Verhaltenssystem), a system that does not have its own meaning, but derives its meaning from a "first order" behavioral system (primäres Verhaltenssystem) which it represents ([1], Vol. 2, p. 309ff.). This distinction makes the theoretical definition of reflective processes of thinking at the level of the representation of cognitive structures possible, because the representation of one behavioral system by another behavioral system turns the operations of one of them into the ideally constituted object of the other.

For the purpose of this study, a *first order representation* will be understood as a direct representation of actual objects and actions in signs and sign transformations, a *second or higher representation*, on the other hand, as the representation of the elements and operations of cognitive constructs. For example, the waiter's marks on the beer coaster for each glass of beer consumed, a custom to be found in some areas of Germany, constitute a first order representation of this quantity, since their use as a means of control does not presuppose any arithmetical mental operations. The figures on the check, on the other hand, constitute a second order representation of the beers consumed, since they cannot be understood and used without knowledge of their arithmetical meaning.

The construction of cognitive structures by means of reflective abstraction is basically supported in the same way by first order representations as by the real actions represented by the sign transformatons of such representations. The handling of first order representations does not presuppose these cognitive constructs. Higher order representations, on the other hand, have this characteristic only with regard to the cognitive constructs represented by them, but not with regard to the real actions on which they are based. For this reason, they can be understood as first order representations at a metalevel. In contrast to the first order representations, they presuppose the constructs represented by them. However, in turn they support the reflective abstraction of the elements of a higher developmental level without on their part presupposing the cognitive constructs of this level. Thus, the distinction between first order and higher order representation explains theoretical relationships between representations and reflective stages in such a way that the reflective processes of thinking acting at the level of the representation of cognitive structures can be expressed theoretically.

The distinction between first order and higher order representation is of particular importance with regard to the development of the number concept as well. According to Piaget's theory, the representations of numbers and numerical operations, encountered in almost all cultures ([41]; [56]), do not play a substantial role in the development of the number concept; and in those instances where they do play a central role—that is, in learning arithmetical procedures—it follows from Piaget's theory that the construction of the cognitive structure of the number concept is not at issue. But according to Bruner's "instrumental conceptualism," as he characterized his view in contrast to Piaget's universalism ([5], p. 319), such representations have a central function. In Bruner's view, all operations to be performed mentally that relate to the numbers belong to the cognitive structure of the number concept, and this structure is embodied in the representations. However, his conception of the three media of representation (action, image, symbol) basically allows only for the static representation of the structures in these media. Consequently, their function can only be that of "amplifiers" of mental operations. By distinguishing between first order and higher order media, on the other hand, the essential difference between representations of numbers or numerical operations of various degrees of generality can be represented theoretically. In this way the function can be determined that the representations have in the process of the reflective abstraction while constructing the number concept.

However, even if we confine ourselves to the limited scope of Piaget's theory, the historiogenesis of the number concept poses problems requiring an analysis of the role of culturally specific representations for the ontogenesis of the number concept. Even if, in accordance with Piaget's assumptions with regard to the elementary number concept, such representations were shown to have no significant influence on the cognitive construct itself, historiogenetic, as opposed to ontogenetic, developmental processes can only be explained in terms of the cultural conditions embodied in such representations. Thus, as indicated above, Piaget assumes the existence of a "prelogical" phase of the historical development of the number concept which has been preserved in the culture of "primitive" peoples, a phase in which the cultural milieu does not offer sufficient conditions for the development of the number concept. This is a view that, based on the results of comparative psychology, is being held to this day ([36]). In this context, the parallel that presumably exists between the historiogenesis of the number concept and the ontogenetic development consists in a correspondence between the gradual evolution of such a cultural milieu and the gradually expanding action-based experiences of the child within a cultural context already providing a variety of possibilities for such action-based experiences.

Thus, however the role of culturally specific representations of cognitive constructs for the development of the cognitive structures of logico-mathematical thinking is conceived within the framework of a psychogenetic theory, Piaget's studies on the genesis of the number concept need to be complemented in two ways by means of comparative and historical cultural analyses:

- First, the systems of actions postulated by Piaget as the basis of the number concept must be identified in their historical and cultural context—that is, they must be described not only with regard to their logical and psychological structure, but also with regard to their concrete realization employing culturally specific means of representation which are required for their transmission by means of the interaction of individuals and their social transmission.
- Second, the psychogenetic functions of assimilation, accommodation, and reflective abstraction, which are presupposed by the construction of the number concept must also be identified in their historical and cultural context—that is, a culturally specific explanation is needed as to which cognitive constructs constitute arithmetical thinking in each case and how these constructs are culturally transmitted by the representations.

Let us, for the present purpose, make use of a term from the constructivist tradition for such systems of actions which are structured arithmetically and embodied in representations and which constitute the culturally specific conditions for the construction of arithmetical concepts in ontogenesis, and call them protoarithmetical systems. In this sense, *protoarithmetic* is not just a precursor of arithmetic, but a component of any culture in which ontogenesis produces cognitive constructs that are arithmetical in nature. With regard to the number concept, the historical analysis of cognitive constructs introduced here aims at deducing the genesis and development of the cognitive structures of arithmetical thinking from the development of protoarithmetical techniques confirmed by historical sources.

PROTOARITHMETIC OF A STONE AGE CULTURE

The first example to be analyzed in this fashion concerns the arithmetical thinking in those cultures which Piaget calls the "primitive" ones. Even to this day there exist primitive peoples in very few remote places on this planet who live in a world that, to a certain extent, can be compared to the living conditions of prehistoric times. These are cultures that have not had any contact with the arithmetical techniques of modern cultures until very recently or maybe even until today. The existence of number words in the languages even of such cultures is considered evidence for the universality of the number concept ([39], [40]). However, this argument is problematic, for it is in the preliterate cultures in particular that we find considerable differences in the elaboration of the counting sequences. To be specific, there are cultures where the sequence of number words ends before it reaches ten. For example, many of the approximately 200 languages of the Australian natives, the aborigines, do not even have a designation for the number "four," although in other respects these languages are neither syntactically nor semantically less complex and expressive than, say, the English language ([20], especially p. 107f.; [21]).

Structurally speaking, the technique of counting is a typical protoarithmetical technique. It serves to control quantities by constructing one-to-one correspondences to subsets of a conventionally defined and culturally transmitted ordered standard set of symbols, generally a lingual standard composed of a potentially infinite sequence of number words. In languages with a very limited sequence of number words—whether the sequence ends with eight or three is of little concern in this case—such a technique for the construction of correspondences capable of including any number of objects is generally lacking. The few so-called "number words" of these languages do not constitute a sequence of number words in the proper sense. Rather, they are qualitative expressions for quantities without reference to a general procedure for the construction of such correspondences.

The construction of correspondences is doubtlessly not a necessary prerequisite for perceiving quantities and for taking them into account in practice as required by the particular situation. All the languages on this earth, even languages without a sequence of number words like the languages of the aborigines, possess numerous qualitative concepts of quantity as well as grammatical forms to make discriminative quantitative statements. What distinguishes concepts like "far," "recently," "lovelier," etc. from arithmetical concepts denoted by number words is their context-specific meaning. Although they represent quantities, they are not, unlike arithmetical concepts, applicable or transferrable to any given object. They acquire their semantic meaning from comparison operations associated with specific objects. Consequently, they often are much better suited for expressing the precise meaning of a quantity specific to a particular context than abstract designations of quantity like number words.

The difference between expressive quantitative concepts, that are, however, dependent on their specific context and cannot be generalized, and semantically sparse but universally applicable number words explains why many languages do not have a sequence of number words in the proper sense. It is an indication that the importance of the construction of clearly defined correspondences for the perception and understanding of quantities is merely secondary. The relative cultural importance of such protoarithmetical constructions determines the extent to which a formalized counting technique has been codified lingually. This was especially true of the Australian aborigines, before their country was taken over by Europeans. Their language and culture was characterized by a rich variety of highly developed techniques for adapting their way of life to the exacting natural conditions they had to endure. The technique of constructing one-to-one correspondences was not a necessary part of them.

Thus, to a certain extent this is a confirmation of Piaget's assumption that the systems of actions on which, according to his theory, the number concept is based and for which we are using the term protoarithmetic, represent a precondition for the development of the number concept. However, this trivial case of the correspondence between ontogenesis and cultural transmission—that is, that neither the protoarithmetical systems of actions postulated by Piaget nor an arithmetical terminology exist in a culture—is on the whole not very revealing. Not until we consider the more frequent case of a preliterate culture with a developed counting technique, do the theoretical problems become evident. It is precisely this precondition that is true, for example, for almost all the cultures of New Guinea and, in particular, for the culture of the Eipo as well whose protoarithmetic will be examined more closely in the following.

The area inhabited by the Eipo is an inhospitable, tropical mountain region in the Western part of New Guinea which belongs to Indonesia. Their settlements are located in the southern valley of the Eipomek river at an elevation of between 4,600 and 7,500 feet. The valley, situated in the central mountains of the island, is surrounded by mountain ranges of up to 9,800 feet. Although the area is only a few degrees south of the equator, the climate is cool, foggy, and extremely rainy throughout the year.

The Eipo are sedentary and engage in horticulture using the simplest of means. In terms of its material characteristics, Eipo culture is comparable to the cultures of the Neolithic age. The Eipo have their own language spoken only by the roughly 800 members of this tribe; it is one of the estimated 730 Papuan languages of New Guinea. Until a group of researchers conducted field research from 1974 to 1976, this culture had remained virtually untouched by external influences [44]; [24]). At this time Eipo culture was an intact Neolithic culture without any contact whatsoever with Western traditions of arithmetic in particular.

Like most of the cultures of New Guinea, the Eipo have a developed technique of counting. They make use of so-called "body counting numerals." In these counting systems, points of the human body serve as the counting standard ([37], especially p. 18; photo documentation of the counting process: [44], p. 124). The "number words," if they can be called that, are identical with the names for these parts of the body. In the case of most of these number words the affixes "-barye" or "-digin" to designate the left or the right side of the body are, with few exceptions, added to the proper name of the body part. The counting process begins with the little finger of the left hand, proceeds via the left arm to the top of the head, then down the right arm again and ends with the little finger of the right hand as the denotation for the number 25.

1	ton	left little finger
2	betinye	left ring finger
3	winilye	left middle finger
4	dumbarye	left index finger
5	fangobarye	left thumb
6	nakobarye	left wrist
7	tekbarye	left lower arm
8	finbarye	left arm bend
9	toubnebarye	left upper arm
10	takobarye	left shoulder
11	koklombarye	left side of the neck
12	obarye	left ear
13	mekbarye	top of the head
14	odigin	right ear

15	koklomdigin	right side of the neck
16	takubdigin	right shoulder
17	toubnedigin	right upper arm
18	findigin	right arm bend
19	tekdigin	right lower arm
20	nakubdigin	right wrist
21	famdigin	right thumb
22	dumdigin	right index finger
23	winilyaba	right middle finger
24	betinyaba	right ring finger
25	seselekyaba	right little finger

After this cycle is finished, a higher order counting process, known to us from all developed counting procedures, commences, and the counting process starts from the beginning. The term "yupe ton," "counted once" (literally "one speech") indicates that the counting process was completed once and then started again; "yupe betinye" means "counted twice" and represents the second completed cycle, etc. This counting technique is, just like our own, basically unlimited and can be expanded any time as needed.

Body counting numerals of this kind can be found in many cultures of New Guinea and in part in the surrounding areas as well ([72]; [46]; [50]; [58]; [97]). They constitute a protoarithmetical cultural technique of this region with a consistent logic across clan and language barriers, but without general norms concerning the correspondence of the parts of the body and the number of body points used. The Oksapmin who live to the east of the Eipo, for example, use 27 body points and count in a similar way; however, they reverse the order by going from right to left. The Yupno who inhabit the northeastern part of the island's massif make use of 33 body points, and their order of counting is completely different in that they alternate between left and right.

Thus, in contrast to the Australian natives, the cultures of New Guinea use and transmit techniques for constructing one-to-one correspondences by means of the body counting systems. Piaget based his assumption that the thinking of primitive peoples is prelogical and that they, in particular, lack a number concept, on the insufficiently developed technique in these cultures and, consequently, on the insufficiently reflected development of the coordination of actions from which the logico-mathematical structures emerge. However, this is obviously not the case here. If we accept Piaget's theory of abstracting the number concept from protoarithmetical systems of actions, then the developed counting techniques of the cultures of New Guinea must be regarded as positive proof that they already possess the cognitive schema of the infinite sequence of natural numbers.

However, in the following we nevertheless want to examine the question as to whether the counting technique of the Eipo is indeed used with the same semantics as the sequence of number words in modern cultures. Unlike Piaget who followed Lévy-Bruhl in this point, Wertheimer, for example, did not generally consider the thinking in preliterate cultures to be prelogical; they were merely based on different cognitive constructs. It must, therefore, be examined whether the cognitive achievements of the Eipo with regard to counting might not be based on context-specific constructs in the sense of Wertheimer's "number analogues" rather than on the abstract number concept. A fictitious example ([99], p. 144) with which Wertheimer sought to illuminate the idea of such context-specific number analogues may serve to set the stage for answering this question:

> Logs are brought in. Main supports or trusses for building a house. Things can be counted. Or: People can go and get the necessary logs carrying the image of the house (the skeleton of the house) in their heads. They have a conception of the grouping of the posts which presents itself quite concretely in the shape of the house.

Wertheimer's fictitious example proves to be most appropriate when compared to the report on the construction of a sacred men's house of the Eipo village community of Munggona in November 1974. Around 40 to 50 men, in large part from the surrounding communities, participated in the undertaking which, in terms of its order of magnitude, is rather a rare event in Eipo society.

The construction of a new men's house is a ritual act. Cutting and hauling the building materials is already part of the ritual. Thus, the ayukumna, the long post beams cut from particularly strong material, are carried in a solemn procession to Munggona where they must remain hidden in the tall grass until the day of the event. During the night of November 26 the old house is dismantled under cover of darkness. Shortly after 6 a.m. the construction of the building begins. This is how it is described in the report:

> It is now 6:40 a.m.—All the supports for the new house are brought by the men to the designated place in a rhythmical, almost ecstatic procession. ... The men position themselves with the supports in the envisioned circular outline of the men's house. The supports must be rammed into the ground up to a depth of 40 centimeters.—Many of the men are eager to touch some of these important building elements. Although there are many men and youths working together, there is no supervision. Nobody gives any instructions or commands. They all work together smoothly, everybody knows his part, although such a structure is not often built. Only now and then, a few of them are seen to consult with each other. ... The rolled up bark of a pandanus tree is brought over. Later it will be used to cover the floor of the finished house. Now the men use the cylinder to determine the proper diameter of the new roundhouse. The work of the numerous participants continues without discernible guidance. ([45], p. 144).

Although the Eipo have a developed technique of counting, the manner in which the construction of the representative building takes place is in complete accordance with Wertheimer's assumption. It would not occur to any of the men involved to count the posts of such a house; but everyone is keenly aware of their function. The men involved in the construction need a number concept about as much as they need to envision, for example, the idea of approximating a circle by means of a regular polygon in order to determine the distance of the posts using a piece of rolled up bark. The anticipation of the function determines the construction and the proper size of the house. In addition, the men forming a circle anticipate the construction of the house represented enactively and transmitted ritually. This ritual component frees the event from the necessity of cognitive planning. The construction is, as it were, embodied in the collective ritual action, in the roles the actors must assume during the construction of the house. The political relationships between the villages, evident in the participation of the communities of men's houses, ensure that the mythical continuity of transmitting the knowledge is maintained even for such rarely used techniques as the construction of a sacred building.

If the Eipo do not use their counting technique in situations such as these, when do they use it? This is a key question with respect to the identification of the cognitive functions of this technique. If the mere existence of a counting sequence is taken as an indication for the existence of a number concept applicable to any kind of situation, the assumption is that the counting technique is not only structurally identical with our way of counting, but that it is generally used in the same way as well. However, the example of Eipo culture goes to show that this need not necessarily be the case.

When do the Eipo count? The answer is: very rarely. While the question *yate barye?* which roughly means "how much?" does exist in the language of the Eipo, an Eipo does not as a rule expect a numerical response to this question and will seldom answer this question in that way. If, in such cases, he is asked to give an accurate count, he is perfectly capable of doing so and will oblige—as a matter of courtesy towards the stranger, but not because he might detect any meaning in it. If an Eipo is asked about the number of guests at a celebration, he will enumerate them one by one. If he is asked about the number of the inhabitants of his village, approximately 200 people, the answer is *weik,* which is the equivalent of "many," and this answer to such a question is in accordance with the way of life of the Eipo.

In a few important matters the Eipo do depend on trade, for instance with regard to the acquisition of unhewn stone axes. However, they do not exchange one piece for another which might require counting; rather, amounts of the goods to be exchanged are balanced as a whole until both sides are satisfied with the result.

Even if the question *yate barye?* "how much?" is answered in numerical terms, this answer does not necessarily imply a numerical sense. If, for example, an inquiry is made as to the number of enemies in an armed conflict, the answer might be *yupe winilye,* "counted twice." Taken literally, the answer would mean that there were exactly 50 people. The answer is, however, meant as an overall assessment of the enemy. In the course of a confrontation, no Eipo in his right mind would count the people involved.

In many cases a numerical answer is common, but the numerical range is limited. An Eipo's answer as to the distance of a particular place might be for instance: "I slept twice." However, in such a situation he will never say, "I slept seven times," because in their wanderings the Eipo never wander beyond a distance requiring four days. Any mention of a higher figure in such a sentence would be objectively meaningless to him.

The Eipo do not apply their counting technique to temporal situations either. While the language of the Eipo has a wealth of qualitative expressions for determining particular times, their concept of time does not display any arithmetical structures.

Many of the specifications of concepts with temporal connotations are dependent on the context. They refer to sequences of actions, not to a time independent of action. *mane li,* literally "time of the marsupials," is the time of the hunt before a big feast. *ninye li,* "time of the people," is the time when the men return from the hunt before a feast, or the time when the guests are coming. Generally speaking, particular points in time are defined in relation to a particular context and not by embedding them in a general schema comparable to our arithmetically defined concept of time. A special suffix, the suffix *-sum,* helps to communicate particular times by means of indicating a certain context. In connection with an expression for an event, a situation or an action *-sum* identifies the coincidence of the situation with this identifying context. *mot sum* is the time of the feast. *bilum-sum male febeyik,* literally "you went *-sum,* arrows they shot." "After you left, they went on the warpath." Here the time of the outbreak of hostilities is determined by the coincidence with the departure. *monob sum* "earlier" fixes an event in relation to another by saying that it does not coincide with it in a specific way.

Some concepts indicating the temporal distance to the present are formed using the suffix *-sum* independently of the body counting system. Curiously enough, these concepts are indifferent in terms of the direction of time regarding the past or the future. They constitute something like the preliminary stage of a second counting sequence with a specific temporal meaning:

ambosum	"one day difference in relation to today"
amkyesum	"two days difference in relation to today"
nesum	"three days difference in relation to today"
katensum	"four days difference in relation to today"
wandaluksum	"five days difference in relation to today"

The sequence ends here. The last of these concepts already has the connotation "soon." Consequently, it is no longer used exclusively for the exact interval of five days. This preliminary stage of a counting sequence explicitly demonstrates the limited extent to which the Eipo consider it necessary to express time in arithmetical terms.

For the Eipo the relationship to the contexts replaces the application of a general schema of time. No Eipo would count years, months, or days by means of body counting numerals. For instance, no Eipo knows his age or that of his fellows. No Eipo knows the number of days in a month or the number of months in a year—not because the Eipo are incapable of counting them, but because that kind of activity would make no sense to them.

In the cultural context of the Eipo this orientation within time by contextual rather than arithmetical means must not be seen as a deficit. Rather, it affords possibilities of temporal coordinations of actions which we no longer have in our culture. Let us take, for example, the preparations for a feast, a socially significant and costly event. If we ask an Eipo, when the planned feast will take place, we receive seemingly imprecise answers, usually about the status of the preparations. At some point the time of the feast has come: *ninye li,* the "time of the people" has arrived. For the European observer it comes as a complete surprise when the guests, even from far away, descend the mountainsides in an impressive demonstration.

How is this coordination possible? It would not be possible, if everyone did not continually synchronize their preparations with those of the others. That is a difficult problem in view of the geographic circumstances, but it is still easier than working without coordination towards a date set in advance. It is precisely the multiplicity of context-dependent plans following their own individual "time scales," comparable perhaps to the critical path method in modern project planning, that makes a complexity of coordination possible. Given the conditions under which the Eipo live, planning in accordance with an absolute time frame would most likely be bound to fail because of its lack of flexibility.

For the Eipo, meaningful counting is a process associated with a few exceptional situations. The strands of the Nassa headbands are counted—they number approximately 13 at most. The more strands such an ornament has, the greater the prestige of the owner. Members of the research project mentioned earlier recall a situation where 64 unhewn stone axes where accurately counted. However, in Eipo culture counting is essentially a method not really needed and therefore not generally used.[2]

The cognitive consequences for the thinking of the Eipo are predictably insignificant. There is some doubt whether the so-called "numbers" used in body counting have any meaning other than their literal meaning as denotations for the body parts pointed to in the counting process. For this reason, the number 25 apparently has two different denotations in the language of the Eipo. The term *seselekyaba* names and connotes the right little finger with which the simple counting action is concluded, *yupe ton,* on the other hand, names and connotes "one speech"—that is, the first completed cycle in the case of a counting action that is repeated more than once.

When an Eipo is counting, he does not determine a number as a result, but the body part to which the procedure familiar to him is leading him. Pointing out the body part becomes a symbolic gesture for a particular quality of reality. In modern cultures this quality is endowed with a separate substratum and with an internal structure, that is to say, with the structure of the number concept. In the case of the Eipo the denotations for the body parts they use in counting do not have such a structure. Perhaps nothing demonstrates this more clearly than the fact that the Eipo do not use their body counting numerals in any way for computing—that is, for the very activity which, for us, is just as much a part of our number concept as counting.

PROTOARITHMETICAL TECHNIQUES IN THE TRANSITION TO A LITERATE CULTURE

The second example to be analyzed here is quite different. This example represents a historical period with far more developed forms of arithmetical thinking. In the following, the arithmetical thinking at the time of the emergence of writing in the Near East and thus the historical transition from a preliterate arithmetic to one represented in writing will be examined.

It is well known that in antiquity Mesopotamia was the center of a system of writing, the so-called "cuneiform," which was used in many parts of the Near East for approximately 3,000 years. The writing material consisted of clay tablets. The characteristic signs of this system of writing composed of wedge-shaped markings were pressed into the surface of these tablets with a wedge-shaped stylus. In the course of its long history, numerous languages were written in this system of writing. After the cuneiform had been deciphered in the 19th century, most of them were thoroughly studied and are by now well known.

The oldest tablets displaying the developed system of cuneiform writing date back to around the middle of the 3rd millennium B.C. The beginnings of this system of writing, however, go back to the turn of the 4th to the 3rd millennium, to the last phase of the so-called Late Uruk period. At this time the culture of Mesopotamia and the surrounding areas differs fundamentally from Neolithic village cultures like the preliterate culture of the Eipo described in the last section. As early as the 4th century B.C. urban centers of a society with a highly developed division of labor and social stratification emerged. Remnants of representative buildings in the city centers attest to the existence of temples and palaces which were the administrative centers of a redistributive barter economy. A sophisticated apparatus of officials organized the deployment of labor and supervised the distribution of the products of labor collected in central storehouses (Chap. 8; [11]; [63]; [64]).

The administrative tasks associated with this type of economy could obviously not be accomplished without administrative aids for the qualitative and quantitative control of the economic resources. Thus, we are familiar with aids predating the invention of writing which served the registration and symbolic representation of economic goods and the designation of who controlled them, for example, standardized containers, sealed containers supplied with bullae, signs with numerical meanings, but most importantly a special kind of clay symbols with simple geometric shapes (sphere, cone, pellet, tetrahedron, cylinder, etc.), which were apparently used, among other things, as counters to record quantitative data. This function of the clay symbols is attested to by the fact that combinations of such clay symbols, apparently for the purpose of preventing the forgery of the encoded information, were sometimes kept in closed and sealed clay spheres some of which exhibit markings on their surfaces. With regard to type and number these markings generally correspond to the clay symbols inside the spheres and can easily be identified as precursors of the numerical signs on the tablets from a later period. Numerous clay tablets that, with the exception of the impressions of seals, contain only such markings, the so-called preliterate "numerical tablets," must also be located in the period before the invention of writing.

The addition of a system of pictographs to these symbolic means of representation around the last century of the 4th millennium is generally considered to be the actual invention of writing. In this way, two different systems of writing emerged in the Near East in quick succession, the so-called "archaic" systems of writing; soon afterwards, a third system of writing, the Egyptian system of writing, emerged—however, the evidence from the early period of this system is sparse. The older of the two archaic systems of writing is the "protocuneiform" of southern Mesopotamia from which the cuneiform evolved. To date approximately 5,600 clay tablets with this type of writing have been excavated. The oldest are tablets from the IVa layer of the ancient city of Uruk, the most important archaeological site yielding protocuneiform tablets.

Proto-Elamite writing is somewhat more recent, documented by some 1,500 texts. Most of the proto-Elamite texts come from Susa, the urban center of a region in the southeast of Mesopotamia. Around 3000 B.C. this system of writing was common in all of contemporary Iran, but was subsequently completely superseded by the rapidly emerging cuneiform.

Each of these two systems of writing contained more than 1,000 different signs with widely standardized notations and conventionally defined meanings. Although there is apparently little resemblance between the actual signs of the two systems, it seems to be certain by now that proto-Elamite writing was directly influenced by the protocuneiform model. The numerical signs are a case in point. The proto-Elamite and the protocuneiform numerical signs are similar in many respects. The characteristics of the differences indicate that the numerical signs of proto-Elamite writing were adopted from the protocuneiform. Thus, the protocuneiform served as a model for proto-Elamite writing and not the other way around.

Writing is generally defined as the representation of the spoken language by a medium other than the fleeting sound of the human voice. However, the archaic texts of the Near East document a developmental stage of writing before it acquired this semiotic function. On the surface this is expressed by the fact that most of the tablets found represent administrative documents and almost exclusively contain numerical records of economic processes (see Fig. 1 and 2; on the preliminary tentative interpretation of the two texts see [64]; Chap. 8; both texts will be published in their entirety with annotations in: [10]). Neither the syntax of these texts nor the semantics of the signs and sign combinations exhibit the structures characteristic for the spoken language. The information is not ordered in a linear way but hierarchically. Its organization does not correspond to the organization of

332 CHAPTER 9

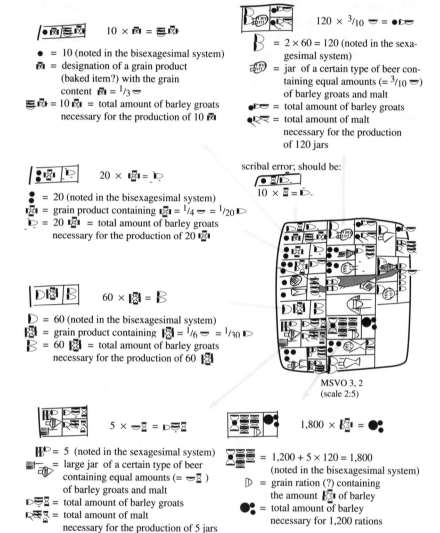

10 × 𒀭 = 𒀭

- • = 10 (noted in the bisexagesimal system)
- 𒀭 = designation of a grain product (baked item?) with the grain content 𒀭 = $1/3$ ⌒
- 𒀭 = 10 𒀭 = total amount of barley groats necessary for the production of 10 𒀭

20 × 𒀭 = 𒀭

- ⁝ = 20 (noted in the bisexagesimal system)
- 𒀭 = grain product containing 𒀭 = $1/4$ ⌒ = $1/20$ ▷
- ▷ = 20 𒀭 = total amount of barley groats necessary for the production of 20 𒀭

60 × 𒀭 = 𒀭

- ▷ = 60 (noted in the bisexagesimal system)
- 𒀭 = grain product containing 𒀭 = $1/6$ ⌒ = $1/30$ ▷
- 𒀭 = 60 𒀭 = total amount of barley groats necessary for the production of 60 𒀭

5 × 𒀭 = 𒀭

- 𒀭 = 5 (noted in the sexagesimal system)
- 𒀭 = large jar of a certain type of beer containing equal amounts (= ⌒𒀭) of barley groats and malt
- 𒀭 = total amount of barley groats
- 𒀭 = total amount of malt necessary for the production of 5 jars

120 × $3/10$ ⌒ = •⌒

- 𒀭 = 2 × 60 = 120 (noted in the sexagesimal system)
- 𒀭 = jar of a certain type of beer containing equal amounts (= $3/10$ ⌒) of barley groats and malt
- •⌒ = total amount of barley groats
- 𒀭 = total amount of malt necessary for the production of 120 jars

scribal error; should be:

10 × 𒀭 = ▷

MSVO 3, 2
(scale 2:5)

1,800 × 𒀭 = •⁝

- 𒀭 = 1,200 + 5 × 120 = 1,800 (noted in the bisexagesimal system)
- ▷ = grain ration (?) containing the amount 𒀭 of barley
- •⁝ = total amount of barley necessary for 1,200 rations

Figure 1: Calculation of the amounts of the ingredients needed in the production of dry cereal products and beer.

ONTOGENESIS AND HISTORIOGENESIS OF THE NUMBER CONCEPT

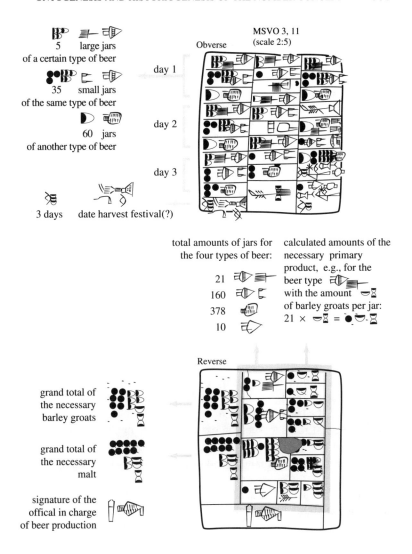

Figure 2: Account on barley groats and malt used in beer production.

language by means of sentences consisting of subject and predicate. The pictographs are often used already abstractly and no longer according to their pictorial meaning; on the other hand, however, the phonetic codification of information is at best only of marginal significance. The semantics of the signs and sign combinations is structured by a relatively small number of categories. The majority of the signs and sign combinations that can be interpreted up to now represent the objects of economic activities, identify the persons involved, or indicate the type of these activities themselves. With regard to their semiotic function, protocuneiform and proto-Elamite writing did not represent media for the codification of language, but for modeling economic processes.

To date the meaning of the signs and sign combinations is only partially understood in detail. The decipherment of proto-Elamite writing proves to be particularly difficult because it was superseded by the subsequent systems of writing. The efforts to understand the protocuneiform texts, on the other hand, have by now yielded numerous conclusive interpretations because of the close relationship between the protocuneiform and cuneiform writing.

The level of knowledge is particularly good with regard to the numerical signs of both systems of writing, because in these cases arithmetical operations documented by the existing texts even offer internal evidence of their numerical meaning, evidence that is independent of the relations to the subsequent development of writing. The numerical relations between the signs can often be unambiguously determined from such arithmetical operations. In addition, the use of numerical signs in the archaic texts was already subject to general rules to such an extent that, in many instances, a statistical analysis of the use of the signs makes it possible to distinguish accidental sign constellations from those that lend themselves to be interpreted as a result of applied rules that can be explicated in semiotic terms. The decipherment of the numerical signs by means of such methods has yielded detailed information about a highly developed system of signs that, during the time immediately following the invention of writing (around 3100 to 2800 B.C.), was used by the officials of the Mesopotamian cities and—in a slightly different form—also by the officials of the proto-Elamite cultural sphere for managing quantities of a magnitude already encompassing hundreds of thousands of counting and measuring units.

For the present purpose the details of this system of signs and the particular features of its protocuneiform and proto-Elamite forms will not be discussed in detail.[3] Rather, in the following the discussion will center on general semiotic characteristics of this system with the aim of gaining insight into the arithmetical thinking at the time of the development of writing.

Up to now the numerical signs of the archaic texts have been interpreted by scholars exclusively from the perspective of their subsequent development as a somewhat clumsy form of representation of an otherwise full-fledged number construct.[4] At first sight the archaic numerical signs do indeed seem to prove that, at the time of the codification of arithmetical thinking by means of the invention of the archaic systems of writing, the number concept had already developed to the same degree as is indicated by later evidence of mathematical thinking in the early civilizations. In particular the signs of the sexagesimal numerical sign system of cuneiform writing containing symbols for the numbers 1, 10, 60, 600, and 3600 are already to be found in the archaic texts in the familiar form of subsequent texts. Beyond that the documents undoubtedly seem to confirm the sophisticated use of the arithmetical operations of addition and multiplication. Between the prearithmetical thinking of the preliterate culture of the Eipo discussed in the previous example and the example discussed here there seems to exist the very gap which Piaget sought to identify when he distinguished the prelogical thinking of "primitive" peoples from the developed logico-mathematical thinking at the ontogenetic developmental stage of formal operations in the "civilized" cultures.

However, a closer look reveals that the numerical signs of the archaic texts display peculiar characteristics which render their interpretation as symbols representing the construct of number along the lines of Piaget's theory and the attribution of abstract numerical values doubtful. The mere fact of the graphical variation of many signs already poses problems. If each of the variants is counted as a separate sign, the result is an overall number of about 60 different numerical signs.[5] It is by no means the case that 60 different numerical values correspond to this large number of different numerical signs; rather, it results from the fact that many of these signs are applied only in specific contexts. It is the result of confounding numerical, metrological, and qualitative meanings which limits the applicability of the individual numerical signs to particular areas of application.

However, even more peculiar than this specificity of archaic numerical signs is their arithmetical ambiguity. Although those signs that, on the one hand, have a simple graphic shape and might, therefore, be considered basic signs have a broader area of application than the graphically complex signs and could therefore be most readily interpreted as representatives of numbers, it is, on the other hand, precisely these "abstract" signs that are used with different numerical meanings depending on the area of application. The archaic numerical signs have no specific connection with particular numerical values with the exception of those graphic variants which are associated with specific areas of application. Rather, their numerical values were also determined by the counting method or metrology of the particular context of their actual application.

Essentially these "abstract" basic signs are the signs of the sexagesimal system which remained in use. Expressed anachronistically in terms of abstract numbers these signs have the values 1, 10, 60, 600, 3600, and 36000. During the archaic period, however, these signs could (with regard to the sign with the value 1) also take the following values in other contexts: the second sign the values 6 and 18 instead of 10, the third sign the value 180 instead of 60, the fourth sign the value 1800 instead of 600, the fifth sign the values 60 and 1080 instead of 3600, finally the sixth sign also the value of 180 instead of 36000. The systematics of archaic systems of numerical signs implied by this will have to be considered in the following.

An ambiguous use of numerical signs is essentially nothing unusual. Any place-value system as, for instance, our modern decimal system presupposes an ambiguous use of the basic signs, for the efficiency of such a numerical representation results precisely from the fact that the same sign is used with different meanings and that it is the "place" within the numerical representation which indicates the numerical value. In contrast to the archaic numerical signs, however, the signs have an unambiguous basic meaning in this case, and the ambiguity is merely the result of the second order use of the signs in a systematically structured, developed system of numerical representation. By contrast, in the case of the archaic numerical signs there is in most cases no definite numerical meaning that has been designated as the basic meaning, and the ambiguity does not result from a systematic structuring of the representation, but from the dependency of the numerical meanings on the context.

The sign for the number one is a notable exception. In all applications this sign and its graphic variants represent either the counting unit or a standard measure which in some way or other plays a significant role in the metrology in question. This semiotic idiosyncrasy of the sign for the counting unit proves to be a natural consequence of its semantic function, if we consider the general principle of the representation of quantities by means of archaic numerical signs. As with all historio-genetically early developmental stages of numerical representations, especially with all numerical sign systems of the early civilizations, the basic numbers are not represented by specific individual signs, but by repetition of the sign for the counting unit, for example the number three by repeating the sign for the number one three times.

Thus, the elementary quantities of the counting process are not represented abstractly through signs but quasi-physically by means of a corresponding number of repeated signs—that is, by means of an isomorphism of the number of the objects counted and the number of the signs representing them. In the simplest case, each of the objects counted is represented by a sign with the value "1."

However, normally, especially where larger quantities are concerned, numerical signs with other values are also used. Usually such a numerical sign is created by substituting a new sign for a certain number of repetitions of the counting unit. This sign then has a certain value that can be expressed in terms of the smaller unit: It is, however, used semiotically just like the original unit. It can, thus, be characterized as a unit with a higher value the repetition of which constitutes a higher order counting process. Thus, the introduction of signs with higher values is based on the same elementary technique for transcending counting limits already encountered in the system of the body counting numerals of the Eipo as a form of counting characterized by how often the counting cycle was repeated.

For example, when recording quantities of discrete objects up to the number nine the sign for the counting unit is repeated accordingly; for the number ten, however, a separate sign is used in lieu of ten repetitions of the symbol for the counting unit. The reason for the change in the notation is obvious. In this case, as in most cultures, it seems that the two hands originally served as a higher value counting unit, and the notational mode merely reflects an anthropomorphically defined counting stage already prefigured in the sequence of number words. This example is however not typical for the higher value units used in the archaic numerical sign systems. As a rule the unit with a higher value is determined by a new metrological unit, for instance by defining the receptacle or natural measure next in size as a standard. For this reason the numerical relationships between the signs of a numerical system may vary a great deal.

The semantic principle of representing quantities by repeating signs that are symbols for measuring and counting units provides evidence for the origin of the peculiar ambiguity of the archaic numerical signs. If the signs initially corresponded to real objects and standard measures and did not represent numbers, they acquired their numerical meanings only as a second order meaning and, depending on the area of application, as an implication of the respective quantitative relationships of the units to each other.

Thus, the ambiguity can be seen as resulting from the fact that initially only the repetitions of numerical signs in their role as first order representations of quantities, not the numerical signs themselves acquired numerical meanings. The so-called numerical signs themselves merely remained signs for the units to be counted.

This interpretation of the archaic numerical signs explains a further peculiarity as well. At first blush, these numerical signs seem to be easily distinguishable from the rest of the characters on account of the graphical differences and the repetitions of signs characteristic of them. The numerical signs were pressed into the clay with round styluses, while the other signs were originally incised and later impressed into the clay with an angular stylus. However, with regard to their semantics the transition between the numerical signs and the other signs was smooth despite these apparent differences. As signs for counting and measuring units indicating a specific context the numerical signs conveyed information about the object of quantification as well. Thus, they often rendered additional characters for the purpose of identifying the object superfluous and occasionally completely took over their function. For example, certain numerical signs belonging to a system of capacity measure for grain often served directly to designate grain products and grain rations of a particular size. Conversely, written signs denoting such grain products often implicitly represented their size as well and were used in numerical calculations just like numerical signs. This is, for instance, the case in the administrative text shown in Figure 1. The smooth transition between numerical signs and other signs also led to the formation of numerous composite signs which represent graphical combinations of numerical signs and other written signs and which were probably without exception deliberately constructed in order to express quantitative and qualitative information by means of a single sign. An example for this assumption is provided by the time notations in the text reproduced in Figure 2.

After these remarks on the semantics of individual signs let us now turn to the global structures into which the archaic numerical signs were integrated. First, the question must be raised whether it was still possible in particular situations to attribute ad hoc meanings to the signs, as in the case of counters, tally notches, and similar symbols of primitive arithmetical means of control, or whether they already represented fixed, standardized, and socially transmitted qualitative and quantitative structures of meaning independent of particular situations. The latter is without a doubt the case. This is already evident in the very way the signs were written. A comparison of the quantitative notations of texts excavated at different archaeological sites indicates that the numerical signs were, just like the other written signs, to a large extent written uniformly according to established conventions extending over a wide area and for a considerable period of time.

The same is true of the semiotic rules that were applied to the numerical symbols: It is especially true of the numerical relationships between the signs that were established by substituting higher value symbols for certain sign repetitions. Apart from a few exceptions in the border area between the numerical signs and other written signs,[6] the quantitative relationships between the signs were by no means determined by empirical relations of size between the counting and measuring units represented by them. Rather, they represent normatively assigned relations between units and integral multiples of them to which the actual measuring units seem to have in fact been adapted. As a matter of fact, the numerical notations composed of several different signs provide further evidence of this relative independence of the numerical symbols from the actual characteristics of the units represented by them. Whenever possible the appropriate higher value units were substituted for sign repetitions. In the archaic texts there are very few examples that deviate from the strict application of this rule (unpublished, but cited in detail in [8]). It seems fair to say that the assumption that the objects and measures represented, for example containers for grain, were also treated in that way—that is, that the content of small containers was poured into larger containers—can be excluded for the period of archaic writing (in contrast to the preliterate period from which many of the so-called "numerical tables" originate). Thus, as in the case of the standardization of the individual signs, the structure of numerical notations, too, indicates a meaning that is to a certain extent independent of the concrete objects. The sign combinations no longer represent actually existing combinations of counting and measuring units, but a standardized decomposition into abstract units solely determined by the standardized relations between them.

In accordance with their origin as representatives of measures the numerical intervals between the numerical signs are generally irregular. Only the signs with high values exhibit a certain tendency towards a regular sexagesimal structure.[7] Thus, according to the way they emerged the archaic numerical signs can be divided into two different types corresponding to two different developmental stages of representation. Some of the numerical signs represent established counting units and standard measures that probably already existed at the time of the emergence of the numerical signs and thus predetermined certain structures of the numerical sign system. These signs call our attention to an older developmental stage which, by the time the archaic texts were written, had already been superseded. At this stage the signs were actually still first order representations of the objects and units to be quantified, and consequently their use did not presuppose any arithmetical thinking whatsoever. In contrast, the numerical signs of the second type are artificial signs constructed according to an arithmetical rule to which an actual unit or a standard measure actually used for measuring need not necessarily have corresponded.

Thus, there is every indication that, at the time the archaic texts were written, the numerical signs were already higher order representations no longer representing objects, but certain cognitive constructs, in particular socially standardized constructs of a metrological nature.

However, not only does the way in which the numerical notations are written indicate the existence of standardized meanings of the numerical signs as early as the archaic period of writing, but it also illustrates the limitations of this standardization. While the sign combination representing a particular quantity was standardized as such, and the graphical arrangement of the signs followed general rules such as, for instance, that the higher value signs were placed to the left of or above the lower value signs, the arrangement of the signs was no longer standardized in every detail. Rather, it was dependent on the spatial conditions on the writing tablet. Thus, the sign repetitions of a numerical notation were not yet integrated into a standardized graphical unit as was largely the case with later cuneiform writing. For this reason they specifically were not yet representations of numbers fused into a uniform sign, but remained combinations of counting and measuring units and were as such first order representations of metrological constructs.

The global structures of the numerical relations between the numerical signs correspond to these metrological constructs. Because of their specificity with regard to context these relations can be assigned to several systems with specific applications that are independent of each other. This is possible if the meaning of the signs is considered and ambiguous signs are taken into account as different signs more than once. In this way, 13 different numerical sign systems could be identified for the protocuneiform numerical signs and eight for the proto-Elamite numerical signs. They were used as strictly distinct systems; however, because of the ambiguity of the basic symbols they are partly made up of the same symbols.[8] If only the arithmetical meanings are taken into consideration and if the graphical variants that do not involve a modification of the numerical values are not counted as separate systems, then there are still six protocuneiform and five proto-Elamite basic systems. These systems largely consist of the same signs. However, depending on the structure of the system, the numerical relations between them differ. The six basic protocuneiform systems include two different counting systems for distinct objects (with a sexagesimal and a bisexagesimal structure),[9] a system for measuring grain, a system for area measures, a calendar system, and a system with an unknown metrological background. Of these basic systems the two counting systems, the measuring system for grain, and the calendar system were incorporated with minor modifications into the proto-Elamite system of writing and complemented by another counting system (with a decimal structure). In general, these systems were linear—that is, the signs can be arranged on a one-dimensional scale with relations between the individual signs expressed in terms of integral multiples. There are, however, exceptions to this rule as well. Thus, the small measures for grain were apparently not generated by multiplication, but by dividing a basic unit into 2, 3, 4, etc. parts,[10] as the relevant numerical sign system exhibits a corresponding nonlinear structure in this range. However, when it was integrated into the proto-Elamite system of writing, this numerical sign system was linearized by restricting the division of this basic unit into 2, 6, 12, and 24 parts.

In general, the areas of application of the systems did not overlap.[11] Different numerical sign systems were never applied to the same category of objects. For this reason in particular there are no indications of conversions from one system to another which would permit an inference as to the existence of more general cognitive constructs.

The areas to which the systems were applied (see [8]) were extraordinarily diverse. Some systems had very narrow applications, for example, two grain systems which, as far as we know to date, were used exclusively for two very specific types of grain—that is, for the grain variety emmer and for malt—whereas others were used frequently and in a wide variety of situations. The areas in which such systems were used often consist of applications that do not seem to have any connection with each other.[12] The sexagesimal system in its basic form was most widely applied. This already indicates its later function as the dominating numerical notational system. A graphic variant of this system (system S' in Fig. 3), on the other hand, only had a few, very specific and at the same time completely divergent areas of application such as the notation of deceased animals in administrative documents on animal husbandry or the notation of the number of jugs of a particular type of beer. Furthermore, there is no apparent reason why a bisexagesimal rather than the sexagesimal system of numerical signs was con-

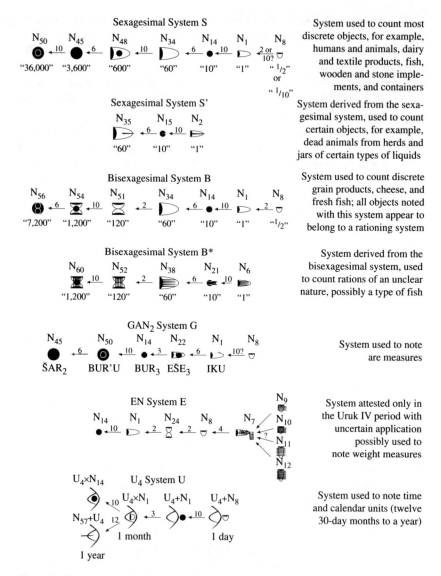

Figure 3: The numerical sign systems of the proto-cuneiform texts from Uruk (the numbers located above the arrows indicate how many respective units were substituted by the next higher unit).

ONTOGENESIS AND HISTORIOGENESIS OF THE NUMBER CONCEPT 343

System used to note capacity measures of grain, in particular barley; the small units also used to designate bisexagesimally counted cereal products

ŠE System Š

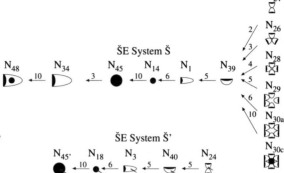

System used to note capacity measures of a certain grain, probably germinated barley (malt) used in beer brewing

ŠE System Š'

$N_{45'}$ $\xleftarrow{10}$ N_{18} $\xleftarrow{6}$ N_3 $\xleftarrow{5}$ N_{40} $\xleftarrow{5}$ N_{24}

System used to note capacity measures of a certain grain, probably various kinds of emmer

ŠE System Š"

N_{46} $\xleftarrow{6?}$ N_{49} $\xleftarrow{10}$ N_{36} $\xleftarrow{3}$ N_{46} $\xleftarrow{10}$ N_{19} $\xleftarrow{6}$ N_4 $\xleftarrow{5}$ N_{41}

System used to note capacity measures of grain, probably barley groats used to make certain grain products

ŠE System Š*

N_{37} $\xleftarrow{3}$ N_{47} $\xleftarrow{10}$ N_{20} $\xleftarrow{6}$ N_5 $\xleftarrow{5}$ N_{42} $\nearrow^{2}$ N_{25}
$\rightarrow^{3}$ N_{27}
$\searrow_{4}$ N_{28*}

System used to note capacity measures of certain products, in particular a milk product, probably dairy fats

DUG_b System Db
$N_1.DUG_b$ $\xleftarrow{10}$ $N_1.SILA_3$

System used to note capacity measures of certain products, probably dairy fats

DUG_c System Dc
$N_1.DUG_c$ $\xleftarrow{2}$ $N_1.KU_3$ $\xleftarrow{5}$ N_2

sistently used in the protocuneiform numerical notation for several types of countable objects, particularly for certain grain products and grain rations, and why a decimal system was used without exception for recording all animals and workers in the proto-Elamite numerical notations. Apart from that, for all forms of the differentiation of the context of application through modifications of the numerical notation there are parallels where the differentiation was achieved instead by adding specific non-numerical signs (as, for instance, in the DUG systems in Fig. 3; for further evidence see [8], p. 129f.). These examples may suffice to show that the global structures of the archaic numerical sign systems and their context-dependent use exhibit obvious deficits of a conscious structuring.

With regard to the way in which the numerical signs were used in the archaic administrative documents, the simple notation of the information on certain amounts of a particular product is by far the most frequent kind of application. In this case the notations represent empirical data on the objects of the particular economic transactions recorded, data which are independent of each other. However, in addition, there are entries with numerical relations indicating that these entries are not independent of each other, but are linked by arithmetical operations.

The most frequent of these operations—which was already mentioned earlier—is the totaling of several individual entries and recording the result as a particular entry for the total amount. This arithmetical operation is substantiated by hundreds of texts (see, for instance, the totals on the reverse of the text reproduced in Fig. 2). The operation is greatly facilitated by the semiotic structure of the archaic numerical signs. As a first order representation of metrological constructs the numerical representation is constructive in the sense that the numerical notation does not merely symbolize the quantity to be represented by repeating the symbols, but is in effect generating it. For this reason the totaling of several notations is exactly the same operation as the mere construction of an individual numerical notation. To construct a numerical notation the archaic scribe had to produce a correspondence between the quantity to be recorded and the numerical signs; he then had to simplify the notation by introducing higher value units in place of unnecessary sign repetitions. Exactly the same operations were needed for combining several individual entries to form a total except that the actual counting or measuring units were replaced by the units of the individual entries to be totaled.

With regard to the arithmetical structure this totaling of entries corresponds to our modern addition. There is, however, an essential semiotic difference. The modern method of adding composite numerical notations is a method for the transformation of signs solely guided by the forms of the signs. This is not the case when archaic entries are totaled.

While, on the one hand, the characteristic of the archaic numerical signs of representing the actual counting units effectively made the totaling much easier, the context dependency of the arithmetical meaning of the signs associated with this characteristic presented, on the other hand, a formidable obstacle for developing a formal method of addition solely based on the forms of the signs. Since the arithmetical meaning of the archaic numerical signs changed depending on the given context, the totaling could not be achieved without recourse to the substantial meaning of the signs.[13]

Aside from the generation of totals additional types of relationships between numerical entries of archaic texts can be identified which can be interpreted as the result of operations. They are, however, comparatively rare. Two types are documented well enough by extant texts to permit their identification as standardized operations. Judged in light of the modern number concept, both cases are special instances of multiplication. In the archaic texts there is, however, no evidence that the two types of operation are related to each other in terms of their computational techniques.

The first of these two operations can be considered as multiplication with an implicit factor. Pairs of two entries as in the text reproduced in Figure 1 provide typical examples. In each of these examples a certain number of units of a grain product is recorded in the first entry and a quantity of grain in the second, apparently the quantity of grain required for production; between the two entries there is a simple numerical ratio which, for different examples referring to the same product, is always the same. Thus, the second entry can be calculated by multiplying the first entry by this numerical ratio.

Despite this, it appears certain that the actual development of such mutually dependent pairs of numerical notations has nothing to do with such a multiplication. This is evidenced by the fact that, as a rule, the entries were recorded in disparate, incompatible numerical sign systems. Thus, a formal method of calculating one value from the other seems virtually impossible. How these values were in fact obtained remains unclear. The signs for the initial value probably were, in accordance with their particular meaning in fixed numerical ratios, exchanged for the signs of the value to be determined. Such a technique would explain in particular why the entries are almost always related to each other by simple numerical ratios.

The second of the two operations has a much more specific area of application. Some of the archaic texts deal with the surveying of fields and, in addition to data on the lengths and widths of fields, contain data about the corresponding areas. These data were obviously in some way derived from the length measures ([64], pp. 55–57). Although the data pertaining to area are incorrect in most of the existing examples, there can be no doubt that these data do not simply represent empirical values. In this case the method of how the values were obtained is not known either. However, because of the incongruity of the measures pertaining to lengths and areas it is unlikely that the areas were determined by a multiplication in the modern sense and only subsequently converted into the area measures. Presumably the method for determining the areas of fields was a specific method developed solely for this purpose. It was symmetrical concerning the data for lengths and widths and thus undoubtedly different from the unsymmetrical method in the case of the relations with implicit factors.

After this summary on the semiotic characteristics of the archaic numerical signs let us once more return briefly to the question of prime interest in the present context—that is, the question as to the cognitive constructs represented by these signs.

As we have seen, the numerical signs of the archaic texts constituted an efficient instrument for controlling quantities of any size likely to occur within the economic framework of that time. If Piaget considered the ability to construct correspondences between different sets of objects as a minimal precondition of the number concept as well as the insight that the quantities thus identified remain invariant against modifications of the spacial arrangement, then this precondition was without a doubt met by the archaic scribes. In archaic numerical notations the construction of correspondences between sets of objects and repetitions of signs was the very principle of the representation of quantities.

Since these sign repetitions were not yet completely standardized as far as their arrangement is concerned, they even provide direct evidence of the elementary abilities to deal with limited quantities. The flexible grouping of the signs, adapted to the individual spatial conditions on the tablet, shows, for example, that the archaic scribe anticipated the configurations of the signs mentally, before actually pressing them into the clay. The archaic scribes had a cognitive ability, undoubtedly applicable to any kind of objects, to identify limited quantities correctly and handle them accurately. In this regard there is no difference between the archaic scribe and the modern accountant.

However, no comparable parallels can be observed, if the global semiotic structures of the sign systems of the archaic scribe and the modern accountant are compared. The representational system of contemporary arithmetic is based on concepts and symbols no longer directly related to the objects of the reality to which they are applied. Rather, they represent the abstract arithmetical constructs and operations of arithmetic. It is in particular based on concepts and symbols for numbers and for the operations of addition and multiplication. Context-dependent quantities and operations, for example, metrological data are usually represented by putting these more general symbols and operations into a particular context, for example, by combining abstract numerical information with expressions for measuring units, with expressions for objects and similar matters. For instance, in the quantitative information "17 grams" the abstract number is embedded in the specific context of the physical category "weight" by adding the designation for the standard measure. Aside from this there are, however, historical relics in modern numerical representational systems as well, such as, for example, antiquated designations for particular quantities or designations referring to particular situations as, for instance, currency symbols like $ or £ which themselves represent particular numerical values. However, such representational forms can be identified as exceptions. First, because the disparate levels of generalization of such different representational media as numerals, measures, and currency symbols make it possible to determine the basic level of abstraction even internally; second, because their arithmetical meanings can always be translated into the more general representational system of abstract numbers and numerical operations. Although the abstract number concept of modern arithmetic is not expressed directly by the modern representational system, the fact that the semantics of all forms of representation refers more or less directly to the abstract construct of the number concept is, nevertheless, clearly reflected in the high degree of generality of numerous semiotic operations.

The search for evidence of a construct of comparable generality within the system of archaic numerical signs is futile, because these numerical signs are characterized by two features that specifically indicate the lack of a context-independent, overarching construct. First, the numerical signs do not represent fixed numerical values; rather, it is the context that determines their numerical meaning. Second, there are no purely syntactic, arithmetical rules, that is to say, no methods for the transformation of symbols resembling computational rules that would suggest general numerical operations.

Systems resembling numbers can only be recognized, when the semantics of the numerical signs is scrutinized—that is, if the ambiguous signs are differentiated according to their meaning. However, the semiotic structures that can be determined in this fashion and that we call here numerical sign systems are area specific. If the concept of metrology is conceived broadly enough to include object-specific counting techniques, then these global semiotic structures must be considered without exception symbolic representations of metrologies; and while these metrologies already conformed to widely applied conventions, their general structures can nevertheless only be explained historiogenetically in terms of their function as means to the symbolic control of the economic processes which were becoming more and more complex.

Because of the deficits of a deliberate structuring of the archaic numerical sign systems it seems in particular problematic to infer the existence of generalized and arithmetized cognitive constructs such as number, volume, area, and time from the existence of fundamental systems that, with regard to the numerical values of the symbols, agree with the systems derived from them. The areas of application of these systems do not represent a meaningful conceptual structuring in terms of abstract concepts any more than the numerical structures. The most striking characteristic concerning the use of the archaic numerical signs is their specificity and their contextual dependency which is especially conspicuous with regard to the graphic variants and the numerous quasi-numerical notations in the border area between the numerical signs and the other written signs. The numerical signs are distinguished from the rest of the signs merely by some particular syntactic rules applied to them. The numerical signs are reiterated by means of sign repetitions used to represent quantities. Certain repetitions of signs are replaced by higher order signs on the basis of standardized numerical relations. And finally the numerical notations constructed in this way and composed of different signs display a standardization of the sequence of the numerical signs according to their actual value in the particular context. These syntactic rules are obviously direct consequences of the standardization of the metrological background of the signs.

The assumption which, in light of later developments, has a certain plausibility, that one of the systems—namely the sexagesimal system—occupied a special role and, as the only one, did not represent an object-specific metrology, but rather a number construct superior to the other systems, is not supported by the use of this system documented in the texts. The sexagesimal system had not yet become particularly prominent in any way. When the protocuneiform numerical sign systems were adopted by the proto-Elamite system of writing, the significance of the sexagesimal system was even reduced, since it was replaced in part by another system, the proto-Elamite decimal system. The sexagesimal system was only one of the counting systems for distinct objects, albeit a system with a particularly broad area of application. The signs of the system are the same as those of the other basic systems; thus, the signs of the sexagesimal system in particular were used in different contexts with different meanings. The sexagesimal system was not particularly important in relationship to the other systems to which, as a symbolic representation of a more general concept, it would have had to be superior. At best the fact that the higher values of these systems, which were presumably added at a later date, tend to have a sexagesimal structure, can be attributed to the influence of the sexagesimal system. There is after all not a single example for the transformation of a metrological notation into a sexagesimal representation which would prove that the sexagesimal system did indeed have a more general status than the other systems. The conclusion, therefore, seems inescapable that the arithmetical techniques documented by the archaic numerical signs were in fact techniques without an integrating number construct.

Indeed, we obtain a much better explanation for the semiotic characteristics of these signs, if we assume that it was not an abstract concept, but the context documented directly by the administrative documents which, as a cognitive construct, constituted the intellectual background of the numerical sign systems as well. As was demonstrated at the very beginning, the archaic systems of writing did not yet constitute representational systems of the spoken word; rather, they were means for modeling economic processes. The numerical sign systems were central to these systems of writing. The numerical signs mainly served to represent quantities by sign repetitions, but their use was area specific and they had a host of connotations so that they also had functions similar to those of the non-numerical signs. Numerical and non-numerical signs were signs with only a slightly different function within a protoarithmetical representational system which, in terms of the numerical signs and for the most part also the other signs used, reflected only the economic and administrative actions of the administrators in the period of the early development of writing.

The hypothesis that this function also determined the cognitive constructs represented by the numerical signs seems to be justified by the fact that there are, in the semantics of the signs, to the extent it can be reconstructed, no indications of a specific conceptuality separating the metrological notations from their context and systematizing them under overarching aspects.[14] Their use in the context of administrative techniques is completely explained if we assume that the use of the non-numerical as well as the numerical signs was based uniformly on a cognitive model of the redistribution of products and of the administrative activities associated with this redistribution which were regulated by these techniques. The archaic scribe understood the numerical notations and the context-dependent transformations he performed with them probably just as they have been transmitted to us in the administrative documents, that is to say, as symbolic representations of products quantified in terms of their particular category and of the transactions performed with them.

This would indicate that, as in the case of the Stone Age culture of the Eipo, the existence of counting and measuring techniques did also not result immediately in the development of an abstract number concept in the early protocuneiform and proto-Elamite literate cultures. The structure of the representation of quantities by sign repetition and by substitutions of higher value units does not require any global cognitive structure other than a cognitive model of the context of application; for by introducing such units, which after all always have the same semiotic structure, all numerical problems are reduced to the same simple operations in the range between one and at most ten (see Fig. 4).

The assumption that archaic arithmetic is simply based on a cognitive model of economic administration and not yet on an abstract number concept is further backed up, if we examine this arithmetic in the context of its historical development. If the numerical notations and operations documented by the protocuneiform and the proto-Elamite administrative texts are compared, on the one hand, with the evidence of preliterate arithmetical techniques (counters and numerical tables) mentioned earlier and, on the other hand, with the arithmetical techniques of the developed cuneiform, then it becomes evident that the specific form of the arithmetic of the first written documents only represents a brief transitional stage in the development of Babylonian arithmetic.

Our knowledge of the preliterate arithmetical techniques is unfortunately still rather incomplete. Interpretations of the function of the counters and of the preliterate numerical notations are for the most part speculative and, what is more, almost universally called into question.[15] However, at least it seems to be certain that a change of the semiotic function of the numerical signs occurred between the preliterate numerical notations and the numerical notations of the first written tablets. Many difficulties in deciphering the numerical meanings of preliterate numerical signs and in interpreting the numerical notations of this time are due to this change of their function.

ONTOGENESIS AND HISTORIOGENESIS OF THE NUMBER CONCEPT

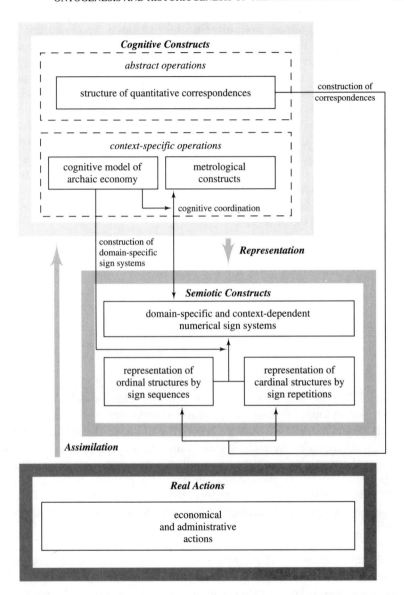

Figure 4: Structure of proto-arithmetical thinking of archaic arithmetic.

The numerical tablets from diverse sites constitute the most important source of information on preliterate numerical notations. It makes sense to compare the numerical notations of these preliterate tablets with the numerical notations of the first written tablets to determine whether the systems of numerical signs of these tablets already existed before the invention of writing. However, such a comparison encounters basic difficulties. In particular the numerical notations of the numerical tablets are more inconsistent than those of the archaic written tablets thus making it difficult to draw general conclusions. Some of the numerical tablets contain notations that can easily be identified as notations in the numerical sign systems of the archaic written tablets. Many of the notations are, however, of a different nature and, in part, do not exhibit any discernible parallels at all with the numerical notations of these tablets.[16]

The inconsistency of the numerical notations on the numerical tablets could be attributed to the fact that the archaelogical finds must be assigned to very different periods. For most of these tablets it cannot even be determined on archaeological grounds whether they stem from the time before the invention of writing at all. There are even indications that the numerical tablets with numerical notations that lend themselves to interpretation are no older than the archaic tablets, but originated at the same time. For if we limit ourselves to the few tablets for which there is archaeological evidence that they do indeed date back to the time before the invention of writing—that is, specifically the finds of Jebel Aruda ([22]) and Habuba Kabira ([87]) as well as the finds of a more recent excavation in Susa ([48]), we no longer find examples of numerical notations which can definitely be interpreted as notations belonging to one of the systems of numerical notations of the archaic texts. Furthermore, the lack of standardization of how these confirmed examples of preliterate numerical notations were written is such that it is not even possible to identify the individual symbols with any certainty; that means that it is virtually impossible to determine the identity of symbols of different numerical notations.

In the case of the tablets from Jebel Aruda such an identification of the individual symbols is not even possible with regard to the texts from the same site. Furthermore, these tablets have more than ten repetitions of signs within a notation. That is to say, that the rule of substituting a higher value unit for repetitions of signs beyond a certain number, observed almost without exception on the archaic written tablets, was not yet a firm rule in the case of these numerical tablets. It is fair to assume that in this case the individual symbols did not yet represent abstract metrological units, but rather the concrete objects and means for measuring.

Since the number of the numerical tablets that can be located in the preliterate period with any certainty is relatively small, the generalization of such individual findings is more than problematical. There are, nevertheless, many indications that the counting units and measures were integrated into those standardized metrological systems which we know from the numerical sign systems of the archaic texts only in the context of the written representation.

As we have seen, these first numerical sign systems had very peculiar characteristics. The latter did not, however, last long. Even the oldest texts of the developed cuneiform, written about 500 years after the invention of writing, the texts of the Fara period, reveal significant changes not only in terms of the paleography and the structure of writing, but above all in terms of the numerical notations as well. Most of the protocuneiform systems of numerical signs are no longer extant at this time. Only three of the archaic numerical sign systems represented in Figure 3 are documented in the texts of the Fara period (see [17]). Only two of these, modified to a greater or lesser extent, subsequently remained in use—that is, the sexagesimal system and the system of area measures. In the Fara period the sexagesimal system is almost the only system for the notation of concrete objects. Although the bisexagesimal system can still be found in its basic form, its use apparently is no longer strictly separated from the sexagesimal system. In later groups of texts it has completely vanished.

The ambiguous archaic signs are replaced by metrological notations consisting for the most part of combinations of a sexagesimal numerical notation and additional signs for measures as well as a host of different kinds of notations for specific purposes. While specific notations continue to be used for the metrological systems, the numerical information and the information about the type of the objects counted and measured are already largely differentiated within these notations. The numerical ambiguity of the numerical signs has been removed.

These changes inevitably led to a higher level of formalization. To a large extent the arithmetical operations could now be performed solely on the basis of the symbolic notations. However, this does not necessarily mean that the cognitive constructs associated with these operations had changed as well. It must remain open when this actually occurred. It is possible that it was the revolutionary change of the symbolic notations following the invention of the sexagesimal place-value system of Babylonian mathematics towards the end of the 3rd millennium (see Chap. 7) which resulted in more general cognitive constructs of arithmetic. The numerous tables with transcriptions of metrological notations into abstract sexagesimal numbers and the systematically tabulated arithmetical operations using these numbers show that a cognitive concept encompassing all numerical notations and operations was indeed associated with this form of numerical notation. Thus, the first firm indications of an inclusive numerical construct appear some 1,000 years after the invention of writing.

ONTOGENESIS AND HISTORIOGENESIS OF THE NUMBER CONCEPT

In closing, the consequences of the previous analyses of protoarithmetical techniques in a neolithic culture and in an early literate culture for evaluating Piaget's psychogenetic theory of the number concept will be discussed. First, a general remark on the significance of historical analyses of this nature for questions concerning developmental psychology is called for.

As a rule, basic assumptions of psychological developmental models are framed at such a general theoretical level that the claims of their validity go far beyond what can be verified or falsified by investigations of ontogenetic processes in the context of developmental psychology. There is a factual reason for this generalization. By necessity, developmental psychology raises problems that encompass ontogenesis and historiogenesis; for the problem of the ontogenetic development of fundamental cognitive structures of human thinking inevitably raises the problem of the nature of these structures. And whatever the answer, it will, whether intentionally or not, include statements on the relationship of such structures to cultural evolution.

As we have seen, this is especially true of Piaget's genetic epistemology as well. In contrast to the tradition of Platonic and a priori interpretations, Piaget's interpretation of the number concept is based on the assumption that the number concept is not an inherited schema of thought but a cognitive construct. The starting point for this assumption was not the historical development, but the development of the child's thinking. In his empirical investigations of the ontogenetic development of elementary concepts, for example, the concept of number, Piaget outlined the construction of the cognitive structures basic to logico-mathematical thinking in the course of ontogenesis by means of processes of reflection on concrete actions and on the schemata of thought and the operations derived from them. The theory of the development of the number concept which, among other things, arose in this fashion identifies systems of actions consisting of correspondence, comparison, and grouping activities as the protoarithmetical conditions for the cognitive construction of the number concept by means of reflective abstraction, systems of actions which, in almost any culture, are part and parcel of the universal mental equipment of sensori-motor schemata acquired by the child.

Piaget's assumption that the number concept is a cognitive construction abstracted from coordinations of actions belongs exactly to those basic psychological assumptions which include answers to historical problems as well. Nobody was probably more aware of this than Piaget himself. With the results of his investigation of the ontogenetic development of the number concept he provided answers not only to narrow problems merely concerned with the ontogenetic development, but he was forced to take sides in the dispute over the essence of numbers dating back to antiquity, a dispute that up to that point had, however, taken place mainly in mathematics and philosophy and not in psychology. As we have seen, Piaget aggressively defended even those implications of his investigations that go beyond the narrow sphere of research in developmental psychology and involve the historiogenesis of the number concept, although he hardly subjected them to a serious examination in light of historical research. The status of such assertions must, however, remain tenuous as long as their tenability is not even examined *ad hoc,* let alone being in agreement with the results of an independent study of the historiogenesis of the number concept.

The tenuous status of such generalizing assertions of developmental psychology permits the conclusion that psychological research ought to limit itself in terms of its problems and methods in such a way as not to even consider far-reaching implications of this nature like those of a structuralist developmental psychology so as not to jeopardize criteria pertaining to validity and verification internal to the discipline. This consideration probably plays a significant role in psychological research; in any case, among the numerous investigations produced in Piaget's wake there are hardly any studies that made the question raised by Piaget concerning the relationship between ontogenesis and historiogenesis of elementary cognitive structures the subject of a thorough analysis. It may, however, also be argued that it is precisely the tenuous status of generalizing statements of developmental psychology which affords the possibility to subject psychological findings to a more far-reaching control than is possible by internal criteria of validity and verification. This might even result in resolving open questions in developmental psychology by making use of external considerations. In view of this, the results of the case studies on arithmetical thinking in an extant preliterate culture and in the historical transition to a literate culture introduced in the previous two sections will be discussed in the following.

As we have seen, it is first of all necessary to supplement the studies on the number concept in developmental psychology: The protoarithmetical systems of actions postulated by Piaget as the basis of this concept must be identified in their historical and cultural context. If Piaget's assumption is correct that the number concept is a result of the reflective abstraction of these protoarithmetical systems of actions, then the latter must be correlated in a meaningful way to the identifiable historiogenetic developmental stages of the number concept. If, for example, such developmental stages did not even exist and the basic logico-mathematical structure of the number concept were universal, then it would have to be possible to prove the existence of a universal protoarithmetic as a cross-cultural component of ontogenesis capable of explaining the ontogenetic development of the number concept in any cultural setting. On the other hand, assuming the existence of historically identifiable developmental stages it could be expected that the historiogenetic developmental surges are preceded or accompanied by a change along the lines postulated by Piaget regarding the protoarithmetical systems of actions characteristic of a particular culture and transmitted by social interactions.

If the demonstration of such relationships between a protoarithmetic identifiable in terms of the cultural context and the development of cognitive constructs of logico-mathematical thinking is considered to be a scientifically productive method at all, then the conclusions of the two case studies must be regarded as a convincing confirmation of Piaget's basic assumptions concerning the development of the number concept.

In the preliterate cultures the evolution of a precise concept of quantity seems to be directly related to the degree to which symbolically represented and socially transmitted standard quantities have been developed for the purpose of constructing one-to-one correspondences and of identifying the results of such constructions—that is, particularly the development of counting techniques. The reason for the lack of a sequence of number words in many languages of the aborigines is, for example, not due to a lesser expressiveness of these languages with regard to certain, universally existing mathematical forms of thinking, but rather to a culturally specific function of the cognitive technique for the construction of unambiguous correspondences. This technique constitutes the foundation on which the semantics of the sequence of number words is based. As long as such a technique is not part of the transmitted means of a culture, the number words possibly already existing in such a culture are not an arithmetical, but a linguistic phenomenon. They are variants of the qualitative designations of quantities to be found in every known language. The example of the Eipo furthermore shows that there is a direct relationship between the actual use of such techniques for constructing correspondences and the degree of the assimilation of real situations to concepts quantifying magnitude and intensity. In this context only those concepts directly related to the actual use of the counting technique using the system of body counting numerals acquire a quantitative meaning.

Presumably the break to be anticipated in the historiogenetic development of the number concept during the transition from preliterate forms of representation to the representation of correspondence activities by means of numerical signs may for all intents and purposes be considered as having been established by the analysis of the systems of numerical signs of the archaic written tablets. In the historical period before the emergence of writing we must take note of an extensive use of preliterate means for constructing correspondences and representing the results in order to cope with the emerging problems of controlling an expanding economy in the early urban centers of the Near East, problems that, in terms of their magnitude alone, cannot be compared to the limited scope of a rural economy. The means used for this purpose such as the archaeologically attested counters and numerical symbols are the typical means with which preliterate cultures control modest quantities if necessary. However, in the cultural environment of the administration of the centralized economy of the cities dominating their surroundings these means are generalized to apply to all objects and means of economic reproduction. The evolution and standardization of metrologies for all economically relevant objects is an indication of this development.

The invention of writing and the evolution of context-specific numerical sign systems representing these metrologies must be understood as a consequence of this development. The numerical and non-numerical signs of this earliest phase of the evolution of writing are the elements for modeling economic transactions and the administrative procedures governing them. The semantic categories represented by them are identical with the generalizations established in the context of administrative actions. In the strict sense of Piaget's theory they represent reflective abstractions from the activities of the archaic scribes. Thus, the composition and decomposition of sets of objects, for example, which constitute the basis for the arithmetical operation of addition were represented merely enactively by the usage of counters in the preliterate period, and the relationship between individual amounts and their total could be represented by such counters only in temporal succession. However, on the archaic written tablets this relationship is represented in spatial simultaneity. Even if, as we have seen, the operations carried out in the course of recording entries did not yet constitute context-independent systems of sign transformations, they already were essentially operations with numerical notations and their relations to each other and no longer merely operations constituting first order representations of concrete actions. Thus, they relate only indirectly to the quantified objects. They are operations at a metalevel, a level of the second order representation of the initial actions, which displays relationships such as the independent existence of the parts and the whole (the individual entries and their total) that cannot be realized at the level of first order representation.

Thus, a fundamental change of the preliterate techniques for constructing correspondences is evident during the period of the early development of writing. It consists, on the one hand, in a more extensive use, on the other hand, in the representation of cognitive constructs of reflective abstraction which constitute a second order system of actions. The result of our reconstruction of the cognitive correlates of the protoarithmetical systems of actions that are fundamental to the development of the number concept corresponds to this historically attested radical change. If this reconstruction is correct, then the archaic systems of numerical signs and the operations performed with them represent a transitory stage of cognitive construction of a relatively short historical duration. In this transitory stage the system of actions of the number concept has already expanded in comparison to the techniques familiar from preliterate cultures. However, at this transitory stage the cognitive constructs still reveal a deficit with regard to abstract operations that are context-independent operations which are nonspecific with respect to their areas of application—that is, a deficit with regard to mental operations referring directly to the representation by numerical signs and only indirectly to the actual economic products represented by these signs. In this transitory period we encounter a protoarithmetic, already highly developed in terms of its capabilities, but without an abstract number concept that would permit the assimilation of the protoarithmetical operations with numerical signs to a homogeneous cognitive construct.

Only shortly afterwards, the semiotic structure of the numerical notations has changed. In particular the peculiar characteristic that values are assigned to the numerical signs only by the context in which they are used has disappeared. They now represent a more abstract numerical construct with a greater degree of independence from the context in which they are used. However, at this point the nature of this construct must remain unanalyzed. Its explication will be reserved for a psychological analysis of the development of the numerical and metrological notations in the administrative documents of the 3rd millennium up to the invention of the abstract sexagesimal place-value system (see Chap. 7).

Thus, the relationship between counting technique and arithmetization in the preliterate culture of the Eipo as well as the relationship between the development of administrative techniques and the cognitive constructs associated with the systems of numerical signs developed by the archaic scribes of the Near East substantiate Piaget's assumption that the number concept is generated by means of reflective abstraction of protoarithmetical systems of actions.

However, in other respects the results of the two case studies make a revision of Piaget's theory of the development of the number concept necessary. As was shown at the beginning, Piaget assumed that, while the number concept is acquired by means of action-based experiences, its structure is not substantially influenced by the content of these experiences. For according to his theory the number concept belongs to that logico-mathematical structure of thinking which is abstracted from a biologically developed form of the coordination of actions and therefore appears universally in ontogenesis as long as the cultural context offers the necessary possibilities for these actions at all. Thus, a historical development of the number concept paralleling the stages of the ontogenetic development was supposed to take place only in the "prescientific" phase. This is the phase between the cultures of primitive peoples where this condition is not met and where thinking is confined to the prelogical stage, and the beginning of the "scientific" phase where the logico-mathematical thinking is fully developed and the developmental stages of scientific thought are connected with ontogenesis merely in functional rather than substantial terms.

The conclusions of the two case studies presented above contradict this theory. Neither the Eipo nor, to an even lesser degree, the archaic scribes thought prelogically as demanded by Piaget's theory. The cultural presence of transmitted counting techniques is a sure sign that the particular condition transcending prelogical thinking which Piaget considered to be the minimum requirement for the number concept has been met: the ontogenetic construction of the invariance of quantity. However, this obviously does not mean that, in these cultures, the remaining operations of the number concept in their general form or even the integrated overall structure of logico-mathematical thinking understood by Piaget as being canonical are constructed ontogenetically as well. In the case of the Eipo there is no arithmetic in our sense at all despite the ability to construct correspondences for the purpose of identifying quantities. While it is true that there are developed arithmetical techniques in the case of the archaic scribes, the mental operations of these techniques remain context-specific operations of metrological constructs. At best they were integrated into a cognitive model of their particular area of application—that is, the centralized economy—but had not been broadened into a general number concept with generalized operations. Historically the abstract number concept is a relatively recent product. In contrast to Piaget's theory it is more likely that it was abstracted from the written representation of those operations of counting and measuring on which it is based than directly from these actions themselves.

Thus, if in Piaget's theory of psychogenesis encompassing ontogenesis and historiogenesis parallels between the sequence of the developmental stages in ontogenesis and historiogenesis are explained in terms of the assumption that the structure of the operations of logico-mathematical thinking is independent of the nature of the actions from which it is abstracted, then the results presented demand a revision of this specific assumption. In light of his assumption that the coordinations of actions are determined by developmental biology, Piaget postulates that, in accordance with the model of ontogenesis, the historiogenesis of the logico-mathematical concepts takes place exclusively in a "prescientific" historical period. The distinction between a "prescientific" period of historiogenesis determined by ontogenesis and a "scientific" period based on a universal structure of logico-mathematical thinking in which ontogenesis and historiogenesis only agree in terms of the cognitive functions of assimilation, accommodation, and reflective abstraction, cannot be verified historically as far as the number concept is concerned.

For it appears that the development of this concept depends on socially generated and historically transmitted actions of economically mediated cooperation and the control of economic processes to a much larger extent than would probably be the case for an epigenetic process determined by developmental biology. It is not the mere existence of certain cognitive operations in a culture, but rather the area of application of techniques based on these cognitive operations in a particular historical context that determines into which universalized cognitive constructs these operations are integrated.

It is true that there are indeed elements of the number concept as conceptualized by Piaget which are to be found universally in almost all cultures. In particular this includes the construction of one-to-one correspondences which occurs as a socially transmitted protoarithmetic pattern of action in nearly all cultures. As a rule the correspondences are established with the aid of standards of correspondence activities. These are for the most part number word sequences. In the case of the Eipo it is, as with many cultures of New Guinea, a system of body counting numerals. However, such protoarithmetical techniques do not necessarily result in the development of the number concept of Piaget's theory. To a certain extent such techniques can form the basis for the control of quantities in a specific context without being integrated into the universalized structures generated by reflective abstraction which, in Piaget's view, define the ontogenetic developmental stages of the concrete and formal operations. Apparently Piaget underestimated the possibilities of generating those performances to a practically satisfactory extent solely on the basis of the mental construction of correspondences and context-specific cognitive constructs for counting and measuring units to be controlled in terms of their quantity, performances which, according to his theory, necessitated a universalized number concept.

If, however, the assumption that the structure of logico-mathematical thinking is epigenetically determined in terms of developmental biology, it is narrowed in Piaget's theory in order to account more fully for the historically demonstrable dependence of cognitive constructs such as the number concept on coordinations of actions by means of socially generated and transmitted protoarithmetical techniques, then the picture of the relationship between ontogenesis and historiogenesis implied by this theory undergoes a basic change. Then the distinction between a period of historiogenesis determined by ontogenesis paralleling it and a period in which the parallelism of ontogenesis and historiogenesis consists merely in similar cognitive functions becomes obsolete. Then the historically specific, symbolic representations of cognitive structures which determine the historical transmission play a substantial role. The reason for this is that they, as higher order representations the meaning of which has to be reconstructed, open up those areas of action to the individual in the process of ontogenesis which guide the process of reflective abstraction to general levels in culturally specific ways. Thus, higher order representations of cognitive structures prove to be the essential factors for transmitting the results of reflection, once they have been achieved, to the totality of the individuals or to professionalized sections of a certain culture.

Piaget's theory must be commended for recognizing that there are parallels between ontogenetic and historiogenetic stages of development such as, for instance, that between the area-specific nature of cognitive constructs at the ontogenetic developmental stage of concrete operations and the area-specific nature of quantifying thinking in the transition to written representation. It must also be admitted that it makes sense to apply the same theoretical categories to the developmental mechanisms of ontogenesis and historiogenesis. However, these relationships do not prove the identity of ontogenetic and historiogenetic developmental processes. Rather, they are the result of the relationship of cognitive constructs and their culturally specific representations which determine their historical transmission. Actually, the apparent parallelism of ontogenesis and historiogenesis is a parallelism of the cognitive constructs of a culture to be realized ontogenetically and their symbolic representation by culturally specific means of historical transmission.

NOTES

[1] See Chapter 1 of the study. In view of Piaget's often repeated assertions regarding the parallelism of these developments it is surprising that only one of the relevant examples is meant to document this parallelism in a work explicitly dedicated to the relationship of ontogenesis and historical development. The study was not published until after Piaget's death. Therefore the question must remain open whether the study represents a reevaluation of the role of psychogenesis with regard to the historical development of concepts or unresolved differences of judgment on the part of the two authors. This is not the place for a detailed discussion of the example that, in this way, became the touchstone for Piaget's theory. It should, however, be pointed out that it is precisely this example which could raise doubts with respect to Piaget's thesis. It appears that the alleged parallelism between conceptual and theoretical structures of ancient and medieval natural philosophy on the one hand and the thinking of the child on the other can only be maintained, if the elaborate conceptual systems and deductive inferences of the Scholastic scholars as well as Galileo and his contemporaries are interpreted in a completely one-sided fashion. If such a parallelism did indeed exist, it would apply only to an exceedingly marginal aspect of scientific development in the course of those centuries. See Damerow, Freudenthal, McLaughlin, and Renn ([14]) on the role of conceptual structures of preclassical mechanics regarding the evolution of the concepts of classical mechanics.

[2] We would, therefore, be well advised to take the interpretation of the marginal role of the counting process based on a generalization of this observation with a grain of salt which was provided by Eibl-Eibesfeldt, Schiefenhövel, and Heeschen ([24], p. 36): "Our knowledge of the self-restraint of the Eipo in situations where they are engaged in exchange and giving, of the studied reserve with which they give away valuables and of the detachment with which they accept them is based in particular on film footage. A dance presents the

appearance of generosity. We did ask ourselves why there are number words in the Eipo language at all. It is superfluous to know how many trees are cut down, because it is the number of boards they yield which counts; to state how many pigs are served is tabu. What gives satisfaction is whether the stomach is agreeably filled, not however the statement how many sweet potatoes someone ate. Paul Blum did discover when the Eipo count: When they secretly spread out the gifts after a dance to examine them. Then the number of taro tubers and unhewn stone axes are established with exactitude. They painstakingly observe whether they receive gifts of comparable value in return for what they themselves gave. Below the surface of the festive atmosphere greed and avarice are seething."

3 The first rather flawed attempts to decipher the archaic numerical signs date back to Scheil ([74]). A certain conclusion of these early endeavors was reached with the works of Deimel, Langdon, and Falkenstein ([17]; [47]; [27], [28]). Not until about 50 years later Vaiman ([88], [89], [91]) and Friberg ([30], [31]) discovered serious shortcomings and mistakes concerning these generally accepted interpretations. With regard to the present status of the decipherment of the numerical signs cf. the writings of the Berlin Project ([8], [9]; [25]; [64]). On the methods used see Damerow and Englund ([8], [9]). On the further development of arithmetic see Chapter 7 and Friberg ([32]).

4 See the standard work by Neugebauer ([60], especially p. 98), subsequently adopted by most authors (e.g., [94], p. 10 and pp. 17–19; [101], p. 33, containing a misinterpretation of tablet VAT 12593 shown as an example, a misinterpretation based on Neugebauer's carelessness in his treatment of the archaic numerical signs; a correct interpretation had already been given by Deimel, the editor of the tablet, as early as 1923 as well as by Neugebauer in his edition of mathematical cuneiform texts in 1935). Not until the writings of Friberg ([32]) and Høyrup ([38]) has a convincing case been made that Babylonian mathematics must be understood in terms of its thousand year history and not vice versa.

5 The problems of the identification of these signs and their differentiation from the remaining non-numerical signs cannot be treated more fully in the present context. See Damerow and Englund ([8]) concerning the numerical signs of the archaic texts from Uruk and Englund and Grégoire ([26]) concerning the number signs of the archaic texts from Jemdet Nasr. A preliminary list of all the numerical signs is to be found in Nissen, Damerow, and Englund ([64]); however, a complete list will probably not be available until all archaic text groups have been edited.

6 For the time being the text reproduced in Figure 1 may serve as evidence for this assertion. It shows the sign of a ration bowl which, as can be determined from the data, denotes the daily ration of a worker; here, however, it represents $\frac{1}{5}$ of a certain measure of grain, although it normally represents $\frac{1}{6}$ of this measure, for example, in the very similar text IM 23426, published by Falkenstein ([28]) with a faulty determination of the numerical values and interpreted correctly only by Friberg ([31]; see also [32]). Further evidence is provided by Englund and Grégoire ([26]) as well as in the forthcoming publication by Damerow and Englund and in the archaic texts from Uruk yet to be published.

7 See the following examples of a sexagesimal structure of the numerical signs with higher values in the numerical sign systems to be discussed in detail in the following (for the present purposes notation anachronistically in abstract numbers). Bisexagesimal system: 120–1200–7200 units (120, times 10, times 6); field areas: 18–180–1080 area units of the measure iku (18, times 10, times 6); measures for grain: 180–1800–10800 units (180, times 10, times 6). The assumption that this is not accidental, but that these are the beginnings of the transfer of the sexagesimal system to the areas of application of other systems is confirmed by the subsequent development. To the area measures further measures in sexagesimal sequence are added. The weight measures, attested only in more recent text groups, are constructed almost entirely with a sexagesimal structure. The measures for grain are completely replaced, but starting with the measure gur they are again sexagesi-

mally structured. On this point compare the archaic numerical sign systems ([8]) with those of the subsequent Fara period [17]).

[8] In this context we must forego a detailed discussion of these systems. The most important protocuneiform systems are shown in Figure 3. For details see Damerow and Englund ([8]). The proto-Elamite systems are discussed in Damerow and Englund ([9]). However, it should be mentioned that the exact number of the systems depends on the extent to which variants in style and context are considered separate systems. In addition, there are some numerical signs and numerical notations for which evidence is so scarce that the systems on which their use is based cannot be determined.

[9] See the protocuneiform systems in Figure 3. When interpreted numerically the sexagesimal system (excepting problematic signs for fractions with context-specific values) includes signs for the numbers 1, 10, 60, 600, 3600, and 36000. Furthermore, in a school exercise there is a combination of signs that was later used to represent the number 216000; a numerical interpretation has, however, not been verified. The bisexagesimal system contains the same signs for the numbers 1, 10, and 60, but continues with signs for the numbers 120, 1200, and 7200. The numerical signs of both systems were never used together in a numerical notation—for example, no numerical notation has ever been found in which the symbol for 600 of the sexagesimal system and the number 120 of the bisexagesimal system were used at the same time. Here the problem of the usual practice of attributing modern, abstract numbers to the numerical signs of archaic texts becomes apparent, since it does not allow for the expression of this difference between two distinct types of numbers.

[10] See the systems for the notation of grain in Figure 3 (right side). A similar structure of system E (see Fig. 3 [left side]) with an unknown area of application must be assumed, but has not been corroborated. Although the creation of measures through division instead of the multiplication of a unit is rare in the early period of the development of writing, it is nevertheless known from other contexts. It is, for instance, exemplified by the weight measures in Ebla.

[11] An exception is a system of two signs with values for 1 and for 10 units (see the system L in Chap. 8, Fig. 6) not shown in Figure 3 which probably did not have its own area of application, but merely served accounting purposes. This system was generally used for recording subentries, presumably in order to distinguish them from the main entries so as not to include them unintentionally in the summations. At first sight, the general system for the notation of grain measures and the special variant presumably used for groats seem to overlap as well (see Fig. 3). However, these apparent overlaps can presumably be attributed to a careless rendering of the notations of the latter system, especially since it exhibits a peculiarity which often results in the graphic similarity of the numerical notations to those of the general system for the notation of grain measures. For in this case the differentiation of the numerical signs of both systems by means of little dots with the stylus is not the usual differentiation by means of graphic variations of the numerical signs, but by means of marking the whole "case" containing the notation. The dots, generally few in number, were evenly distributed over the numerical notation and at times even over the whole "case." For this reason it probably eluded earlier investigators that this constitutes a separate system of numerical notations.

[12] In this context a fundamental problem of the notation we chose for the numerical relations between the signs by assigning them to systems becomes evident. This representation suggests overarching, arithmetically constructed cognitive constructs along the lines of our modern physical quantities. However, such constructs need not have existed at all. The systems are after all formed merely by metrological relations between counting and measuring units and not necessarily by overarching arithmetical operations and structures as well. The integration into a system is really only justified when comparable integrating forms of representation from the particular period itself confirm the existence of a corresponding cognitive construct or if, at least, generalized arithmetical operations implicitly

establish a systematic connection between the units. With regard to the numerical notations of the archaic period, however, this condition is only partially satisfied. Up to now no explicit representations of metrological systems from the archaic period have been found that, like for instance the metrological lists of later periods (see the examples in Chap. 7, Fig. 5, and in [64], p. 148f.), indicate a deliberate treatment of notational rules for representing metrologies. The existence of overarching arithmetical operations to be discussed in the following depends exclusively on the context of application. The semiotic rules which make up the archaic numerical sign systems and which we reconstruct as syntactic rules for the use of the signs may well have been unconscious consequences of the semantics of the signs for the archaic writer—that is, consequences of context-specific procedures for measuring and comparing quantities. The problem which follows from this for the interpretation of the cognitive background of the systems of numerical signs is illustrated in particular by the grain systems covering a particularly broad numerical range. In these systems the applications of the signs for the smallest units have almost nothing in common with the applications of the largest units. The latter almost exclusively serve to record quantities of grain and their summation (see, e.g., the reverse of the tablet reproduced in Fig. 2). In contrast, the small units are almost exclusively used for the notation of grain rations and grain products (as in the text reproduced in Fig. 1). Only a few examples indicate that they, just like the rest of the signs of these systems, could also be used for representing quantities of grain in the usual manner by sign repetition. Summations of such signs are exceedingly rare. There are even indications for the existence of an independent type of notation for the signs with higher values if they were used for a similar purpose (combination with the symbol $NINDA_2$, see the system N in Chap. 8, Fig. 6), as well as the survey of the supporting evidence in Damerow and Englund ([8], p. 138). Therefore it appears somewhat unlikely that there was always a cognitive metrological construct corresponding to the numerical sign systems we constructed.

[13] The central obstacle is indeed the arithmetical ambiguity of the signs. This becomes especially obvious when we compare very simple summations in different systems. Thus, the text W20672,2 includes the addition of one sign N_{14} and seven signs N_1 resulting in two signs N_{14} and two signs N_1 (for the denotations of the numerical signs see the diagrams in Fig. 3). Since the sign N_1 represents the counting unit, the sign N_{14} consequently has the value 10 (addition in the sexagesimal system). In text W15897,c21 exactly the same signs are used, but with a different numerical relation. Twelve signs N_1 are totaled into two signs N_{14}; in this case this sign therefore has the value 6 (addition in a system of grain measures); cf. the illustration of the two texts in Chapter 8 or in Damerow ([7], p. 147). The operation to be performed with the signs is based on the values of the signs dependent on the content of the texts and can in no way be inferred from the shape of the signs, the format of the tablet, or any other formal indicator.

[14] In most languages number words used in a variety of contexts constitute the elementary precondition for such a conceptuality. The degree to which counting sequences composed of such number words had developed in the Near East at the time of the emergence of writing is unfortunately entirely unknown. The existence of diverse protocuneiform and proto-Elamite numerical sign systems for distinct objects even raises doubts as to whether the number words of the counting techniques already revealed a developmental level comparable to these systems of numerical signs. In its strictly sexagesimal structure the sexagesimal system corresponds to the number word sequence of the Sumerian language (see [70]), which, at this time, is believed to have been the language of the inhabitants of the southern part of Mesopotamia. It cannot, however, be concluded from this that this numerical sign system evolved as a representation of the number word sequence. Throughout the 3rd millennium numerical signs were almost exclusively used in the texts; virtually no number words were used. Our knowledge of the Sumerian sequence of number words is obtained from lexical lists written down more than a 1,000 years after the archaic

tablets, if we disregard the evidence for the number words from one to ten which was only recently discovered in conjunction with the excavation of the archives of Ebla (see [23]), where the distance to the development of writing only amounts to a few hundred years. It is entirely impossible to decide whether the Sumerian number word sequence had evolved to the same extent as the sexagesimal system of numerical signs before the development of this system for which there is evidence up to the sign for the number 36000 in the texts of the more recent layer Uruk III and evidence up to the symbol for the number 3600 in the texts of the older layer Uruk IV. In particular it cannot be determined in light of this evidence whether the number word sequence had a sexagesimal structure even before the evolution of writing. Only with the number 3600 does the sexagesimal system reach the second power of the basic number 60. Etymologically the number word sequence only shows a former counting limit at 60 with the word geš, a counting limit of unknown etymological origin, and a counting limit at 3600 with the word šár (= totality, world). Thus, the number word geš is the last genuine individual name of this number sequence. The word geš-u (= sixty-ten) for the number 600 simply expresses the counting of the units of 60. Apart from that the number word sequence conforms so closely to the written numerical notations that it seems only logical to assume that this number word sequence, emulating the cuneiform numerals, was artificially created much later. What is more, it might have been created by scribes at a time when Sumerian was no longer spoken. As far as the sexagesimal numerical sign system in the archaic texts is concerned, the close association with the construction of the number word sequence can, therefore, be merely stated. It can be assumed with a certain degree of probability that, in the period before the development of writing, the šár still represented a counting limit which, at the time it acquired its written representation, might have been identified with the second power of the counting unit 60; it was probably exceeded in the usual fashion by counting higher counting units. In the number word sequence there is no evidence of the bisexagesimal system of the archaic texts which is identical with the sexagesimal system up to the sign for the number 60, unless the break in the law governing the generation of number words at 120 is taken as evidence for a previous counting limit. The addition of a decimal system primarily for animals and workers at the time the numerical sign systems were adopted by the proto-Elamite system of writing could possibly be viewed as an indication of the preexistence of a (semitic?) decimal number word sequence in an area with pastoral traditions; this is, however, purely speculative. It must be stated that there is an insoluble discrepancy between the as such not implausible assumption of developed number word sequences at the time of the emergence of writing and the archaic numerical sign systems that do not admit of the identification of a universal counting sequence.

[15] The interpretation of the cognitive background of these symbols and signs did indeed become the subject of a long-standing scientific controversy. Since this controversy has not been acknowledged to date either by psychology or by the history of mathematics, it is in order to cite the most important contributions at this point: Oppenheim ([65]), Schmandt-Besserat ([75], [76], [77], [78], [79], [80], [81], [82], [83], [84]), Le Brun and Vallat ([49]), Vallat ([92], [93]), Liebermann ([53]), Powell ([71]), Shendge ([85]), Amiet ([2]).

[16] The numerical tablet W14148 which unquestionably contains the notation of a field area as well as the tablet reproduced in Damerow, Englund, and Nissen ([11], ill. 11f) on which a quantity of grain was most certainly recorded may serve as evidence for the assertion that the recorded products can be identified on some tablets ([8], Plate 59). The numerical tablets of the illustrations 110, 111, 113, and 114 in Nissen, Damerow, and Englund ([64]), on the other hand, do not exhibit any similarities to the numerical notations of the archaic texts and, therefore, can neither be interpreted qualitatively nor quantitatively.

BIBLIOGRAPHY

[1] Aebli, H.: *Denken: Das Ordnen des Tuns.* 2 vols., Stuttgart: Klett-Cotta, 1980/1981.
[2] Amiet, P.: *L'Age des Echanges Inter-Iraniens.* Paris: Réunion des musées nationaux, 1986.
[3] Berry, J. W. & Dasen, P. R. (Eds.): *Culture and Cognition: Readings in Cross-Cultural Psychology.* London: Methuen, 1974.
[4] Brainerd, C. J.: *The Origins of the Number Concept.* New York: Praeger, 1979.
[5] Bruner, J. S. et al.: *Studies in Cognitive Growth: A Collaboration at the Center for Cognitive Studies.* New York: Wiley, 1966.
[6] Cole, M. & Scribner, S.: *Culture and Thought: A Psychological Instruction.* New York: Wiley, 1974.
[7] Damerow, P.: "Individual Development and Cultural Evolution of Arithmetical Thinking." In Strauss, S. (Ed.): *Ontogeny, Phylogeny and Historical Development.* Norwood, NJ: Ablex, 1988, pp. 125–152.
[8] Damerow, P. & Englund, R. K.: "Die Zahlzeichensysteme der Archaischen Texte aus Uruk." In Green, M. W. & Nissen, H. J. (Eds.): *Zeichenliste der Archaischen Texte aus Uruk.* Berlin: Mann, 1987, pp. 117–166.
[9] Damerow, P. & Englund, R. K.: *The Proto-Elamite Texts from Tepe Yahya.* Cambridge, MA: Harvard University Press, 1989.
[10] Damerow, P. & Englund, R. K.: *The Proto-Cuneiform Texts from the Erlenmeyer Collection.* Berlin: Mann, in prep.
[11] Damerow, P., Englund, R. K. & Nissen, H. J.: "Die Entstehung der Schrift." In *Spektrum der Wissenschaft, February,* 1988a, pp. 74–85.
[12] Damerow, P., Englund, R. K. & Nissen, H. J.: "Die ersten Zahldarstellungen und die Entwicklung des Zahlbegriffs." In *Spektrum der Wissenschaft, 3,* 1988b, pp. 46–55.
[13] Damerow, P., Englund, R. K. & Nissen, H. J.: "Zur rechnergestützten Bearbeitung der archaischen Texte aus Mesopotamien (3200–3000 v. Chr.)." In *Mitteilungen der Deutschen Orientgesellschaft, 121,* 1989, pp. 139–152.
[14] Damerow, P., Freudenthal, H., McLaughlin, P. & Renn, J.: *Exploring the Limits of Preclassical Mechanics.* New York: Springer, 1991.
[15] Dasen, P. R.: "Are Cognitive Processes Universal? A Contribution to Cross-Cultural Piagetian Psychology." In Warren, N. (Ed.): *Studies in Cross-Cultural Psychology.* London: Academic Press, 1977, pp. 155–201.
[16] Dasen, P. R.: "'Strong' and 'Weak' Universals: Sensori-Motor Intelligence and Concrete Operations." In Lloyd, B. & Gay, J. (Eds.): *Universals of Human Thought.* Cambridge: Cambridge University Press, 1981, pp. 137–156.
[17] Deimel, A.: *Liste der archaischen Keilschriftzeichen.* Leipzig: Hinrichs, 1922.
[18] Deimel, A.: *Schultexte aus Fara.* Leipzig: Hinrichs, 1923.
[19] Dittmann, R.: "Seals, Sealings and Tablets. Thought on the Changing Pattern of Administrative Control from the Late-Uruk to the Proto-Elamite Period at Susa." In Finkbeiner, U. & Röllig, W. (Eds.): *Gamdat Nasr. Period or Regional Style.* Wiesbaden: Reichert, 1986, pp. 332–336.
[20] Dixon, R. M. W.: *The Languages of Australia.* Cambridge: Cambridge University Press, 1980.
[21] Dixon, R. M. W. & Blake, B. J. (Eds.): *Handbook of Australian Languages.* Amsterdam: Benjamins, 1979ff.
[22] Driel, G. van: "Tablets from Jebel Aruda." In Driel, G. van (Ed.): *Zikir Šumin.* Leiden: Brill, 1982, pp. 12–25.
[23] Edzard, D. O.: "Sumerisch 1 bis 10 in Ebla." In *Studi Eblaiti, 3,* 1980, pp. 121–127 and Fig. 26a, b.

[24] Eibl-Eibesfeldt, I., Schiefenhövel, W. & Heeschen, V.: *Kommunikation bei den Eipo. Eine humanethologische Bestandsaufnahme.* Berlin: Reimer, 1989.
[25] Englund, R. K.: "Administrative Timekeeping in Ancient Mesopotamia." In *Journal of the Economic and Social History of the Orient, 31,* 1988, pp. 121–185.
[26] Englund, R. K. & Grégoire, J.-P.: *The Proto-Cuneiform Texts from Jemdet Nasr.* Berlin: Mann, 1991.
[27] Falkenstein, A.: *Archaische Texte aus Uruk.* Leipzig: Harrassowitz, 1936.
[28] Falkenstein, A.: "Archaische Texte des Iraq-Museums in Bagdad." In *Orientalische Literaturzeitung, 40,* 1937, pp. 401–410.
[29] Frege, G.: *Foundations of Arithmetic: A Logico-Mathematical Enquiry Into the Concept of Number.* Trans. by J. L. Austin. Evanston, IL: Northwestern University Press, 1968.
[30] Friberg, J.: *A Method for the Decipherment, Through Mathematical and Metrological Analysis, of Proto-Sumerian and Proto-Elamite Semi-pictographic Inscriptions.* Göteborg: Chalmers Technische Hochschule, 1978.
[31] Friberg, J.: *Metrological Relations in a Group of Semi-Pictographic Tablets of the Jemdet Nasr Type, Probably from Uruk-Warka.* Göteborg: Chalmers Technische Hochschule, 1979.
[32] Friberg, J.: "Mathematik." In Edzard, D. O. (Ed.): *Reallexikon der Assyriologie und Vorderasiatischen Archäologie.* Berlin: de Gruyter, 1990, pp. 531–585.
[33] Fuson, K. C.: *Children's Counting and Concepts of Number.* New York: Springer, 1988.
[34] Gelman, R. & Gallistel, C. R.: *The Child's Understanding of Number.* Cambridge, MA: Harvard University Press, 1978.
[35] Gericke, H.: *Geschichte des Zahlbegriffs.* Mannheim: Bibliographisches Institut, 1970.
[36] Hallpike, C. R.: *The Foundations of Primitive Thought.* Oxford: Clarendon, 1979.
[37] Heeschen, V. & Schiefenhövel, W.: *Wörterbuch der Eipo-Sprache.* Berlin: Reimer, 1983.
[38] Høyrup, J.: "Algebra and Naive Geometry." In *Altorientalische Forschungen, 17,* 1990, pp. 27–69 and 262–354.
[39] Hurford, J. R.: *The Linguistic Theory of Numerals.* Cambridge: Cambridge University Press, 1975.
[40] Hurford, J. R.: *Language and Number.* Oxford: Basil Blackwell, 1987.
[41] Ifrah, G.: *Histoire Universelle des Chiffres.* Paris: Seghers, 1981.
[42] Kant, I.: *Critique of Pure Reason.* Trans. by F. M. Müller. 2nd ed., New York: Macmillan, 1927.
[43] Kant, I.:"Prolegomena to any Future Metaphysics That Will be Able to Come Forward as Science." In Ellington, J. W. (Ed.): *The Philosophy of Material Nature.* Indianapolis, IN: Hackett, 1985, pp. 1–121.
[44] Koch, G.: *Malingdam.* Berlin: Reimer, 1984.
[45] Koch, G. & Schiefenhövel, W.: *Eipo (West-Neuguinea, Zentrales Hochland), Neubau des sakralen Männerhauses in Munggona.* Sonderserie 7, Nr. 9, Film E 2475. Göttingen: Institut für Wissenschaftlichen Film, 1987.
[46] Lancy, D. F.: *Cross-Cultural Studies in Cognition and Mathematics.* New York: Academic Press, 1983.
[47] Langdon, S.: *Pictographic Inscriptions from Jemdet Nasr.* Oxford: Oxford University Press, 1928.
[48] Le Brun, A.: "Suse, Chantier 'Acropole I'." In *Paléorient, 4,* 1978, pp. 177–192.
[49] Le Brun, A. & Vallat, F.: "L'Origine de l'Ecriture à Suse." In *Cahiers de la Délégation archéologiques française en Iran, 8,* 1978, pp. 11–60.

[50] Lean, G. A.: *Counting Systems of Papua New Guinea* (Draft Edition). Papua New Guinea: Department of Mathematics, Papua New Guinea University of Technology, 1985.
[51] Lévy-Bruhl, L.: *Primitive Mentality*. Trans. by L. A. Clare. New York: MacMillan, 1923.
[52] Lévy-Bruhl, L.: *How Natives Think*. Trans. by L. A. Clare. New York: MacMillan, 1926.
[53] Liebermann, S. J.: "Of Clay Pebbles, Hollow Clay Balls, and Writing: A Sumerian View." In *American Journal of Archaeology, 84,* 1980, pp. 339–358.
[54] Lorenzen, P.: *Einführung in die operative Logik und Mathematik*. Berlin: Springer, 1955.
[55] Lorenzen, P.: *Metamathematik*. Mannheim: Bibliographisches Institut, 1962.
[56] Menninger, K.: *Number Words & Number Symbols: A Cultural History of Numbers*. Trans. by P. Broneer. New York: Dover, 1992.
[57] Mill, J. S.: *System of Logic*. Charlottesville, VA: Ibis, 1986.
[58] Mimica, J.: *Intimations of Infinity*. Oxford: Berg, 1988.
[59] Minsky, M. L.: *The Society of Mind*. New York: Simon & Schuster, 1985.
[60] Neugebauer, O.: *Vorgriechische Mathematik*. Berlin: Springer, 1934.
[61] Neugebauer, O.: *Mathematische Keilschrifttexte*. Berlin: Springer, 1935.
[62] Neumann, D.: *The Origin of Arithmetical Skills: A Phenomenographic Approach*. Göteborg Studies in Educational Sciences 62. Göteborg: Acta Universitatis Gotheburgensis, 1987.
[63] Nissen, H. J.: *The Early History of Ancient Near East, 9000–2000 B.C.* Chicago, IL: University of Chicago Press, 1988.
[64] Nissen, H. J., Damerow, P. & Englund, R. K.: *Archaic Bookkeeping: Early Writing and Techniques of Economic Administration in the Ancient Near East*. Chicago, IL: University of Chicago Press, 1993.
[65] Oppenheim, A. L.: "On an Operational Device in Mesopotamian Bureaucracy." In *Journal of Near Eastern Studies, 18,* 1959, pp. 121–128.
[66] Piaget, J.: *Introduction à l'Epistémologie Génétique*. 3 vols., Paris: Presses Universitaires de France, 1950.
[67] Piaget, J.: *The Child's Conception of Number*. Trans. by C. Gattegno and F. M. Hodgson. London: Routledge & Kegan Paul, 1952.
[68] Piaget, J.: *Biology and Knowledge: An Essay on the Relations Between Organic Regulations and Cognitive Processes*. Chicago, IL: University of Chicago Press, 1971.
[69] Piaget, J. & Garcia, R.: *Psychogenesis and the History of Science*. New York: Columbia University Press, 1989.
[70] Powell, M. A.: *Sumerian Numeration and Metrology*. Ann Arbor, MI: University Microfilms, 1973.
[71] Powell, M. A.: "Three Problems in the History of Cuneiform Writing: Origins, Direction fo Script, Literacy." In *Visible Language, 15,* 1981, pp. 419–440.
[72] Saxe, G. B.: "Culture and the Development of Numerical Cognition: Studies among the Oksapmin of Papua New Guinea." In Brainerd, C. J. (Ed.): *The Development of Logical and Mathematical Cognition*. New York: Springer, 1982, pp. 157–176.
[73] Saxe, G. B., Guberman, S. R. & Gearhart, M.: "Social Processes in Early Number Development." In *Monographs of the Society for Research in Child Development, 52* (2, Serial No. 216), 1987, pp. 1–162.
[74] Scheil, V.: "Essai de Déchiffrement des Textes protó-Elamites: Système de Numération protó-Elamite." In Scheil, V. (Ed.): *Documents en Ecriture protó-Elamite*. Paris: Geuthner, 1905, pp. 115–118.
[75] Schmandt-Besserat, D.: "An Archaic Recording System and the Origin of Writing." In *Syro-Mesopotamian Studies, 1,* 1977, pp. 31–70.

[76] Schmandt-Besserat, D.: "The Earliest Precursor of Writing." In *Scientific American, 238* (June), 1978, pp. 38–47.
[77] Schmandt-Besserat, D.: "Reckoning Before Writing." In *Archaeology, 32* (3), 1979, pp. 22–31.
[78] Schmandt-Besserat, D.: "The Envelopes that Bear the First Writing." In *Technology and Culture, 21,* 1980, pp. 357–385.
[79] Schmandt-Besserat, D.: "From Tokens to Tablets: A Re-Evaluation of the So-Called 'Numerical Tablets.'" In *Visible Language, 15,* 1981, pp. 321–344.
[80] Schmandt-Besserat, D.: "Before Numerals." In *Visible Language, 18,* 1984a, pp. 48–60.
[81] Schmandt-Besserat, D.: "Proto-Literate Counting: The Archaeological Evidence." In *Recueil de travaux et comm. de ... Proche-orient ancien, 2,* 1984b, pp. 21–24.
[82] Schmandt-Besserat, D.: "Tonmarken und Bilderschrift." In *Das Altertum, 31,* 1985, pp. 76–82.
[83] Schmandt-Besserat, D.: "Tokens at Susa." In *Oriens Antiquus, 25,* 1986, pp. 93–125.
[84] Schmandt-Besserat, D.: "Tokens at Uruk." In *Baghdader Mitteilungen, 19,* 1988, pp. 1–176.
[85] Shendge, M. J.: "The Use of Seals and the Invention of Writing." In *Journal of the Economic and Social History of the Orient, 26,* 1983, pp. 113–136.
[86] Strauss, S. (Ed.): *Ontogeny, Phylogeny and Historical Development.* Norwood, NJ: Ablex, 1988.
[87] Strommenger, E.: *Habuba Kabira. Eine Stadt vor 5000 Jahren.* Mainz: Zabern, 1980.
[88] Vaiman, A.: "O svazi protoelamskoj pis'mennosti s protošumerskoj" (Eine vergleichende Untersuchung der protoelamischen und der protosumerischen Schrift). In *Vestnik Drevnej Istorii, 3,* 1972, pp. 124–133.
[89] Vaiman, A.: "Über die Protosumerische Schrift." In *Acta Antiqua Hungaricae, 22,* 1974a, pp. 15–27.
[90] Vaiman, A.: "Protošumerskie sistemy mer i sceta." In *Trudy XIII. mezdunarodny Kongr. po Istorii Nauki,* 1974b, pp. 6–11.
[91] Vaiman, A.: "Beiträge zur Entzifferung der Archaischen Schriften Vorderasiens." In *Baghdader Mitteilungen, 20,* 1989, 92–138.
[92] Vallat, F.: "Le Matériel Epigraphique des Couches 18 a 14 de l'Acropole." In *Paléorient, 4,* 1978, pp. 193–195.
[93] Vallat, F.: "The Most Ancient Scripts of Iran: The Current Situation." In *World Archaeology, 17,* 1986, pp. 335–347.
[94] Vogel, K.: *Vorgriechische Mathematik. Part 2: Die Mathematik der Babylonier.* Hannover: Schroedel, 1959.
[95] Vygotsky, L. S.: "Thinking and Speech." In Rieber, R. W. & Carton, A. S. (Eds.): *The Collected Works of L. S. Vygotsky,* Vol.1. New York: Plenum, 1987, pp. 37–285.
[96] Wartofsky, M. W.: "From Genetic Epistemology to Historical Epistemology: Kant, Marx, and Piaget." In Liben, L. S. (Ed.): *Piaget and the Foundations of Knowledge.* Hillsdale, NJ: Erlbaum, 1983, pp. 1–17.
[97] Wassmann, J.: *Das Ideal des gebeugten Menschen.* Basel: Habilitationsschrift, 1991.
[98] Welti, E. L.: *Die Philosophie des Strikten Finitismus.* Bern: Lang, 1986.
[99] Wertheimer, M.: "Über das Denken der Naturvölker, Zahlen und Zahlgebilde." In Wertheimer, M. (Ed.): *Drei Abhandlungen zur Gestalttheorie.* Erlangen: Palm & Enke, 1925, pp. 106–163.
[100] Wolff, M.: *Der Begriff des Widerspruchs.* Königstein/Ts.: Hain, 1981.
[101] Wussing, H.: *Mathematik in der Antike. Mathematik in der Periode der Sklavenhaltergesellschaft.* Leipzig: Teubner, 1962.

CHAPTER 10

ABSTRACTION AND REPRESENTATION

[1988]

It is a common view that abstraction means refraining from using the information available on a given real object, and instead isolating certain properties and dealing with these independently. But this concept of abstraction reveals itself as unsatisfactory if it is used to conceptualize the development of mathematical thinking. Abstraction in this sense does not explain that outcome of new knowledge which obviously does result from mathematical thinking. Furthermore, this concept of abstraction makes it impossible or at least difficult to understand why certain abstractions turn out to be very useful but the huge mass which might be produced by arbitrarily isolating properties of mathematical objects would only result in nonsense.

To give an example I will refer to a well-known anecdote. Gauss when he was still a school boy is said to have solved the problem of adding the numbers 1 to 100 by reordering the sequence of operations from $1 + 2 + 3 + \ldots$ to $(1 + 100) + (2 + 99) + (3 + 98) + \ldots$ thus reducing the problem to the simple multiplication $50 \times 101 = 5050$. He disregarded certain features of the problem at hand and abstracted a general structure which directly led to the rule for solving the problem. If we conceive of abstraction simply as a process of isolating certain properties, we express only a very unimportant aspect of such a problem solution. To understand abstraction essentially means understanding *what* has to be abstracted rather than merely knowing *how* it has to take place. To understand the abstraction leading to an elegant solution of the problem means understanding how the solution can really be found.

Abstraction is a *constructive* process which is essential for mathematics teaching and learning. There is a fundamental difference between the teaching of mathematics and the teaching of subjects like physics or biology. A biology teacher who demonstrates something on a real plant is in fact trying to teach something about plants. A mathematics teacher, however, who demonstrates the concept of a triangle on some real object usually does not at all intend to deal with real objects. Instead he is trying to communicate a mental construct which is not identical with anything in the real world. In biology teaching, the picture of a tree is an iconic model of real trees, whereas in mathematics teaching a triangle drawn on the blackboard is a model of an *abstract idea.*

This aspect of mathematics teaching results in specific difficulties. It is a principal problem in the didactics of mathematics how to invoke abstract thinking and to guide abstraction toward the mathematical concepts which should be taught.

In my opinion abstraction is in itself an activity directed towards some end or goal, since it depends on "reflection" (see Chap. 1 and 3). Abstraction results in the construction of cognitive structures. By a cognitive structure I mean a system of relations between mental objects which is determined by mental actions building the relations between these elements of the structure. To recognize a real object and to identify it as something already known is here interpreted as assimilation to a cognitive structure. This assimilation relates the recognized object to earlier experiences by interpreting it in the framework of a system of knowledge which itself is logically structured. Cognitive structures are built up in the process of ontogenetic development. They are the result of mental constructions which transform real actions into corresponding mental actions. Thus, cognitive structures are mental reflections of real actions and abstraction is but a special function of reflection as a self-determined and self-determining process.

The cognitive structures underlying logical and mathematical thinking can be regarded as one special type of such structures. According to Piaget they are constituted by "operations"—that is, reversible mental actions organized in closed systems. It is a fundamental idea in Piaget's work on the development of logical and mathematical thinking that ideal objects such as abstract mathematical ideas have their origin in the closure of such systems. Once a system is closed the properties of the constituted ideal object are no longer dependent on empirical experience. Rather they are determined by the structure as a whole and can be elaborated by the operations of logical and mathematical thinking.

Piaget's analysis of conservation of quantity is an example of reconstructing the process of "reflective abstraction" building up an abstract mathematical idea ([4]). Conservation of quantity is a precondition of the number concept. Any use of numbers is meaningful only if there is something—called quantity—which is unchangeable by transformations such as spatial displacements. Quantity is an abstract entity to which numbers can be attached. According to Piaget, it is a mental construct based on a system of reversible correspondence operations between sets. If the reversibility of these operations is achieved, conservation of quantity is no longer merely a matter of empirical judgment but under certain conditions an implication with logical necessity.

A powerful means of abstraction, I should like to claim, is *representation* (see Chap. 2). Representation is based on a very fundamental and general function of the mind which is called by Piaget the symbolic function. The symbolic function is the ability to conceive something as representing something else. Real objects functioning as symbols can represent abstract ideas and real objects as well, and can thus be used as external tools for performing mental operations. They constitute complex representations of cognitive structures.

Representations of cognitive structures can be distinguished according to the media of their representation. Practical intelligence in the way this term is used by Piaget is based on enactive representations. Goal-directed activities are represented by schemes of actions such as gestures and routines which support intuitive thinking. Iconic representations are representations of objects and actions, for example by images and graphs, using similarities and correspondencies between abstract structures and their visualizations. Iconic representations can be conceived as special types of material representation. Material representations more generally are representations of material qualities by prototypes, material models, and standards. In the area of social cognition persons may function as models of social behavior. However, the most important kind of representation is symbolic representation—that is, representation by free, appropriate, or traditionally determined convention without any material similarities between symbols and symbolized objects and actions.

In the following I will introduce as a further distinction that between first order and second and higher order representations. Contrarily to the above-mentioned distinctions which are made according to media of representation, the distinction between first order and higher order representation refers to the role in the developmental process in which a cognitive structure is built up.

First order representations are representations of real objects by symbols or models which permit the performance of essentially the same actions or operations with these symbols as can be performed with the real objects themselves. However, handling the symbols is easier because it is free of certain limitations of the real-world situation. Of course handling symbols cannot be a substitute for real action, but it can initiate the same cognitive structures. Consequently first order representations are constructive and at the same time more abstract than the represented objects and actions. They may serve as cognitive tools for abstracting a cognitive structure because the underlying actions do not depend on the preexistence of this cognitive structure. Adequate handling of the symbols or models of first order representations is possible as soon as the underlying real actions can be performed.

Some simple examples of first order representation are:
- representation of the ordinal structure of numbers by number words and similar counting devices;
- representation of the cardinal structure of numbers by counters, by constructive number notations such as notations by means of the old Roman, Babylonian, or Egyptian numerals, or by constructive didactic aids as they are used in the elementary teaching of arithmetic;
- representation of geometrical relations by figures constructed with straightedge and compasses—that is, the constructive basis of the geometry of Euclid;
- representation of the arithmetical structure of the concept of area by cut-and-paste operations and standard area measures.

Second and higher order representations are representations of mental objects by symbols and symbol transformation rules which correspond to mental operations belonging to the cognitive structures constituting the mental objects. Higher order representations are not constructive in general, and in particular they are not constructive on the level of the underlying cognitive structures. Adequate use of these representations depends on the preexistence of the underlying cognitive structures because it is generally not possible to reconstruct these structures from the symbol transformation rules. However, higher order representations are constructive on the meta-level of these structures. Therefore higher order representations can serve as tools for a guided reflection which constitutes meta-structures and hence can make mental operations more efficient. (See, e.g., [3] on the development from Pythagorean arithmetic to deductive arithmetic in Euclid's *Elements*.)

If higher order representations are to be used they have to be linked somehow to the real objects which are indirectly represented. In other words, these objects have to be interpreted by assimilation to the underlying cognitive structure. Adequate application of higher order representations depends on their meaningful use, whereas first order representations are automatically applied adequately by using the symbols as symbols for the real objects which are subject to the real actions corresponding to the symbol transformation rules.

Some simple examples of higher order representations are:
- representation of numbers by conventional designations like number words (one, two, three, ...) or nonconstructive numerical signs (1, 2, 3, ...);
- representation of the number concept by the word "number" or by variables;
- representation of spatial objects and relations by concepts and theorems of Euclid's *Elements*.

In the process of cognitive development, first order representations may change their function and turn into second order representations. The sequence of number words, for instance, develops from a first order representation of an ordinal structure into a second order representation of cardinality by means of correspondence relations constructed by the counting activity, and finally into a higher order representation of the concept of number.

The development of Euclidean geometry from Euclid's *Elements* to Hilbert's *Foundations of Geometry* can be interpreted in a similar way. Representation by geometrical concepts is constructive only in so far as deduction is independent of real figures. In Euclid's *Elements* first order representation by geometrical construction is still an essential part of the system, whereas Hilbert's geometry represents a fully developed system of meta-knowledge.

In the following, I shall derive some consequences from the concept of representation which I have introduced. It is a fundamental problem of any theory of mathematical education that up to now there is no consensus about the answer to the question, to what extent mathematical thinking is universal and to what extent it depends on cultural conditions. The concept of representation enables us to solve at least part of this problem, whatever the nature of the cognitive structures underlying logical and mathematical thinking is, there can be no doubt that their *representations* are culturally dependent. Whatever the nature of the construction process which leads in ontogenetic development to these cognitive structures, mathematical learning includes in addition to this process the reconstruction of the meaning of culturally determined representations.

The conditions of this reconstruction are very different in the case of first order representations and in that of higher order representations.

The reconstruction of the meaning of a first order representation is to a certain extent independent of the cognitive structures which the learner has already built up, and also independent of the interaction between teacher and learner which connects the learner's cognitive preconditions in some way with the teacher's didactic aims.

The meaning of higher order representations, on the other hand, cannot adequately be reconstructed solely by performing actions with these representations according to the specific rules on the representation level, because these representations are constructive only on a metalevel of mathematical thinking.

Accordingly, any theory of the individual development of mathematical thinking has to explain three different processes which are connected with the development of mathematical abilities:

1. The process of building up the fundamental cognitive structures of mathematical thinking.

    As long as we make no assumption about the role of representations in this process of construction nothing can be inferred about this process from what I have written so far. I shall return to this point shortly.

2. The process of reconstruction of the meaning of first order representations by the learner.

    This process is closely linked to the manner of content presentation and to the nature and manner of use of learning materials. One might see the conception of this process as a theoretical explanation of the "opportunity-to-learn" variable, which has turned out to be such a good predictor of mathematical ability in recent achievement studies. On the other hand, such an explanation also results in strong criteria for the operationalization of this variable in order to raise its content validity.

3. The learner's process of reconstruction of the meaning of higher order representations.

    This process has two aspects, the meaning reconstruction is primarily based on classroom interaction. This guides the integration of specific information about the representation, which is the subject of teaching, into the learner's knowledge system. But in so far as a higher order representation is constructive it is nothing else than a first order representation of a meta-concept of the cognitive structure underlying the representation.

The first of these three processes connected with the development of mathematical abilities is probably decisive for the level of mathematical thinking which can be achieved by the learner. So it is tempting to introduce the strong assumption that the construction of cognitive structures is essential-

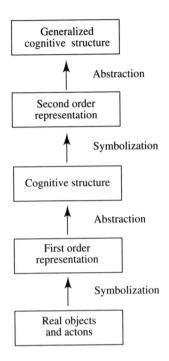

Figure 1: Abstraction and representation.

ly determined by first order representations, so that the first and the second processes are identical.

It should be stressed that at present there is no consensus possible about this assumption, and especially it contradicts Piaget's influential theory ([5]; [2]). In fact this assumption implies an answer to the open question to what extent mathematical thinking is universal. If representations determine the construction of the cognitive structures of mathematical thinking, and if representations are dependent on culture, then abstraction cannot be adequately explained solely by psychology or epistemology, and mathematical thinking is essentially a cultural phenomenon.

On the other hand, this assumption provides us, in my opinion, with a powerful theoretical instrument to obtain good explanations both of the historical development of mathematical thinking and of the individual development of mathematical thinking as well.

What are the essentials of the resulting theoretical model? Any process of mathematical abstraction starts with first order representations such as counters, standards, drawings or elements of language such as words for intuitive quantities, geometrical objects and relations, and so on. Thinking about real objects and actions without using such representations as tools simply means what it means. Two cows are two cows and not one general cow multiplied by the number two as a mental object. But representations are in themselves more general. Two counters for two cows may function as two counters for two sheep if they are used in another context. But still counters are counters and not representations of numbers. They are concrete objects representing a possible abstraction, one which is not arbitrary but reflects something general in a scheme of action.

Operating with first order representations instead of performing the underlying real actions isolates the scheme of action from its real context and guides the construction of a corresponding cognitive structure. This cognitive structure constitutes a mental object which is no longer dependent on direct empirical experience. The first order representation develops into a link between the abstract idea and its adequate applications to real objects.

The abstract idea may be represented by a symbol or a symbol system, thus generating a second order representation. This representation reflects the level of thinking involved and guides the abstraction of a metalevel, which again is not arbitrary but reflects something general in the operations on the level of thinking about real objects.

At the same time such second order representations are first order representations on the level of thinking, because only this can make them useful tools for performing mental operations. Thus, they guide the construction of a structure of meta-cognition.

Any development of mathematical thinking is an iteration (and possibly a combination) of such cycles of reflective abstraction.

One remark should be added. It is an immediate consequence of the relative independence of higher order representations from the initial first order representations that certain elements of the structure of meta-cognition can no longer be applied to reality within the framework of the initial first order representation. This explains the possible contradictions between formal reasoning and meaningful interpretations on a lower level of abstraction, which are well known both in the history of mathematics and in individual mathematics learning. Instances are:

ABSTRACTION AND REPRESENTATION 379

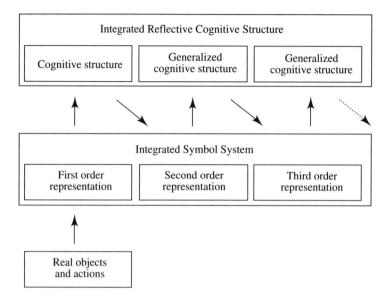

Figure 2: The reflective structure of abstraction.

- accepting negative, irrational, or imaginary numbers as numbers in the sense of the representation of objects by counters;
- accepting non-Euclidean geometries as geometries in the sense of the geometry generated by space representation in constructions with straightedge and compasses.

The theoretical model briefly outlined above is a general model of the development of mathematical thinking which results from the proposed distinction between first order and higher order representation and from the assumption, that representations guide the process of abstraction leading to the cognitive structures of mathematical thinking. This model can easily be specified for the very different processes of historical development and individual learning.

The *historical development* of mathematical thinking is essentially based on three different functions of representations (see Chap. 7 and 12).
- They may be used as tools in applications of mathematical thinking useful in certain areas of intellectual work. This function assures that there will be serious efforts to perform the difficult task of qualifying certain

people to perform mathematical thinking as it is represented by these tools.
- They may be used simply to train students in their adequate use. In this function the tools are not determined by any special application. Such a use of the tools is often accompanied by testing them to limits beyond any practical application.
- They may be used simply to develop their implications or to solve unsolved problems, thus generating new mathematical knowledge and constructing new higher order representations.

Once we have introduced these distinctions it is easy to apply the proposed model to explain the different functions of representations in the historical development of mathematical thinking.

Finally, let us look at the *individual development* of mathematical thinking. There seem to be parallels between individual and historical development. Indeed, the individual development of mathematical thinking in a particular historical situation is essentially a process of mental construction aimed at just those cognitive structures which are the outcome of the process of the historical development up to the given situation (see, e.g., [1]; Chap. 6).

But according to the proposed theoretical model, both processes have to be explained quite differently. Even if we disregard for a moment the simple fact that the individual development of mathematical thinking is an intraindividual process, whereas the historical development of mathematical thinking is an interindividual process based on the social reproduction of knowledge by individual development—so to speak a developmental process of a higher order—then still there remains an essential difference:

Progress in the historical development of mathematical thinking is always based on first order—and that means constructive—representations which themselves have been created in a situation in which the cognitive structure, whose construction they make possible, does not yet exist in the historical situation concerned (such progress can, of course, stem from higher order representations, but only in so far as they function as first order representations on a metalevel).

Progress in individual development, on the contrary, is up to a high level of achievement based completely on higher order representations of historically already existing cognitive structures, or on first order representations which are artificially created as teaching devices to guide abstraction to preexisting cognitive structures (see Chap. 6). The individual development of mathematical thinking consists therefore essentially of two processes which are not necessarily favored by equal conditions and may even contradict each other. On the one hand, mathematical meaning is constructed by using each representation offered by teaching as if it were simply a first order representation, thus using all available sources in their constructive potential. On the other hand, the constructed structures are used to interpret the social interaction about representations in order to reconstruct the cognitive structures which define mathematical thinking in a given culture.

BIBLIOGRAPHY

[1] Damerow, P.: "Historische Analysen zur Entstehung des arithmetischen Denkens." In Steiner, H.-G. & Winter, H.: *Mathematikdidaktik, Bildungsgeschichte, Wissenschaftsgeschichte. Untersuchungen zum Mathematikunterricht,* Vol. 12. Köln: Aulis, 1985, pp. 26–29.
[2] Damerow, P.: "Individual Development and Cultural Evolution of Arithmetical Thinking." In Strauss, S. (Ed.): *Ontogeny, Phylogeny and Historical Development.* Norwood: Ablex, 1988, pp. 125–152.
[3] Lefèvre, W.: "Rechensteine und Sprache." In Damerow, P. & Lefèvre, W.: *Rechenstein, Experiment, Sprache: Historische Fallstudien zur Entstehung der exakten Wissenschaften.* Stuttgart: Klett-Cotta, 1981, pp. 115–169.
[4] Piaget, J.: *The Child's Conception of Number.* Trans. by C. Gattegno and F. M. Hodgson. London: Routledge & Kegan Paul, 1952.
[5] Piaget, J.: *Biology and Knowledge: An Essay on the Relations Between Oganic Regulations and Cognitive Processes.* Chicago, IL: University of Chicago Press, 1971.

CHAPTER 11

THE CONCEPT OF LABOR IN HISTORICAL MATERIALISM
AND THE THEORY OF
SOCIO-HISTORICAL DEVELOPMENT
[1977]

LABOR IN HUMAN HISTORY

Historical materialism is a theory of the most general developmental laws of human society. Historical materialism is based on dialectical materialism: The development of society is understood as being based on the assumption of the material unity of nature and society and considered to be determined by the material life processes of society—that is, by the development of the mode of production in which the change of matter between nature and society takes place. The mode of production of a particular social system includes the relations between human beings and nature in the production process, that is, the forces of production, and the relations of human beings to each other in the production process, that is, the relations of production. The forces of production and the relations of production form a contradictory, evolving unity in which the forces of production in particular constitute the revolutionary moment. Thus, historical materialism primarily conceives of the development of the forces of production as the driving force of socio-historical development.

This basic theoretical structure of historical materialism gives particular significance to the concept of labor. The relations between human beings and nature are established through labor. According to historical materialism, the socio-historical development begins with labor, and it is labor that moves this development forward. Within the context of a materialist theory the concept must correspond to material reality—that is, in historical materialism the concept of labor to the actual status of labor with regard to the development of society. In historical materialism the theoretical development of the most general laws of history proceeds from the concept of labor. A correct definition of the concept of labor on the basis of dialectical materialism is, therefore, essential for the theoretical rigor of historical materialism.

The concept of labor was not introduced originally by historical materialism, human history already made this concept available as "rational abstraction." As rational abstraction the concept of labor emphasizes those aspects of the relation between human beings and nature common to all periods of human history.

However, this abstract concept of labor resulting from rational abstraction does not reflect real labor, since it includes only the distilled general aspects. Real labor is concrete—that is, it always is simultaneously a particular kind of labor and a singular performance of labor. Furthermore, real labor exists in different particular forms. If the concept of labor is to conceive of real labor adequately, then the concrete "totality" of all its constituents must be included in the concept. Therefore, historical materialism must develop the abstract concept of labor available to it as rational abstraction into the concrete totality given as the object of cognition by the real types of labor and their relations to each other ([4], pp.100–111).

This mental reproduction of real labor, the rising from the abstract to the concrete concept of labor, must not be mistaken for the real development of labor in the course of human history. Like any abstract concept, the abstract concept of labor, encountered by historical materialism, is not just a rational abstraction, the result of an incomplete act of cognition in general; rather, this rational abstraction itself is specifically based on the developing relation of the subject of cognition to its object and thus acquires a specific meaning in its relation to real labor. A particular form of concrete labor corresponds to the abstract concept of work, where labor is really determined almost completely by its general constituents rather than being fused with its particularity and dominated by it. This, however, becomes only true under the conditions of bourgeois society, where labor acquired the form of the developed totality of real categories of labor and where individuals move with ease from one type of labor to the next. Under these conditions the possibility arises that labor as real labor can no longer be imagined only in its particular form. Only in this society "labor as such" becomes "practically true" as the sole driving force of social development. The thinking course of cognition, the rising from the abstract to the concrete concept of labor is—interpreted in this sense—diametrically opposed to the real development of labor in the course of human history.

In this inversion a general principle of dialectical materialism becomes apparent: The more developed stage provides the key for understanding the previous stages of development. The development of the lower species of animals can be understood in light of the higher species, likewise early forms of social organization in light of later, more developed forms. By conceptually developing the abstract concept of labor, given as rational abstraction, into the concrete totality of the dialectical concept of labor on the basis of the knowledge of the developed form of social organization, historical materialism provides the key to understanding the real history by revealing the driving forces of its development. On the other hand, however, this movement of thought cannot replace the investigation of the real history. Historical materialism in turn is dependent on the investigation of the real history, because real history alone is capable of providing the criteria for the development of the abstract concept of labor into the concrete totality. This mutual dependency of real history and the theory of history which must be included in the historico-materialist concept of labor is reflected within the theory in terms of the relation of historical and logical conditions; and, due to the inherent self-reference of this relation, this occurs in two ways: as the relation of experience and scientific thinking in the process of cognition and as the relation of historical and logical conditions with regard to the concrete object of cognition reproduced by thinking.

Abstract thinking cannot adequately represent this relationship. The rational abstractions, from which scientific thinking proceeds and by means of which experience enters scientific thinking as nothing more than a "chaotic conception of the whole," are rigid, fixed thougth structures. In relation to a given social state reproducing itself in reality and, therefore, identical with itself, these structures appear to exist in fixed relations to each other which can be determined analytically by examining the object. Or, conversely: society appears as an abstract connectedness of the concepts obtained by rational abstractions, as a system constituted by formal relations. Thinking that remains at this stage and identifies the abstract thought structures with reality, obviously misses the specific character of the historical object. To this way of thinking historical changes present themselves as changes of relations existing between abstract concepts, changes that, just like the relations and their very existence, cannot be explained in terms of these concepts. Historical changes appear as the abstract negation of logical relations and not as the result of a specific logic of the object.

By comparison, developing the abstract concept of an historical object into a concrete totality means understanding the changing relations of the concrete connectedness that seem to have been added externally to the abstract concept, as internal moments of the concept itself. The concrete concept is the developing, contradictory unity of the general and the particular. The general reproduces itself under the particular conditions of an historical period in particular ways. In this respect the logical conditions are based on the historical conditions of the period in question. However, the reproduction of the general changes the particular conditions, and in that sense the logical conditions are relatively independent of the historical conditions and cannot be reduced to them. In this sense, Marx distinguishes between the "historical presuppositions" of an historical object reproducing itself which belong to "the history of its formation" and the conditions of its reproduction, which are the same conditions, but "now appear as results of its own realization, reality, as *posited by it—not as conditions of its arisings, but as results of its presence*" ([4], p. 460). The possibility of understanding the object of thinking as an historical object is based on this unity of the general and the particular, of logical and historical conditions, to make inferences from the logic of its reproduction to its development and its passing. The logical conditions "offer the key to the understanding of the past," because the historical conditions of their becoming appear in them "as *merely historical*, i.e. suspended presuppositions." At the same time, the logical conditions "appear as engaged in *suspending themselves* and hence in positing the *historic presuppositions* for a new state of society" ([4], p. 461).

Applied to the concept of labor of historical materialism this means: Even if this concept understands labor as the driving force of human history, history as a process of self-creation of the human being by means of labor, this does not imply that history can be derived from the concept of labor and even less from the abstract concept of labor that is a product of the abstract labor of bourgeois society. In every historical period labor is to be understood as a unity of general and particular labor. Marx writes:

In all forms of society there is one specific kind of production which predominates over the rest, whose relations thus assign rank and influence to the others. It is a general illumination which bathes all the other colours and modifies their particularity. It is a particular ether which determines the specific gravity of every being which has materialized within it. ([4], p. 106f.)

Understanding labor as a totality means unfolding the dialectic of the labor process in terms of the contradictions between the general and the particular aspects of the production of any form of society and developing the driving forces of the historical process in terms of the mutual turning of historical and logical conditions into each other.

BIOLOGICAL EVOLUTION AND HISTORICAL DEVELOPMENT

Let me give an exceptional example which shows the benefit of these reflections with regard to understanding the real historical process by demonstrating the inadequacy of methodological conceptions that proceed from a merely external connection of the general and the particular in the historical process constituted by the cognitive process, that is, the example of the beginning of human history itself, the evolution of human beings out of the context of nature (see especially [2], [3], [5], [6]). Any conception that attempts to deduce the particular in history from human nature in general, however it may be conceived, winds up with the theoretical dilemma of no longer being able to understand human nature itself historically. The transition from the animal to the human being becomes a supernatural act of creation which constitutes a transcendental condition of labor as the driving force of human history. In this view, nature and history are confronted with each other as two ultimately separate spheres related to each other merely externally. Conversely, any conception that seeks to deduce the general nature of human beings from their particular history by means of induction is obscuring the qualitative difference between biological and socio-historical development. The differentiation between animals and humans in terms of the historical development becomes a problem of definition. It does not appear possible to provide a theoretical justification for the distinction between nature and history.

At first sight, the conceptual differentiation between animals and humans does not seem to present a problem. No one would seriously confuse humans and animals. Due to the laws of reproduction animal species reproduce biologically as naturally existing, distinct classes of individuals. They are differentiated from each other and specifically from the particular species of human beings. By emphasizing the common characteristics of the individuals of a particular species and the characteristics that distinguish it from the individuals of other species by means of rational abstraction, useful definitions of the species can be obtained which, although it may occasionally be doubtful whether a given individual belongs to a particular species or not, provide absolutely incontrovertible criteria for the distinction between living animals and human beings in particular. Thus, the particular characteristics that have been emphasized with regard to human beings are: their upright walk, the use and production of tools, consciousness and planned conscious activity, communication by means of language, cooperation by means of the division of labor.

The abstract definition of human beings as a species with the characteristics mentioned above is more than sufficient for distinguishing animals and humans in everyday life. Essentially it merely provides a conceptual representation of how we distinguish animals and humans in our sensual perception. Nevertheless, on closer inspection this definition turns out to have considerable flaws.

First of all, taken by themselves, the characteristics are not at all conclusive and at most identify gradual distinctions between human beings and animals. Is it not true that the spider, too, makes a tool with its web? Is it not true that the lion, too, stalks a herd of zebras with extraordinary intentness, albeit, admittedly, not with the same consciousness of its plan as a human being? There is no doubt that animals, too, communicate with each other by means of sounds and gestures, although the subjects of communication may be limited. Furthermore, highly developed animals often possess, too, developed social structures; what is it then that distinguishes the behavioral differences caused by these structures in the pursuit of a collective goal of the animal group from the cooperation among humans based on the division of labor? And chickens, after all, also walk on two legs! Thus, the high degree of certainty in distinguishing human beings from animals does not result from the conclusiveness of the distinguishing criteria to be gained by means of rational abstraction, but rather from their context. This necessarily renders the status of this context, its relationship to the particular characteristics obtained by means of rational abstractions, problematic. It becomes evident that the all too limited attempts of reducing the particular to the general by means of deduction and the general to the particular by means of induction is a result of the methodological inadequacy of a kind of thinking that is limited to abstraction.

Second, the distinction between humans and animals is unproblematic only because the intermediary developmental links that, in the course of the development of the species, connect animals and humans, the hominids, are extinct. In the stage of the transition from animal to human being the criteria for distinguishing them, gained by comparing the developed human being with animals, fail because they do not necessarily occur together. Furthermore, they appear in a peculiar inversion of their logic, an inversion that is an indication of the reasons for the insufficient conclusiveness of individual criteria occurring in isolation. While, for example, the tool as a means of instrumental action on the part of human beings is defined precisely by the fact that it is produced for a specific purpose, it seems, on the other hand, to be true of the "tools" of the transitional stage between animal and human being that, for a long time, they were, as objects incidentally found to be of use, systematically collected and used as tools, before they were produced from raw materials for a particular purpose. Archaeological finds suggest a continuous transition from natural objects to human tools. The rubble-stone tools of the australopithecines, for example, consist of rocks with roughly hewn edges that can hardly be distinguished from pieces of rocks shaped by accident. The assumption that they were systematically used as tools is based on the fact that they were frequently found in conjunction with fossils. It has, however, not been confirmed and it is extremely questionable whether these early stone "tools" exhibit traces of having been purposefully shaped at all. Similarly, this inversion of the logic of specific characteristics of the human species in the context of the stage of the transition from animal to human being is apparent with regard to all the characteristics mentioned earlier. The communication that is specifically human takes place by means of language. However, in the stage of the transition from animal to human being language develops as a result of communicative processes. The socialization of human beings takes place in the context of labor. By contrast, in the stage of the transition from animal to human being labor apparently develops on the basis of existing social structures that permit cooperation based on a division of labor. The inadequate conclusiveness of the criteria distinguishing between human beings and animals, perceived in isolation, is due to the fact that they do not emerge as late as the transition from animals to humans. Rather, they merely undergo a specific transformation associated with an inversion of their inner logic: conditions of their evolution are transformed into the consequences of their existence.

Finally, third, the distinction between animals and human beings by means of a definition that is based on the rational abstraction of specific characteristics and differences between characteristics does not provide a reconstruction of the process of self-differentiation of the phylogenetic development. However different the significance of the human characteristics may be in the process of evolution, the abstraction from this very evolution, when they are compared externally, levels these differences and provides every characteristic with the same status of being an external characteristic. If the characteristics that distinguish human beings and animals such as walking upright, the use and production of tools, cooperation in the context of the division of labor, planned conscious activity, social organization, and linguistic communication are abstracted from their developmental context, then the developments inherent in them can only be understood as the emergence of particularizations. The abstractly general definition of these characteristics remains unaffected and is therefore not capable of giving a conceptual explication of their inherent developmental potentialities. The development from animal to human being is reduced to its external appearance, as a transition to new characteristics.

Historical materialism understands the transition from animal to human being as self-creation of the human being in the context of concrete labor and the distinguishing characteristics as moments of a developing totality of inherent attributes. In the developmental context human beings are understood as simultaneously being animals and humans, not, however, in the sense of a particularization but in the sense of development; the animal existence of human beings has been suspended in their human existence. They reproduce the specifically human character of their animal characteristics by means of labor. If human beings are isolated from the specific context of social reproduction—this is the truth of legendary reports about lives such as that of Kaspar Hauser—they reproduce themselves merely in their animal existence. Within the context of social labor, on the other hand, the individual reproduction of human beings assumes the particular form that generates the characteristics unique to human beings and that rational abstraction can only acknowledge as a fact. If the example of the transition from animal to human being is to serve as a demonstration of the explanatory power of historical materialism, then concrete evidence must be provided for this theoretical reconstruction.

The guiding question of such an examination is how the transition from biological evolution to socio-historical development took place. The animal must already be understood as a concrete developing totality, if the transition to a development by means of the labor process is not to appear as a supernatural act of creation. Even the animal must not be perceived only in terms of abstract definitions and divided into species on the basis of external characteristics. Rather, it must be seen in terms of the real mediation by means of which it relates to itself and to the other—that is, in the context of its individual reproduction and the reproduction of its species, in its individual development, and in its evolution as a species. In this sense, the animal as a concrete totality has to be conceived as a unity of the concrete general and the concrete singular. The specific form of development characteristic of the animal is determined by the specific form of mediation between the concrete general of the biological organization and the reproduction of the individual and the species on the one hand and the concrete singular individuals and their specific history given particular individual living conditions on the other hand.

The form of mediation unique to the animal between concrete general and concrete singular is determined by two conditions: by the genetic determination of the individual by the characteristics of the species—the genetic reproduction of the genotype—and by the ability of the individual to survive by means of individual adaptation to the environment—the variation of the phenotype. The concrete general of the species is preserved in the individual as a determination by the genotype. While it exists only in the concrete singular form of the phenotype, it is, however, largely independent of it. The singular determines the general in only two ways: by mutation or by selection. If the genetic material is changed by individual conditions, the genotype of the individual in question is changed abruptly (mutation) independent of other conditions. If the determination by the genotype prevents the individual adaptation necessary for the survival of the individuals, then the species perishes together with the individuals (selection). Biological evolution is, therefore, determined by the mutants with a selective advantage. It lacks that extended reproduction of the general characteristic of socio-historical development, which takes place in the context of labor and determines the dynamics of this development.

The task of reconstructing the process of anthropogenesis on the basis of the historico-materialistic thesis that anthropogenesis is the result of self-creation by means of concrete labor, can be stated as follows: First, it must be demonstrated that the development of the moments of the labor process as the historical precondition of anthropogenesis was actually possible within the biological developmental context; second, it must be shown how, given the biological developmental conditions of the stage of the transition from animal to human being, the historical preconditions of anthropogenesis are transformed into the logical conditions posited by the labor process, thereby revolutionizing the evolutionary process and combining the unique characteristics of human beings to form the logical relationship in which they presuppose each other like the chicken and the egg.

In light of these problems the inadequate conclusiveness of the characteristics distinguishing human beings from animals as stated above proves to be an indication for the fact that biological developmental laws paved the way for the development of the human labor process. To be more precise: communication and cooperation among animals, the capacity of animals for individual learning and instrumental action and, above all, the beginnings for creating traditions comprising goal-oriented instrumental actions facilitated by communication in developed animal communities, create the preconditions for the development of the use and production of tools in the context of biological developmental conditions. The assumption that such a (biologically determined) form of the use and production of tools is the precondition for anthropogenesis and not its result is empirically supported by the fact that primitive forms of the use and production of tools can already be found in hominids—that is, long before the biological evolution of *homo sapiens* was completed. Thus, the hypothesis seems admissible and plausible that, due to the use and production of tools which only had minor evolutionary potential because of the dominance of biological conditions whose forms often remained constant for many thousands of years, the enlargement of the brain, the unique characteristic of the transition to *homo sapiens,* proved to be a mutant with a selective advantage in the stage of the transition from animal to human being in which the hominids must be included. It was this mutant that allowed the evolutionary potential, a potential inherent in the use and production of tools, to turn into an actual possibility. For the problem before us concerning a historico-materialistic reconstruction of the inception of history as such, this in our opinion is the most important benefit taking the connection between the social and the natural preconditions of labor revealed in Marx's analysis of the labor process as a starting point.

Primeval society which evolved under the biological developmental conditions was characterized by the fact that the mediation of the individual with the as yet rudimentary general of the labor process by means of communication and cooperation within the community was not result, but precondition of the labor process. In Hegel's terminology as used by Marx: The mediation of the moments of the labor process is not yet posited by social reproduction and thus a logical relationship of this process. Rather, it is merely presupposed. "... the *presupposition is itself mediated;* i.e. a communal production, communality, is presupposed as the basis of production. The labour of the individual is posited from the outset as social labour." ([4], p. 172)

The more deeply we go back into history, the more does the individual, and hence also the producing individual, appear as dependent, as belonging to a greater whole: in a still quite natural way in the family and in the family expanded into the clan *[Stamm];* then later in the various forms of communal society arising out of the antitheses and fusions of the clans. ([4], p. 84)

We can extend this idea to the stage of the transition from animal to human being. The characteristics pertaining to the species of hominids were still determined exclusively by biology, and the development of the individual was determined to a far greater extent by the reproductive conditions of the biologically defined species than by the potentials of extended reproduction and socialization inherent in the labor process. The development of traditions within animal communities constituted the mediating precondition for the use and production of tools by means of which general labor was reproduced in the specific form determined by the particular stage of development in the context of individual labor of the hominid. This mediation had to assure that the phylogenetically acquired techniques and the primitive forms of the division and organization of labor could be ontogenetically reproduced even without genetic fixation in the genotype of the biologically defined species; for only in this way the use and production of tools could become the characteristic of the species of hominids without being fixed genetically. This is precisely what distinguishes the hominid from the animal. For in the animal the use of tools in a particular form emerges as a characteristic of the species only if it is fixed genetically in this form and thus incapable of historical development.

How is the evolutionary process revolutionized in the context of the stage of the transition from animal to human being so that, in addition to the biological developmental context, a socio-historical evolutionary process emerges which increasingly defines the characteristics of the species? The reproduction of animals is characterized by the development of the individual from birth to death, by its physical reproduction in interaction with nature, and by the identical reproduction of the characteristics of the species by means of procreation and its genetic determination. It is, by comparison, unique to the labor process characterized by the use and production of tools that, in the form of the means of production produced, the means of labor, exemplarily represented by the tools, it possesses a material result exceeding the cycle of the identical reproduction of the characteristics of the species in the individual, a result that accumulates in an environment of implements, thereby constituting the basis for the expanded reproduction of the socio-historical development. Inasmuch as the expanding environment of implements does not remain external to the labor process to which it owes its existence, but in turn itself releases its inherent possibilities, the process of accumulation is not a linear process but a process expanding and accelerating exponentially; a drastic example is provided by a comparison of the evolutionary process of the rubble-stone production of the hominids comprising several million years with the process of evolution from the first developed civilizations to modern industrial society that only took a few thousand years. This acceleration of accumulation must not be understood merely quantitatively. Rather, it includes essential qualitative changes based on the reflexive character of the tools. For, inasmuch as the tool, as objectified labor, represents the general, specifically the general of the object of labor, that is, modified nature, as well as the general of the subject of the labor process, that is, the techniques, the organization, and the division of labor, which have become characteristic of the human species, individual development takes place under constantly changing starting conditions in the environment of implements. The ontogenetic reproduction of the characteristics of the species in the individual is no longer identical reproduction, but becomes education. This means: To the extent that ontogenetic reproduction refers to the characteristics of the species, the use and the production of tools, it is from the outset an essential factor of the socio-historical development and at the same time subject to it. The early forms of education are characterized by the identity of the tools of labor and the means of education; education and participation in the labor process are identical. This "original unity of labor and education" ([1], p. 26) transforms the individual results of the labor process into elements characteristic of the species, thus transforming historical preconditions into logical moments of a reproductive context which constitutes a new form of mediation between the concrete singular individual of the species and the concrete general of the species which is characteristic of human beings. It is this which we call socio-historical development.

By proceeding from the specific aspects of the human labor process as a process of extended reproduction the benefit for the problem discussed above, that is, the problem of a materialistic reconstruction of the very beginning of history, is as follows: The development of the human being as a concrete totality, as a unity of species being and individual, can be understood as a process of the self-creation of the human being by means of labor.

BIBLIOGRAPHY

[1] Alt, R.: *Vorlesungen über die Erziehung auf frühen Stufen der Menschheitsentwicklung.* Berlin: Volk und Wissen, 1956.
[2] Holzkamp, K.: *Sinnliche Erkenntnis. Historischer Ursprung und gesellschaftliche Funktion der Wahrnehmung.* Frankfurt a.M.: Athenaeum Fischer, 1973.
[3] Leontjew, A. N.: *Problems of Mental Development.* Washington, DC: U.S. Joint Publications Research Service, 1964.
[4] Marx, K.: *Grundrisse: Foundations of the Critique of Political Economy* (Rough Draft). Harmondsworth: Penguin Books, 1973.
[5] Schurig, V.: *Naturgeschichte des Psychischen.* 2 vols., Frankfurt a.M.: Campus, 1975.
[6] Schurig, V.: *Die Entstehung des Bewußtseins.* Frankfurt a.M.: Campus, 1976.

CHAPTER 12

TOOLS OF SCIENCE
coauthored by Wolfgang Lefèvre
[1981]

LABOR AND COGNITION

The question as to the point when accumulated knowledge is to be interpreted as evidence of "scientific" knowledge and which part of it is to be designated as "scientific" knowledge, receives a variety of answers depending on the definition of what is science. The question presumes that, in the course of history, the nature of human cognition changed which makes it necessary to distinguish between prescientific and scientific forms of knowledge. Knowledge is usually called "prescientific" if it is directly derived from practical experiences and maintains its association with practical applications within the process of labor—regardless of how the dividing line between prescientific and scientific knowledge is drawn. Thus, the problem of how to define science leads to the historical question of how cognition developed in the context of labor so that particular forms of mental labor eventually emerged which can be distinguished from their precursors as "science."

Labor in any of its forms is distinguished from other human activities by its function with regard to the material reproduction of society. It is the tools that define the possibilities of material production and thus determine the particular cultural stage of society.

Tools determine whether goals anticipated mentally can be realized. In that sense tools are never just what they actually are. Rather, they represent the potential of realizing intellectually anticipated goals, which is to say, they represent ideas as real possibilities. Their application mediates between possibility and reality.

The use of tools primarily serves the purpose for which they were produced. But tools are more general than particular purposes, and the accumulated experience acquired in the course of their use leads to knowledge about possibilities capable of being realized and about relationships between goals and means under various conditions of realization. Thus, the primary form in which knowledge about natural and social relationships arising from the labor process is represented is the form of rules for the appropriate use of tools. To the extent that the tools are not merely used for realizing invariable goals, but can in turn influence given goals, the potential of the tools is increasingly probed and elaborated. Traditions of sophisticated styles and designs by skilled artisans which can no longer be explained in terms of utility and which were often considered as the origin of geometry, attest to this detachment of the tools from the purposes and, thus, to the fact that the tools are more general than any given particular purpose.

The experiences gained by making use of the tools can be passed on by means of language in the form of rules based on these experiences. They no longer need to be reproduced again and again as primary experiences acquired in the process of using tools. In this way the cognitive outcome of the use of tools becomes a cognitive precondition of a more advanced use of these tools. Individual experience has become social knowledge.

On the basis that social knowledge has become the precondition for the use of tools, the use of tools was split up into an immaterial and a material activity. The practical execution is now preceded by theoretical planning. In this context planning means choice, combination, and coordination of tools and the determination of the sequence of the activities to be accomplished. However, this immaterial activity itself is dependent on material preconditions. The very fact of the transmission of social knowledge requires the concrete tool of language, and planning, too, makes use of concrete means: of rules represented by language, of drawings, on a more developed level of arithmetical means, of writing, etc. Thus, the labor process initiates the development of a rich inventory of material tools of mental activity long before the emergence of the sciences.

THE EMERGENCE OF SCIENCE

The disposal of alien labor, made possible by the emergence of a class society and the development of the state, makes new demands on the planning of work. Projects only to be accomplished collectively such as the construction of large buildings like temples, storehouses, fortifications of cities, fortifications of territories by means of systems of protective walls, the construction and maintenance of irrigation and drainage systems, the task of securing the livelihood of workers engaged in the construction of large projects, the equipment and maintenance of a standing army, etc., require the coordination of a host of planning tasks hitherto unknown. The planning of work becomes a separate, specialized kind of labor and as such the exclusive domain of particular people. Thus, the emergence of class society and the state was associated with a new kind of the division of labor: the separation of physical and mental labor.

Any division of labor has the effect that the specific potentials of the special tools for particular types of work are further developed. This is also true of the professionalization of mental activities. It leads to the differentiation and further development of the tools and ultimately to a further specialization and professionalization within the domain of mental labor on the basis of this differentiation of the tools of mental activities.

Thus, in the texts of the archives of Mesopotamian cities, for example, numerous occupational designations can already be found for specialized mental activities. These are mostly names for different organizational and supervisory positions of the palace and temple administration, names for the heads of the different groups of artisans, for the heads of the grain storehouse, for the collectors of sacrificial offerings and taxes as well as for various higher administrative offices with differentiated, hierarchical official functions. The emergence of hierarchies indicates the development of mental activities that coordinate other mental activities and, therefore, refer only indirectly to the planning of physical work activities. Moreover, there are occupational designations indicating the further differentiation and development of the material tools of mental labor: designations for the weigher, the surveyor, the royal judges (administering a system of justice with codified laws), and hierarchically arranged designations for the scribes.

More momentous than these general effects of introducing a new division of labor are the particular consequences of the specific division of labor into physical and mental labor. Mental labor is an immaterial activity dealing with a mental object and employing material tools. The object appears only as the meaning of signs, symbols, or other material representations, and the mental activity is the transformation of such meanings by objectively manipulating their material representations. Thus, in the special case of mental labor, the progress of the development of new tools primarily consists in the development of new means of the representation of the objects of planning activities—that is, of representing the means and objects of collective labor. There is an extraordinary example of such progress: The development of early signs and pictographic symbols for words into the objective representation of the spoken word in writing is the first significant result of the separation of physical and mental labor.

The differentiation of the tools of mental labor gave rise to all those means which, in addition to the development of writing, constitute the prerequisites for the development of science. First and foremost among these are the tools of arithmetic, drawing, and observation instruments, measuring tools and standards as well as codifications of systematized knowledge in tables and organizational schemata, text books, and "manuals."

What is true of any tool also applies to these tools, namely, that they are more general than the purposes for which they were developed as instruments of the planning of work. And as with using any tool under conditions that permit a feedback of experiences gained in the process of using it with regard to the stated goals, a process is set in motion when these tools of mental labor are used which was called above the detachment of the tools from the purposes. To the extent that the possibilities are explored that can be achieved by making use of these tools, the more general character of the tools emerges in rules for a more advanced use of them, a usage which can no longer be understood in terms of the goals which established the utility of these tools in the context of the planning of work. The outcome of such an advanced use of the tools of mental labor is of a purely immaterial nature despite the fact that it was achieved with material tools and, therefore, is objective; and the activity must be understood as cognition for the sake of cognition. If such activities, for whatever reason, become a permanent factor of mental labor, then the results of this process that have been transmitted exhibit the characteristics of scientific knowledge. They are the outcome of an activity solely aimed at the pursuit of knowledge without any concern for practical purposes.

This reconstruction of the development of social knowledge leads to the definition: We may speak of science if the goal of a certain social activity consists in elaborating the potentials of the material tools of mental labor, which are otherwise used in the planning of work, apart from such goals solely for the purpose of gaining knowledge about the possible outcome.

THE ROLE OF THE TOOLS OF SCIENTIFIC WORK

Thus, the analysis of the question of how the tools determine scientific knowledge must proceed from a fundamental presupposition. The material tools of planning activities represent certain abstractions in the sense that they represent a general moment of reality—isolated from the context in which it actually exists—which can always be realized by using these tools for constructing a corresponding material representation. Thus, spatial configurations can, for example, be represented as drawings using straightedge and compasses. Quantities can be represented by signs and sign combinations using various techniques of numerical representation, and numerical relations between such quantities can be represented by transformations of sign combinations that depend on the particular technique of representation involved.

By using such tools of planning activities, science proceeds from abstractions which are already realized in the planning component of labor. However, in contrast to planning activities science constructs abstract entities as objects of cognition from these abstractions, entities which are immaterial albeit always capable of being realized. Science, therefore, is not free in forming its abstractions; in this activity it is restricted by material preconditions, more precisely, by the specific tools at its disposal that provide cognition with abstractions which are capable of realization.

However, since science—unlike the activity of planning—does not conceive of the abstractions represented by the tools of mental labor as representations of the material conditions of the process of labor, but limits itself to the realization of abstractions as material representations of abstract entities, the connection between the immaterial objects of science and the nonscientific purposes of mental and physical labor must be restored in a separate act of "application." Thus, application does not mediate between the ideal and the material as independent domains; rather, it only offsets forms of isolation that are the basis of the unfolding of science as a particular type of labor.

An excellent way to get an idea of the manner in which planning activities provide science with abstractions is by studying words and their meaning. The activity of planning makes use of everyday language, but, beyond that, introduces new terms based on abstractions. Such abstract terms, for example, are words referring to actions performed with the material tools of mental activity—for instance, terms for arithmetical operations such as techniques of addition. However, in the context of their use in planning activities their abstract meaning (transformation of meanings) and their concrete meaning (use of material tools) are not differentiated. If, however, as is characteristic of science, such abstract terms refer to the immaterial objects of mental activity which emerge in the context of elaborating the potentials of the tools, they only have abstract meaning. The material representations of these objects always are merely specific realizations of a more general idea—for instance, they are words for abstract objects like number, symmetry, or proportion.

The result can be summarized as follows: The material tools of scientific labor define a scope of objective possibilities that represent the framework for developing scientific abstractions. This means that, although the mental objects of science are abstract entities, they are determined concretely due to their connection with the material tools which define the scope of development of abstractions. It must be emphasized that in each case the tools in question are particular tools defining a particular scope.

If this is disregarded in the research in the field of the history of science, serious misinterpretations can hardly be avoided. Thus, it can, for example, be stated that the Egyptians did fractions by using only unit fractions and sums of unit fractions. If this fact is then interpreted as the Egyptians representing fractions with a numerator other than one by such sums of unit fractions, then this amounts to a misinterpretation. Concepts such as fraction, numerator, and denominator arose from abstractions based on tools available at a later time. Instead of reconstructing Egyptian abstractions by analyzing the potentials of their tools, such an interpretation attributes abstractions to Egyptian arithmetic that were made later and, thus, an idea completely alien to Egyptian thinking. The mistake of such an anachronistic attribution of abstractions is due to inattention to the fact that abstractions are determined by the material tools available at a particular time. These tools open up the possibilities of a given scientific abstraction and simultaneously determine its limits.

The most important tool for the representation of scientific abstractions, for the representation of abstract entities, is language. Due to their representation in the language abstract entities become the object of scientific activity even without the actual presence of the material tools by which they are determined. Thus, science increasingly distinguishes itself from planning activities by a specific use of language. This specific use consists in exploring properties of the mental objects—as, for instance, the divisibility properties of numbers—by manipulating their representations in language thus enriching the meanings of the terms denoting the mental objects under consideration.

A new stage emerges if, on this basis, the relationship between the mental object and its representation in language in turn itself becomes an object of scientific activity. The most general indication of this stage is the act of "defining," that is to say, of intentionally transforming the representation of the mental object in language. Now science no longer operates just with words that represent abstract entities which are constructed by specific abstractions, but by means of the definition the mental object of scientific activity turns into a composite concept represented by language. At this stage, the scientific concepts are composite wholes—wholes of different abstract entities—expressing qualities and transformation rules of mutually dependent mental objects of cognition. The definition of concepts provides the fundamental precondition for any form of conclusive deduction of knowledge and thus for the development of scientific theories. Therefore, the act of defining, expressing the specific use of language as a scientific tool, serves as a preeminent indication of the theoretical proof as a scientific method which is often considered as defining science in the narrow sense.

CONTINUITY AND DISCONTINUITY OF THE DEVELOPMENT OF SCIENCE

The function of the material tools of scientific work as outlined above entails that only the reconstruction of the development of these tools permits research in the field of the history of science to arrive at strong explanations for the development of ideas in the history of science. The development of science depends on the development of its material tools.

We already pointed out some aspects of this development. It is, first, the differentiation of existing tools in the course of the increasing division of labor and their use under changing conditions or with different goals. It is, second, the combination and integration of existing tools. It is, third, the development and the transmission of sophisticated intellectual preconditions for the advanced use of existing tools. And it is, finally, primarily the invention of new tools, depending of course on certain material and intellectual preconditions. However, the key of understanding the growth of scientific knowledge consists in the fact that the knowledge to be gained by using a new tool exceeds the cognitive preconditions of its invention. The reason for this is due to the fact that the tools of science like tools in general are material tools: When using a material tool, more can always be learned than the knowledge invested in its invention.

The transmission of the material tools of scientific knowledge from one generation to the next one is the basis for the continuity of scientific thought. However, the very transmission of knowledge has an evolutionary aspect in so far as it involves the accumulation of experience. Thus, continuity is essentially a continuity of development, even if there are situations where the transmission is restricted to the identical reproduction of the tools and the cognitive preconditions of their use or even comes to a halt. If discontinuities in the development of science are to be explained, the possibility of such an interruption of the transmission might have to be taken into account. It is, however, important to note that even within the overall continuity of this development a distinction must be made between evolutionary and revolutionary forms of development that must be understood as continuous and discontinuous moments of the development of science. The continuous development of existing tools broadens the preconditions for the invention of new tools. Once such an invention has occurred, however, new preconditions for further development are instantly provided.

Such an event, for instance, is the invention of the table of reciprocals which helped to solve the inherent problems of the representation of fractions as sums of unit fractions and to establish the efficiency and the superiority concerning the methods of calculation of early Babylonian arithmetic as compared to early Egyptian arithmetic. The momentous consequences were demonstrated in Chapter 7: Numbers independent of any representation by signs depending on means of measurement became the mental object of arithmetic. Of similar significance was the development of a tool that allowed the relation between the mental object and its linguistic representation to become the object of scientific activity, that is to say, the development of the linguistic tool of definition in Greek antiquity, a tool which opened up possibilities for the conclusive deduction of knowledge and, thus, for the development of scientific theories. As long as the acquisition of this new tool does not become the object of investigation, the creation of scientific mathematics by the Pythagoreans must appear as an inexplicable achievement of the Greek mind. And, finally, without the introduction of the experiment as a tool that was systematically employed to mediate between theoretical inference and experience, the genesis of modern science cannot be adequately understood.

A remarkable example of discontinuity within the context of overall continuity of the development of science is the combination and integration of tools of scientific activity which were developed in different situations but for similar purposes so that they relate to the same mental objects. Such an integration allows the specific structures of these objects which result from the use of particular material tools to become discernible with the effect that the scientific occupation with them is raised to a more general level. The most important aspect of the sciences during the Hellenistic period, which have been relatively neglected by the history of science in comparison to the sciences of classical Greece, possibly consists in the fact that they achieved such an integration of the tools of scientific activity with momentous global consequences for the development of science.

The combination and integration of the material tools of scientific activity—developed and accumulated by the early civilizations of the Near East and the eastern Mediterranean—during the Hellenistic period was the beginning of a uniform global history of scientific development. It constituted the foundation of further integrative developments—in particular the integration of developments in India by the Arabs and the assimilation of Chinese achievements which probably took place at a later time and is not yet adequately understood. The acquisition of the results of this integration by the Occident, however complicated and circuitous its course might have been, was an essential prerequisite of modern science. Copernicus cannot be imagined without Ptolemy any more than Galileo without Archimedes and Vieta without Diophantus. Beginning with the Hellenistic period, the different, historically determined stages of development of the material tools of mental labor are no longer standing side by side without any relationship to each other, but are embedded in a common developmental framework. They generate each other. Ever since, a global historical continuity of the history of science has been based on the transmission of the integrated wealth of the tools of the early civilizations.

To be sure, in each case the actual level of the development of the productive forces and, based on them, the attendant relations of production with their implications for the forms of social life and of culture represent different conditions and prerequisites for different societies which determine the extent to which they can acquire and pass on the available tools, the extent to which they are able to make use of the possibilities provided by them, or whether they might even be capable of further developing these resources. To this extent the developmental connection of the material tools of mental labor does not establish an autonomous logic of scientific development. In each case these tools are manifestations of an historically determined civilization, and the elaboration and development of their inherent possibilities is determined by the conformity with the intrinsic laws of the development of the social systems involved.

The historian of science often conceives of this historical determination as merely being the effect of accidental external conditions of a history of science which in principle is determined by immanent laws of development. This belief is based on the abstraction that establishes the history of science as a special branch of history. The abstraction is appropriate to the extent that the development of the sciences cannot be derived directly from intrinsic laws of the development of social systems, for the intrinsic laws of history determine the development of the sciences only indirectly via their material tools and the developmental connection between them. Objectivity, continuity, and a conformity to the intrinsic laws of the developmental possibilities of the sciences, the reconstruction of which constitutes the responsibility of an explanatory history of science, are based on this developmental connection of the material tools of mental labor. Only such a history of science is capable of avoiding a relativistic historicism.

When the relative autonomy of the history of science is understood in this way, the inner logic of the development of scientific thinking and external conditions do not confront each other in the insurmountable contrariety of thought following intrinsic laws and the incidentals of historical circumstances. Rather, the conformity of scientific development to intrinsic laws only appears then as a relative autonomy of a particular type of labor in the overall process of social reproduction. The history of science, no more than the sciences themselves, can disregard the mediation between this particular type of labor and other types of physical or mental labor. As much as it is a fundamental prerequisite of science to investigate the possibilities of the means of mental activity independently of the specific practical goals that generated these means, its continued development—as demonstrated by the subsequent stagnation of the sciences in the slave societies of antiquity and during European feudalism—is just as fundamentally dependent on efficient forms of overcoming their isolation by means of the mutual reconciliation between scientific labor and the goals and problems of social reproduction, on the reconciliation between cognition and action under conditions of the social separation of physical and mental labor.

LIST OF ORIGINAL PUBLICATIONS

The chapters of this book are based on the following original publications:

Chapter 1
Damerow, P.: "Handlung und Erkenntnis in der genetischen Erkenntnistheorie Piagets und in der Hegelschen 'Logik'." In Beyer, W. R. (Ed.): Hegel-Jahrbuch 1977/1978. Köln: Pahl-Rugenstein, 1979, pp. 136–160.

Chapter 2
Damerow, P.: "Repräsentanz und Bedeutung." In Furth, P. (Ed.): *Arbeit und Reflexion: Zur materialistischen Theorie der Dialektik—Perspektiven der Hegelschen Logik.* Köln: Pahl-Rugenstein, 1980, pp. 188–232.

Chapter 3
Damerow, P.: "Anmerkungen zum Begriff 'Abstrakt'." In Steiner, H.-G. (Ed.): *Mathematik—Philosophie—Bildung.* Köln: Aulis Verlag Deubner, 1982, pp. 219–279.

Chapter 4
Damerow, P.: "Was ist mathematische Leistung und wie entstehen die Leistungsunterschiede im Mathematikunterricht?" In *Neue Sammlung, 20* (3), 1980, pp. 253–273.

Chapter 5
Damerow, P.: "Mathematikunterricht und Gesellschaft." In Heymann, H. W. (Ed.): *Mathematikunterricht zwischen Tradition und neuen Impulsen.* Köln: Aulis Verlag Deubner, 1984, pp. 9–48.

Chapter 6
Damerow, P.: "Vorläufige Bemerkungen über das Verhältnis rechendidaktischer Prinzipien zur Frühgeschichte der Arithmetik." In *mathematica didactica, 4* (3), 1981, pp. 131–153.

Chapter 7
Damerow, P.: "Die Entstehung des arithmetischen Denkens." In Damerow, P. & Lefèvre, W. (Eds.): *Rechenstein, Experiment, Sprache: Historische Fallstudien zur Entstehung der exakten Wissenschaften.* Stuttgart: Klett-Cotta, 1981, pp. 11–113.

Chapter 8
Damerow, P., Englund, R. K. & Nissen, H. J.: "Die ersten Zahldarstellungen und die Entwicklung des Zahlbegriffs." In *Spektrum der Wissenschaft, March,* 1988, pp. 46–55.

Chapter 9
Damerow, P.: "Zum Verhältnis von Ontogenese und Historiogenese des Zahlbegriffs." In Edelstein, W. & Hoppe-Graff, S. (Eds.): *Die Konstruktion kognitiver Strukturen: Perspektiven einer konstruktivistischen Entwicklungstheorie.* Bern: Huber, 1993, pp. 195–259.

Chapter 10
Damerow, P.: "Abstraction and Representation." In Steiner, H.-G. & Vermandel, A. (Eds.): *Foundations and Methodology of the Discipline Mathematics Education (Didactics of Mathematics).* Bielefeld and Antwerpen: University of Bielefeld and University of Antwerpen, 1988, pp. 248–261. (Conference report.)

Chapter 11
Excerpt of: Damerow, P., Furth, P., Heidtmann, B. & Lefèvre, W.: "Probleme der materialistischen Dialektik." In *Sozialistische Politik,* 9 (4), 1977, pp. 5–40. (Excerpt: "Materialistische Dialektik und Theorie der gesellschaftlich-historischen Entwicklung," pp. 19–28.)

Chapter 12
Damerow, P. & Lefèvre, W.: "Arbeitsmittel der Wissenschaft." In Damerow, P. & Lefèvre, W. (Eds.): *Rechenstein, Experiment, Sprache: Historische Fallstudien zur Entstehung der exakten Wissenschaften.* Stuttgart: Klett-Cotta, 1981, pp. 223–233.

NAME INDEX

Abrahamsen, A. A. 67
Aebli, H. 318, 367
Alt, R. 394
Amiet, P. 202, 366–367
Archimedes 403
Aristotle 16, 42, 72–73, 75, 83, 86, 127, 312
Arnauld, A. 86
Arnold, W. 148
Awdijew, W. I. 271

Bailey, D. 145
Balfour, G. 68
Baron, J. 68
Bauer, J. 236, 268–271
Bauersfeld, H. 86
Behnke, H. 145
Bekemeier, B. 145–146
Bell, E. T. 145
Bernal, J. D. 67
Berry, J. W. 367
Blackstock, E. G. 67
Blake, B. J. 367
Bloom, B. S. 96–97, 109–110
Blum, P. 363
Boersma, F. J. 68
Bölts, H. 110, 144–145
Borger, R. 269
Bos, H. 148
Bottero, J. 271
Bourbaki, N. 12, 65, 67, 138
Braine, M. D. 56
Brainerd, C. J. 57, 67, 69, 367
Bruins, E. M. 271
Bruner, J. S. 60, 65–67, 152, 168–170, 317, 319, 367
Brunschvicg, L. 14

Cantor, M. 145
Cardano, G. 127

Cassin, E. 271
Chace, A. B. 145, 170, 268, 271
Chomsky, N. 52, 66–67
Cicero 125, 145
Cole, M. 367
Copernicus, N. 403

Dalin, P. 145–146
Damerow, P. 146, 170, 297, 362–367, 369, 381
Darwin, Ch. 12
Dasen, P. R. 367
Davis, R. B. 142, 146
Dawydow, W. 67
Deimel, A. 163, 170, 226, 229, 271, 363, 367
Delaporte, L. 271
Descartes, R. 11, 66–68, 128
Dieudonné, J. 138–139, 146
Diogenes Laertius 125, 146
Diophantus 403
Dittmann, R. 367
Dixon, R. M. W. 367
Dolch, J. 146
Donaldson, M. 68
Driel, G. van 367

Edzard, D. O. 367
Eibl-Eibesfeldt, I. 362, 368
Eilers, G. 135
Eisenlohr, A. 170, 268, 270–271
Elwitz, U. 146
Englund, R. K. 170, 222, 275, 277, 281–282, 287–288, 297, 300, 363–369
Epicurus 125–126
Erdei, L. 13, 27
Estes, K. W. 68

407

Euclid 51, 66–68, 73, 101, 112, 116, 123, 138–139, 180, 192, 207–208, 306, 374–375, 379
Euler, L. 128–129

Falkenstein, A. 170, 269, 271, 278, 363, 368
Felscher, W. 140
Fermat, P. de 128
Fletcher, T. J. 146
Förtsch, W. 233, 236, 268–271
Frege, G. 302, 307, 368
Freudenthal, G. 362, 367
Friberg, J. 278, 284, 297, 363, 368
Furth, H. G. 170
Furth, P. 27, 271
Fuson, K. C. 368
Fuye, A. de la 270–271

Galanter, E. 68
Galilei, G. 11, 128, 130, 362, 403
Gallistel, C. R. 368
Garcia, R. 311–312, 314, 369
Gauss, C. F. 371
Gearhart, M. 369
Gelman, R. 368
Gericke, H. 368
Gillings, R. J. 170, 271
Goethe, J. W. von 10
Goetze, A. 271
Goldmann, L. 68
Golenishev, W. 176
Gollin, E. S. 68
Grassmann, H. 133
Grégoire, J.-P. 363, 368
Griffiths, H. B. 146
Guberman, S. R. 369
Günther, S. 146
Gutzmer, A. 146

Habermas, J. 55, 68
Hallpike, C. R. 368
Halmos, P. R. 146
Hansen, P. A. 130
Harasym, C. R. 68
Hauser, K. 390

Heeschen, V. 362, 368
Hegel, G. W. F. 1–2, 4–13, 15, 19–24, 27, 55–56, 76–77, 86, 301–302, 392
Heidtmann, B. 55, 68
Heinrich, K. 146
Hemeling, J. 120, 146
Herodotus 189, 268, 271
Hesse, O. 130
Hilbert, D. 375
Hilprecht, H. V. 249, 270, 272
Holzkamp, K. 63, 68, 394
Hopf, D. 78, 86
Hörner, W. 146
Howson, A. G. 146
Høyrup, J. 146, 254, 257, 272, 363, 368
Humboldt, W. von 54
Hume, D. 75, 86
Hurford, J. R. 368

Ifrah, G. 368
Ihnow, G. 146
Inhelder, B. 69, 171
Inhetveen, H. 147
Inhetveen, R. 170
Isocrates 125

Jacobi, C. G. J. 129–130, 135
Jahnke, H. N. 27, 147
Jänicke, E. 170
Joachimsthal, F. 130
Juschkewitsch, A. P. 68, 170, 272

Kant, I. 4, 9, 21, 43, 59, 76, 86, 299, 301–302, 304, 306–307, 368
Keitel, C. 146–147, 170
Keller, M. 68
Kepler, J. 128
Kilpatrick, J. 146
King, W. L. 67
Klein, F. 131, 136–138, 143, 147
Klengel, H. 272
Kline, M. 140, 147
Kling, J. K. 68
Knöss, C. 146
Koch, G. 368
Kohlberg, L. 68

NAME INDEX

Kolmogorov, A. N. 140
Krüsi, H. 169
Kummer, E. E. 133

Lagrange, J. L. 128–129
Lamberg-Karlovsky, C. C. 272
Lamberg-Karlovsky, M. 272
Lancy, D. F. 368
Langdon, S. 363, 368
Laplace, P. S. 128–129
Lawson, G. 68
Le Brun, A. 366, 368
Lean, G. A. 369
Lefèvre, W. 146, 170, 381, 395
Legendre, A. M. 306
Leibniz, G. W. 128
Lenné, H. 144, 147
Leont'ev, A. N.; *see* Leontjew
Leontjew, A. N. 394
Lévy-Bruhl, L. 313, 325, 369
Lewy, H. 269, 272
Lichnerowicz, A. 140
Liebermann, S. J. 366, 369
Lorenzen, P. 170, 302, 369
Lorey, W. 147
Luhmann, N. 55, 68
Lumsden, E. A. 68

Mager, R. F. 110
Maguire, T. O. 68
Maratsos, M. P. 68
Margueron, J.-C. 272
Marrou, H. I. 147
Marx, K. 55, 386, 392, 394
McLaughlin, P. 362, 367
Mead, G. H. 66
Meer, P. E. van der 272
Mehrtens, H. 148
Menninger, K. 171, 272, 369
Mies, T. 27
Mill, J. S. 369
Miller, G. A. 68
Miller, S. A. 68
Mimica, J. 369
Minsky, M. L. 369
Monge, G. 128

Moody, M. 68
Münzinger, W. 144, 148

Neander, J. 144, 148
Needham, J. 171, 272
Nepos, C. 133
Neugebauer, O. 68, 171, 268–269, 272, 363, 369
Neumann, D. 369
Newton, I. 11, 128, 312
Nies, J. B. 268, 270, 272
Nissen, H. J. 275, 297, 300, 363, 366–367, 369

Oppenheim, A. L. 201–202, 272, 366, 369
Oresme, N. 127
Otte, M. 27, 147, 170

Pahl, F. 148
Palermo, D. S. 68
Peet, T. E. 268, 272
Pestalozzi, H. 149–150, 169, 171
Pfeiffer, H. 148
Piaget, J. 1–16, 18–27, 29, 39, 54–62, 64–65, 69, 76, 79, 84, 86, 99, 103, 105, 109–110, 151–152, 167–169, 171, 299–322, 324–325, 335, 346, 354–357, 359–362, 369, 372–373, 377, 381
Plato 43, 73, 124–126, 148, 301, 306, 354
Plücker, J. 133
Pontriagin, L. S. 140
Postlethwaite, N. 110
Powell, M. A. 171, 269, 272, 366, 369
Praetorius, M. J. 114
Pribram, K. H. 68
Projektgruppe Bildungsbericht 148
Ptolemy, C. 117, 403

Renn, J. 362, 367
Rhind, A. H. 176
Richelot, F. 135
Riegel, K. F. 56, 69
Rieke, E. 147
Riley, C. A. 69
Roeder, P. M. 110

Rosenhain, J. G. 130
Rothenberg, B. B. 69
Rottländer, R. C. A. 272
Ruben, P. 57, 67, 69, 171
Russell, B. 41–42, 57–58, 69, 307
Rutten, M. 271

Sachs, A. J. 269, 271–272
Saxe, G. B. 369
Schadler, M. 68
Scheil, V. 161, 171, 279, 363, 369
Schiefenhövel, W. 300, 362, 368
Schlömilch, O. 135
Schmand-Besserat, D. 166
Schmandt-Besserat, D. 164, 171, 202, 212, 218, 272, 366, 369–370
Schneider, I. 148
Schneider, N. 272
Schubring, G. 27
Schulze, J. 133
Schurig, V. 69, 394
Schwenter, D. 114, 144
Scribner, S. 367
Seeger, F. 147, 170
Seidel, L. 130
Sellnow, I. 272
Sewell, B. 148
Shendge, M. J. 366, 370
Siegel, L. S. 57, 68–69
Skemp, R. R. 110
Smedslund, J. 56
Snellius, W. 115
Steiner, G. 69
Steiner, H.-G. 133, 148
Strauss, S. 370
Strommenger, E. 272, 370
Struve, V. V. 176, 273
Szendrei, J. 148

Tartaglia, N. 128
Thureau-Dangin, F. 162, 170
Timerding, H. E. 133, 148
Townsend, D. J. 69
Trabasso, T. 69
Treumann, K. 110
Triarius 126

Tropfke, J. 273
Türk, W. von 149–150, 171

Vaiman, A. 278, 286, 363, 370
Vallat, F. 366, 368, 370
Varga, T. 140
Vercoutter, J. 271
Vieta, F. 128, 403
Vogel, K. 69, 171, 273, 370
Vygotsky, L. S. 54, 65, 69, 317, 370

Wagenschein, M. 148
Wartofsky, M. W. 370
Wassmann, J. 370
Weber, H. 136
Weierstrass, K. 133
Weiner, S. L. 69
Weissbach, F. H. 273
Welti, E. L. 370
Wertheimer, M. 293, 313, 325–326, 370
Westbury, I. 142, 148
Whitehead, A. N. 69, 307
Wiese, L. 134–135, 148
Wilson, E. G. 69
Wittenberg, A. I. 148
Wittmann, E. 148
Wolff, M. 370
Wussing, H. 70, 148, 273, 370

Youniss, J. 70
Yushkevitch, A. P.; see Juschkewitsch

Zech, F. 148
Zermelo 42, 58
Zimmer, J. 146

SUBJECT INDEX

a priori 19, 43, 59–60, 170, 301, 304, 307–308, 354
abstraction
    Aristotelian 16, 72–73, 75, 83
    formal 5, 7–8, 26
    mathematical 76, 378
    rational 384–385, 387–390
    reflective 2–10, 12–27, 57, 84, 109, 304–305, 308, 312, 314, 318–320, 355–356, 357–362, 372, 378–379
    scientific 400
ambiguity of numerals
    Babylonian numerals
        in archaic texts 223, 229, 280–283, 291, 335–338, 340–341, 347, 353, 365
        in the 3rd millennium B.C. 238
        in the Old Babylonian period 204–205, 210, 246, 270
    Egyptian numerals
        representation of fractions by unit fractions 198–199
    relative representation of number 192
anticipation 13, 16–17, 19, 30, 33, 47, 100, 106, 125, 196–197, 262, 326, 346
    of goals 36–40, 42–43, 45, 47, 50, 395
antinomy 41, 57
    Russell's antinomy 41–42, 57–58
approximation 208, 237, 258, 260–261
arithmetic
    subject of the quadrivium 127
arithmetical operations
    addition with archaic sign systems 164
    addition with constructive-additive sign systems 156–158, 195
    algorithms 92

algorithms in Babylonian arithmetic 206, 251–258
    depending on calculating devices 61–62, 154–166
    division in Egyptian arithmetic 178–180, 187, 196
    division in Mesopotamia in the 3rd millennium B.C. 237
    division in the Old Babylonian period 253–258
    Egyptian auxiliary number algorithm 183–186
    in Mesopotamian archaic texts 344–346
    multiplication implicit in Mesopotamian archaic texts 286–288, 345–346
    multiplication in Egyptian arithmetic 177–178, 187, 196
    multiplication in Mesopotamia in the 3rd millennium B.C. 235–237
    multiplication in the Old Babylonian period 246–248, 251–254
    multiplication with Chinese rod numerals 155
    multiplication with Pestalozzi's table of units 149–150
    Old Babylonian standard arithmetical table 251–254
    origin of addition 357–358
    prevalence of additive operations in Egyptian arithmetic 186–188
    reconstruction from archaic numerical notations 279–291
    structure of Babylonian arithmetic 207–211
astronomy
    Babylonian 203, 261
    criticism of the Epicureans 125–126
    in Pythagorean mathematics 123
    subject of the quadrivium 127

classification 2, 41, 219
closed systems 34, 36, 80
   of actions 304
   of numbers 208
   of operations 372
   of thought structures 307
cognitive universal 29, 39, 41, 59, 112, 173–174, 295, 301, 306–311, 355–356, 359–361, 375–377; see also a priori
conceptual development
   in Hegel's philosophy 12
   in mathematics 116, 170
   in the history of science 309
   of children 1
concretization 34, 77, 83–85
consciousness 3, 5, 13, 15–16, 18, 22–26, 55, 64, 99, 108, 126, 314, 387–390
contradiction 25, 34, 41, 57–58, 67, 92–93, 104, 303, 378, 381, 383, 386
   Aristotle's law 42

definition
   as a scientific technique 77, 84, 400–402
   invention of the definition technique 123, 296
   of a problem 43
   of species 387–390
   of the concept of ability 89, 101, 109
   of the concept of abstraction 71
   of the concept of labor 383
   of the concept of number 123, 268, 296
   of the concept of science 395, 398
dialectical concept of abstraction 76, 84
dialectical logic 1; see also logic of reflection
dialectical materialism 56, 383–386
differentiation 14
   and integration 15, 17
   of animal and man 387
   of counters 218–219
   of interpretations 85
   of object and activity 6–7, 23–25
   of signs 215, 363, 364
   of the application of signs 344
   of the number concept 306–307
   of the sign function 228
   of tools 401
   of tools of mental labor 397–398
   of weight and volume 45

empirical certainty 30–31, 36, 56, 306
equations
   in Babylonian mathematics 206, 260
   in Chinese mathematics 62, 156, 175
   in Egyptian mathematics 267
   used by Descartes 67
Euclidean algorithm 180
expansion
   metalinguistic 58
   of a behavioral schema 24
   of a cognitive structure 4
      extensive 33–35
      reflective 35, 41, 43, 105
   of a counting technique 324
   of a language 57
   of an arithmetical schema 196
   of an arithmetical technique 216
   of an interpretation 85
   of experiences 320
   of geometry 66
   of the number concept 45, 61, 306, 314, 358

formalization 12, 34, 101, 353

genetic epistemology 1–4, 8, 10–13, 19–20, 23, 58–61, 299, 302, 309
geometry
   criticism of the Epicureans 125–126
   development of 375
   Euclid must go 138
   Euclidean 51, 73, 101, 116, 374
   historical development of 66–67
   in Pythagorean mathematics 123
   invented by the Egyptians 189
   non-Euclidean 379
   origin of 396
   subject of the quadrivium 127

SUBJECT INDEX

harmonics
    criticism of the Epicureans 125–126
    in Pythagorean mathematics 123
    subject of the quadrivium 127
heuristic 34, 92, 106–107, 185, 196–197, 256, 264
historical materialism 383–386

inferences 39, 42, 58, 75, 123, 305, 362, 365, 402
    inductive 30
intelligence 2, 5, 9, 13, 19, 24, 61, 65, 96, 305, 307, 373
    sensori-motor 13, 20–22, 24–25, 61
internalization 42, 44, 65, 99, 150–151, 153, 167
intuition 4–6, 21, 59, 61, 152, 169, 301, 373

linguistic competence 46–54
logic
    dialectical 1–2, 10–13, 20– 21, 56
    objective 4, 19–21
    of being 4–6, 20–21, 23–24
    of essence 4–6, 20–21
    of reflection 4, 23
    of relations 17
    subjective 19
logical necessity 31, 56, 73, 306–307, 373
logico-mathematical structures 2, 4, 7–9, 12, 16, 19, 20, 25–27, 29, 42, 45, 55–56, 61, 151–152, 175, 306, 309–312, 315–317, 320, 324, 335, 354–356, 359–361

means of labor 8, 386, 390, 393–394
mechanical
    accounting 164–166, 213, 262
    calculation 192, 195–199, 219, 235, 262
    generation of geometric figures 66, 115
    use of formulas 122
    use of representations 51, 100, 279

number
    abstract 162, 247, 275, 291–296, 347, 350, 358–360
    algebraic 307

area numbers 228, 230–231, 234–236, 246–250
concept 40, 45, 61–63, 74
    of conservation of quantity as a precondition 372
    of historical development 291–296
    of ideal objects in the Old Babylonian period 260
    lacking in early civilizations 123, 215, 230, 268, 275
    of ontogenesis and historiogenesis 299–362
    of Pestalozzi's understanding 150
    of representation 375
    of representation by numerals 230
counting numbers 228, 230–231, 233, 236, 238–239, 246–247
fractions
    acceptance as numbers 306
    in Chinese mathematics 61–62
    in early metrology 215, 227–231
    in Egyptian mathematics 62
    in the Babylonian sexagesimal place-value system 208–209
grain numbers 228–230, 234–236, 238
head numbers in Babylonian multiplication tables 247
imaginary 379
    Piaget's view 307–308
irrational 379
    Piaget's view 307–308
natural
    closed system 104
    Platonic idea 306
    relation to logic 307
negative 62, 104, 379
    in Chinese mathematics 156, 175
    Piaget's view 307–308
number analogues 293–295, 313, 325
rational
    sexagesimal representation 208–209
relative representation 192, 210
unit fractions
    in Babylonian mathematics 263–265

414 SUBJECT INDEX

in Egyptian mathematics 177–187,
   197–199, 263–265, 400
origin of the Babylonian sexagesimal place-value system 239–248
numerals
   Babylonian metrological sign systems
      161–164, 219–242
   Babylonian place-value system 160–
      161, 204–206, 242–251
   body-counting numerals in New Guinea
      323–329
   Chinese rod numerals 155–156, 173–
      176
   early development of 261–265
   Egyptian hieratic numerals 157–158,
      193
   Egyptian hieroglyphic numerals 157–
      159, 193–198
   Mesopotamian archaic numerical signs
      219–223, 276–293, 334–354
   Mesopotamian clay counters 164–166,
      212–219
   number words 211–212, 321–322, 365–
      366
   structures of sign systems 190–193

operation
   abstract 294, 358
   concrete 2–3, 24, 61, 152, 313, 361–362
   formal 56, 61, 312, 335, 361

place-value system 45, 61–62, 336
   decimal in China 156, 175
   development of 191, 239
   invention in Babylonia 242–258, 353
   invention in China 157
   none in Egypt 157
   sexagesimal in Mesopotamia 160–161,
      204–207, 210
   sexagesimal structure 207–210
problem solving 29–31, 35, 37, 39–40, 43,
      50, 56, 91, 99–100, 103, 106–
      107, 117, 121, 144, 260, 264
proof
   and definition 401
   and logical necessity 31

by verification 184, 259
formalization of proofs 101
in Babylonian mathematics 259
in Egyptian mathematics 184
in Greek mathematics 51, 66

reflective substratum 305
representation
   material 29, 51–52, 167–169, 373, 397–
      400
   mechanical use of 51, 100
rules
   correspondence rule 280
   formal 66, 90, 198, 212, 294
   generative 34–35, 38, 41–52, 80
   grammatical 52, 211
   rule system 32, 34–35, 38, 41–52, 62,
      80
   semiotic 285, 339, 365
   seriation rule 280
   substitution rule 280
   symbol transformation 374
   syntactic 348, 365
   transformation 401

seriation 2, 17, 59, 167, 304
seriation experiment 29–38, 56
seriation of signs 191, 195, 198, 280

tools 8, 10, 263, 316, 395–404
   and consciousness 18
   in anthropogenesis 44, 387–394
   of mathematical astronomy 261
   of measurement 214–215, 227–231,
      235, 245–246, 248, 250, 260,
      265
   of mental activity 373–374, 378–380,
      396–397
   of mental labor 397–398, 404
   of science 398–404
   tables as tools 242, 251–252
   writing as a tool 189
trial and error 17, 36–37, 42, 257

Boston Studies in the Philosophy of Science

123. P. Duhem: *The Origins of Statics*. Translated from French by G.F. Leneaux, V.N. Vagliente and G.H. Wagner. With an Introduction by S.L. Jaki. 1991
ISBN 0-7923-0898-0
124. H. Kamerlingh Onnes: *Through Measurement to Knowledge*. The Selected Papers, 1853-1926. Edited and with an Introduction by K. Gavroglu and Y. Goudaroulis. 1991
ISBN 0-7923-0825-5
125. M. Čapek: *The New Aspects of Time: Its Continuity and Novelties*. Selected Papers in the Philosophy of Science. 1991
ISBN 0-7923-0911-1
126. S. Unguru (ed.): *Physics, Cosmology and Astronomy, 1300-1700*. Tension and Accommodation. 1991
ISBN 0-7923-1022-5
127. Z. Bechler: *Newton's Physics on the Conceptual Structure of the Scientific Revolution*. 1991
ISBN 0-7923-1054-3
128. É. Meyerson: *Explanation in the Sciences*. Translated from French by M-A. Siple and D.A. Siple. 1991
ISBN 0-7923-1129-9
129. A.I. Tauber (ed.): *Organism and the Origins of Self*. 1991
ISBN 0-7923-1185-X
130. F.J. Varela and J-P. Dupuy (eds.): *Understanding Origins*. Contemporary Views on the Origin of Life, Mind and Society. 1992
ISBN 0-7923-1251-1
131. G.L. Pandit: *Methodological Variance*. Essays in Epistemological Ontology and the Methodology of Science. 1991
ISBN 0-7923-1263-5
132. G. Munévar (ed.): *Beyond Reason*. Essays on the Philosophy of Paul Feyerabend. 1991
ISBN 0-7923-1272-4
133. T.E. Uebel (ed.): *Rediscovering the Forgotten Vienna Circle*. Austrian Studies on Otto Neurath and the Vienna Circle. Partly translated from German. 1991
ISBN 0-7923-1276-7
134. W.R. Woodward and R.S. Cohen (eds.): *World Views and Scientific Discipline Formation*. Science Studies in the [former] German Democratic Republic. Partly translated from German by W.R. Woodward. 1991
ISBN 0-7923-1286-4
135. P. Zambelli: *The Speculum Astronomiae and Its Enigma*. Astrology, Theology and Science in Albertus Magnus and His Contemporaries. 1992
ISBN 0-7923-1380-1
136. P. Petitjean, C. Jami and A.M. Moulin (eds.): *Science and Empires*. Historical Studies about Scientific Development and European Expansion.
ISBN 0-7923-1518-9
137. W.A. Wallace: *Galileo's Logic of Discovery and Proof*. The Background, Content, and Use of His Appropriated Treatises on Aristotle's *Posterior Analytics*. 1992
ISBN 0-7923-1577-4
138. W.A. Wallace: *Galileo's Logical Treatises*. A Translation, with Notes and Commentary, of His Appropriated Latin Questions on Aristotle's *Posterior Analytics*. 1992
ISBN 0-7923-1578-2
Set (137 + 138) ISBN 0-7923-1579-0

Boston Studies in the Philosophy of Science

139. M.J. Nye, J.L. Richards and R.H. Stuewer (eds.): *The Invention of Physical Science*. Intersections of Mathematics, Theology and Natural Philosophy since the Seventeenth Century. Essays in Honor of Erwin N. Hiebert. 1992
 ISBN 0-7923-1753-X
140. G. Corsi, M.L. dalla Chiara and G.C. Ghirardi (eds.): *Bridging the Gap: Philosophy, Mathematics and Physics*. Lectures on the Foundations of Science. 1992 ISBN 0-7923-1761-0
141. C.-H. Lin and D. Fu (eds.): *Philosophy and Conceptual History of Science in Taiwan*. 1992 ISBN 0-7923-1766-1
142. S. Sarkar (ed.): *The Founders of Evolutionary Genetics*. A Centenary Reappraisal. 1992 ISBN 0-7923-1777-7
143. J. Blackmore (ed.): *Ernst Mach – A Deeper Look*. Documents and New Perspectives. 1992 ISBN 0-7923-1853-6
144. P. Kroes and M. Bakker (eds.): *Technological Development and Science in the Industrial Age*. New Perspectives on the Science–Technology Relationship. 1992 ISBN 0-7923-1898-6
145. S. Amsterdamski: *Between History and Method*. Disputes about the Rationality of Science. 1992 ISBN 0-7923-1941-9
146. E. Ullmann-Margalit (ed.): *The Scientific Enterprise*. The Bar-Hillel Colloquium: Studies in History, Philosophy, and Sociology of Science, Volume 4. 1992 ISBN 0-7923-1992-3
147. L. Embree (ed.): *Metaarchaeology*. Reflections by Archaeologists and Philosophers. 1992 ISBN 0-7923-2023-9
148. S. French and H. Kamminga (eds.): *Correspondence, Invariance and Heuristics*. Essays in Honour of Heinz Post. 1993 ISBN 0-7923-2085-9
149. M. Bunzl: *The Context of Explanation*. 1993 ISBN 0-7923-2153-7
150. I.B. Cohen (ed.): *The Natural Sciences and the Social Sciences*. Some Critical and Historical Perspectives. 1994 ISBN 0-7923-2223-1
151. K. Gavroglu, Y. Christianidis and E. Nicolaidis (eds.): *Trends in the Historiography of Science*. 1994 ISBN 0-7923-2255-X
152. S. Poggi and M. Bossi (eds.): *Romanticism in Science*. Science in Europe, 1790–1840. 1994 ISBN 0-7923-2336-X
153. J. Faye and H.J. Folse (eds.): *Niels Bohr and Contemporary Philosophy*. 1994
 ISBN 0-7923-2378-5
154. C.C. Gould and R.S. Cohen (eds.): *Artifacts, Representations, and Social Practice*. Essays for Marx W. Wartofsky. 1994 ISBN 0-7923-2481-1
155. R.E. Butts: *Historical Pragmatics*. Philosophical Essays. 1993
 ISBN 0-7923-2498-6
156. R. Rashed: *The Development of Arabic Mathematics: Between Arithmetic and Algebra*. Translated from French by A.F.W. Armstrong. 1994
 ISBN 0-7923-2565-6

Boston Studies in the Philosophy of Science

157. I. Szumilewicz-Lachman (ed.): *Zygmunt Zawirski: His Life and Work.* With Selected Writings on Time, Logic and the Methodology of Science. Translations by Feliks Lachman. Ed. by R.S. Cohen, with the assistance of B. Bergo. 1994 ISBN 0-7923-2566-4
158. S.N. Haq: *Names, Natures and Things.* The Alchemist Jābir ibn Ḥayyān and His *Kitāb al-Aḥjār* (Book of Stones). 1994 ISBN 0-7923-2587-7
159. P. Plaass: *Kant's Theory of Natural Science.* Translation, Analytic Introduction and Commentary by Alfred E. and Maria G. Miller. 1994
 ISBN 0-7923-2750-0
160. J. Misiek (ed.): *The Problem of Rationality in Science and its Philosophy.* On Popper vs. Polanyi. The Polish Conferences 1988–89. 1995
 ISBN 0-7923-2925-2
161. I.C. Jarvie and N. Laor (eds.): *Critical Rationalism, Metaphysics and Science.* Essays for Joseph Agassi, Volume I. 1995 ISBN 0-7923-2960-0
162. I.C. Jarvie and N. Laor (eds.): *Critical Rationalism, the Social Sciences and the Humanities.* Essays for Joseph Agassi, Volume II. 1995 ISBN 0-7923-2961-9
 Set (161–162) ISBN 0-7923-2962-7
163. K. Gavroglu, J. Stachel and M.W. Wartofsky (eds.): *Physics, Philosophy, and the Scientific Community.* Essays in the Philosophy and History of the Natural Sciences and Mathematics. In Honor of Robert S. Cohen. 1995
 ISBN 0-7923-2988-0
164. K. Gavroglu, J. Stachel and M.W. Wartofsky (eds.): *Science, Politics and Social Practice.* Essays on Marxism and Science, Philosophy of Culture and the Social Sciences. In Honor of Robert S. Cohen. 1995 ISBN 0-7923-2989-9
165. K. Gavroglu, J. Stachel and M.W. Wartofsky (eds.): *Science, Mind and Art.* Essays on Science and the Humanistic Understanding in Art, Epistemology, Religion and Ethics. Essays in Honor of Robert S. Cohen. 1995
 ISBN 0-7923-2990-2
 Set (163–165) ISBN 0-7923-2991-0
166. K.H. Wolff: *Transformation in the Writing.* A Case of Surrender-and-Catch. 1995 ISBN 0-7923-3178-8
167. A.J. Kox and D.M. Siegel (eds.): *No Truth Except in the Details.* Essays in Honor of Martin J. Klein. 1995 ISBN 0-7923-3195-8
168. J. Blackmore: *Ludwig Boltzmann, His Later Life and Philosophy, 1900–1906.* Book One: A Documentary History. 1995 ISBN 0-7923-3231-8
169. R.S. Cohen, R. Hilpinen and Q. Renzong (eds.): *Realism and Anti-Realism in the Philosophy of Science.* Beijing International Conference, 1992. 1995
 ISBN 0-7923-3233-4
170. I. Kuçuradi and R.S. Cohen (eds.): *The Concept of Knowledge.* The Ankara Seminar. 1995 ISBN 0-7923-3241-5

Boston Studies in the Philosophy of Science

171. M.A. Grodin (ed.): *Meta Medical Ethics*: The Philosophical Foundations of Bioethics. 1995　ISBN 0-7923-3344-6
172. S. Ramirez and R.S. Cohen (eds.): *Mexican Studies in the History and Philosophy of Science*. 1995　ISBN 0-7923-3462-0
173. C. Dilworth: *The Metaphysics of Science*. An Account of Modern Science in Terms of Principles, Laws and Theories. 1995　ISBN 0-7923-3693-3
174. J. Blackmore: *Ludwig Boltzmann, His Later Life and Philosophy, 1900–1906* Book Two: The Philosopher. 1995　ISBN 0-7923-3464-7
175. P. Damerow: *Abstraction and Representation*. Essays on the Cultural Evolution of Thinking. 1996　ISBN 0-7923-3816-2
176. G. Tarozzi (ed.): *Karl Popper, Philosopher of Science*.　(in prep.)
177. M. Marion and R.S. Cohen (eds.): *Québec Studies in the Philosophy of Science*. Part I: Logic, Mathematics, Physics and History of Science. Essays in Honor of Hugues Leblanc. 1995　ISBN 0-7923-3559-7
178. M. Marion and R.S. Cohen (eds.): *Québec Studies in the Philosophy of Science*. Part II: Biology, Psychology, Cognitive Science and Economics. Essays in Honor of Hugues Leblanc. 1996　ISBN 0-7923-3560-0
　Set (177–178) ISBN 0-7923-3561-9
179. F. Dainian and R.S. Cohen (eds.): *Chinese Studies in the History and Philosophy of Science and Technology*. 1995　ISBN 0-7923-3463-9
180. P. Forman and J.M. Sánchez-Ron (eds.): *National Military Establishments and the Advancement of Science and Technology*. Studies in 20th Century History. 1995　ISBN 0-7923-3541-4
181. E.J. Post: *Quantum Reprogramming*. Ensembles and Single Systems: A Two-Tier Approach to Quantum Mechanics. 1995　ISBN 0-7923-3565-1

Also of interest:
R.S. Cohen and M.W. Wartofsky (eds.): *A Portrait of Twenty-Five Years Boston Colloquia for the Philosophy of Science, 1960-1985*. 1985　ISBN Pb 90-277-1971-3

Previous volumes are still available.

KLUWER ACADEMIC PUBLISHERS – DORDRECHT / BOSTON / LONDON